5G NR: The Next Generation Wireless Access Technology

5G NR: The Next Generation Wireless Access Technology

Erik Dahlman

Stefan Parkvall

Johan Sköld

ACADEMIC PRESS

An imprint of Elsevier

Academic Press is an imprint of Elsevier
125 London Wall, London EC2Y 5AS, United Kingdom
525 B Street, Suite 1650, San Diego, CA 92101, United States
50 Hampshire Street, 5th Floor, Cambridge, MA 02139, United States
The Boulevard, Langford Lane, Kidlington, Oxford OX5 1GB, United Kingdom

Notices
Knowledge and best practice in this field are constantly changing. As new research and experience broaden
our understanding, changes in research methods, professional practices, or medical treatment may become
necessary.

Practitioners and researchers must always rely on their own experience and knowledge in evaluating and
using any information, methods, compounds, or experiments described herein. In using such information or
methods they should be mindful of their own safety and the safety of others, including parties for whom they
have a professional responsibility.

To the fullest extent of the law, neither the Publisher nor the authors, contributors, or editors, assume any
liability for any injury and/or damage to persons or property as a matter of products liability, negligence or
otherwise, or from any use or operation of any methods, products, instructions, or ideas contained in the
material herein.

British Library Cataloguing-in-Publication Data
A catalogue record for this book is available from the British Library

Library of Congress Cataloging-in-Publication Data
A catalog record for this book is available from the Library of Congress

ISBN: 978-0-12-814323-0

For Information on all Academic Press publications
visit our website at https://www.elsevier.com/books-and-journals

Publisher: Mara Conner
Acquisition Editor: Tim Pitts
Editorial Project Manager: Joshua Mearns
Production Project Manager: Kamesh Ramajogi
Cover Designer: Greg Harris

Typeset by MPS Limited, Chennai, India

Contents

Preface

Long-Term Evolution (LTE) has become the most successful wireless mobile broadband technology across the world, serving billions of users. Mobile broadband is, and will continue to be, an important part of future cellular communication, but future wireless networks are to a large extent also about a significantly wider range of use cases and a correspondingly wider range of requirements. Although LTE is a very capable technology, still evolving and expected to be used for many years to come, a new 5G radio access known as New Radio (NR) has been standardized to meet future requirements.

This book describes NR, developed in 3GPP (*Third Generation Partnership Project*) as of late Spring 2018.

Chapter 1 provides a brief introduction, followed by a description of the standardization process and relevant organizations such as the aforementioned 3GPP and ITU in Chapter 2. The frequency bands available for mobile communication are covered in Chapter 3 together with a discussion on the process for finding new frequency bands.

An overview of LTE and its evolution is found in Chapter 4. Although the focus of the book is NR, a brief overview of LTE as a background to the coming chapters is relevant. One reason is that both LTE and NR are developed by 3GPP and hence have a common background and share several technology components. Many of the design choices in NR are also based on experience from LTE. Furthermore, LTE continues to evolve in parallel with NR and is an important component in 5G radio access.

Chapter 5 provides an overview of NR. It can be read on its own to get a high-level understanding of NR, or as an introduction to the subsequent chapters.

Chapter 6 outlines the overall protocol structure in NR, followed by a description of the overall time−frequency structure of NR in Chapter 7.

Multiantenna processing and beam-forming are integral parts of NR. The channel sounding tools to support these functions are outlined in Chapter 8, followed by the overall transport-channel processing in Chapter 9 and the associated control signaling in Chapter 10. How the functions are used to support different multi-antenna schemes and beamforming functions is the topic of Chapters 11 and 12.

Retransmission functionality and scheduling are the topics of Chapters 13 and 14, followed by power control in Chapter 15 and initial access in Chapter 16.

Coexistence and interworking with LTE is an essential part of NR, especially in the nonstandalone version which relies on LTE for mobility and initial access, and is covered in Chapter 17.

Radio-frequency (RF) requirements, taking into account spectrum flexibility across large frequency ranges and multistandard radio equipment, are the topic of Chapter 18. Chapter 19 discusses the RF implementation aspects for higher frequency bands in the mm-wave range.

Finally, Chapter 20 concludes the book with an outlook to future NR releases.

Acknowledgments

We thank all our colleagues at Ericsson for assisting in this project by helping with contributions to the book, giving suggestions and comments on the contents, and taking part in the huge team effort of developing NR and the next generation of radio access for 5G.

The standardization process involves people from all parts of the world, and we acknowledge the efforts of our colleagues in the wireless industry in general and in 3GPP RAN in particular. Without their work and contributions to the standardization, this book would not have been possible.

Finally, we are immensely grateful to our families for bearing with us and supporting us during the long process of writing this book.

Abbreviations and Acronyms

3GPP	Third Generation Partnership Project
5GCN	5G Core Network
AAS	Active Antenna System
ACIR	Adjacent Channel Interference Ratio
ACK	Acknowledgment (in ARQ protocols)
ACLR	Adjacent Channel Leakage Ratio
ACS	Adjacent Channel Selectivity
ADC	Analog-to-Digital Converter
AF	Application Function
AGC	Automatic Gain Control
AM	Acknowledged Mode (RLC configuration)
AM	Amplitude Modulation
AMF	Access and Mobility Management Function
A-MPR	Additional Maximum Power Reduction
AMPS	Advanced Mobile Phone System
ARI	Acknowledgment Resource Indicator
ARIB	Association of Radio Industries and Businesses
ARQ	Automatic Repeat-reQuest
AS	Access Stratum
ATIS	Alliance for Telecommunications Industry Solutions
AUSF	Authentication Server Function
AWGN	Additive White Gaussian Noise
BC	Band Category
BCCH	Broadcast Control Channel
BCH	Broadcast Channel
BiCMOS	Bipolar Complementary Metal Oxide Semiconductor
BPSK	Binary Phase-Shift Keying
BS	Base Station
BW	Bandwidth
BWP	Bandwidth part
CA	Carrier aggregation
CACLR	Cumulative Adjacent Channel Leakage Ratio
CBG	Codeblock group
CBGFI	CBG flush information
CBGTI	CBG transmit indicator
CC	Component Carrier
CCCH	Common Control Channel
CCE	Control Channel Element
CCSA	China Communications Standards Association
CDM	Code Division Multiplexing
CDMA	Code-Division Multiple Access
CEPT	European Conference of Postal and Telecommunications Administration
CITEL	Inter-American Telecommunication Commission

C-MTC	Critical Machine-Type Communications
CMOS	Complementary Metal Oxide Semiconductor
CN	Core Network
CoMP	Coordinated Multi-Point Transmission/Reception
COREST	Control resource set
CP	Cyclic Prefix
CP	Compression Point
CQI	Channel-Quality Indicator
CRB	Common resource block
CRC	Cyclic Redundancy Check
C-RNTI	Cell Radio-Network Temporary Identifier
CS	Capability Set (for MSR base stations)
CSI	Channel-State Information
CSI-IM	CSI Interference Measurement
CSI-RS	CSI Reference Signals
CS-RNTI	Configured scheduling RNTI
CW	Continuous Wave
D2D	Device-to-Device
DAC	Digital-to-Analog Converter
DAI	Downlink Assignment Index
D-AMPS	Digital AMPS
DC	Dual Connectivity
DC	Direct Current
DCCH	Dedicated Control Channel
DCH	Dedicated Channel
DCI	Downlink Control Information
DFT	Discrete Fourier Transform
DFTS-OFDM	DFT-Spread OFDM (DFT-precoded OFDM, see also SC-FDMA)
DL	Downlink
DL-SCH	Downlink Shared Channel
DM-RS	Demodulation Reference Signal
DR	Dynamic Range
DRX	Discontinuous Reception
DTX	Discontinuous Transmission
EDGE	Enhanced Data Rates for GSM Evolution, Enhanced Data Rates for Global Evolution
ECC	Electronic Communications Committee (of CEPT)
eIMTA	Enhanced Interference Mitigation and Traffic Adaptation
EIRP	Effective Isotropic Radiated Power
EIS	Equivalent Isotropic Sensitivity
eMBB	enhanced MBB
EMF	Electromagnetic Field
eNB	eNodeB
EN-DC	E-UTRA NR Dual-Connectivity
eNodeB	E-UTRAN NodeB
EPC	Evolved Packet Core
ETSI	European Telecommunications Standards Institute
E-UTRA	Evolved UTRA

EVM	Error Vector Magnitude
FCC	Federal Communications Commission
FDD	Frequency Division Duplex
FDM	Frequency Division Multiplexing
FET	Field-Effect Transistor
FDMA	Frequency-Division Multiple Access
FFT	Fast Fourier Transform
FoM	Figure-of-Merit
FPLMTS	Future Public Land Mobile Telecommunications Systems
FR1	Frequency Range 1
FR2	Frequency Range 2
GaAs	Gallium Arsenide
GaN	Gallium Nitride
GERAN	GSM/EDGE Radio Access Network
gNB	gNodeB
gNodeB	generalized NodeB
GSA	Global mobile Suppliers Association
GSM	Global System for Mobile Communications
GSMA	GSM Association
HARQ	Hybrid ARQ
HBT	Heterojunction Bipolar Transistor
HEMT	High Electron-Mobility Transistor
HSPA	High-Speed Packet Access
IC	Integrated Circuit
ICNIRP	International Commission on Non-Ionizing Radiation
ICS	In-Channel Selectivity
IEEE	Institute of Electrical and Electronics Engineers
IFFT	Inverse Fast Fourier Transform
IL	Insertion Loss
IMD	Inter Modulation Distortion
IMT-2000	International Mobile Telecommunications 2000 (ITU's name for the family of 3G standards)
IMT-2020	International Mobile Telecommunications 2020 (ITU's name for the family of 5G standards)
IMT-Advanced	International Mobile Telecommunications Advanced (ITU's name for the family of 4G standards)
InGaP	Indium Gallium Phosphide
IOT	Internet of Things
IP	Internet Protocol
IP3	3rd order Intercept Point
IR	Incremental Redundancy
ITRS	International Telecom Roadmap for Semiconductors
ITU	International Telecommunications Union
ITU-R	International Telecommunications Union-Radiocommunications Sector
KPI	Key Performance Indicator
L1-RSRP	Layer 1 Reference Signal Receiver Power
LC	Inductor(L)-Capacitor

LAA	License-Assisted Access
LCID	Logical Channel Index
LDPC	Low-Density Parity Check Code
LO	Local Oscillator
LNA	Low-Noise Amplifier
LTCC	Low Temperature Co-fired Ceramic
LTE	Long-Term Evolution
MAC	Medium Access Control
MAC-CE	MAC control element
MAN	Metropolitan Area Network
MBB	Mobile Broadband
MB-MSR	Multi-Band Multi Standard Radio (base station)
MCG	Master Cell Group
MCS	Modulation and Coding Scheme
MIB	Master Information Block
MMIC	Monolithic Microwave Integrated Circuit
MIMO	Multiple-Input Multiple-Output
mMTC	massive Machine Type Communication
MPR	Maximum Power Reduction
MSR	Multi-Standard Radio
MTC	Machine-Type Communication
MU-MIMO	Multi-User MIMO
NAK	Negative Acknowledgment (in ARQ protocols)
NB-IoT	Narrow-Band Internet-of-Things
NDI	New-Data Indicator
NEF	Network exposure function
NF	Noise Figure
NG	The interface between the gNB and the 5G CN
NG-c	The control-plane part of NG
NGMN	Next Generation Mobile Networks
NG-u	The user-plane part of NG
NMT	Nordisk MobilTelefon (Nordic Mobile Telephony)
NodeB	NodeB, a logical node handling transmission/reception in multiple cells. Commonly, but not necessarily, corresponding to a base station
NOMA	Nonorthogonal Multiple Access
NR	New Radio
NRF	NR repository function
NS	Network Signaling
NZP-CSI-RS	Non-zero-power CSI-RS
OBUE	Operating Band Unwanted Emissions
OCC	Orthogonal Cover Code
OFDM	Orthogonal Frequency-Division Multiplexing
OOB	Out-Of-Band (emissions)
OSDD	OTA Sensitivity Direction Declarations
OTA	Over-The-Air
PA	Power Amplifier
PAE	Power-Added Efficiency
PAPR	Peak-to-Average Power Ratio

PAR	Peak-to-Average Ratio (same as PAPR)
PBCH	Physical Broadcast Channel
PCB	Printed Circuit Board
PCCH	Paging Control Channel
PCF	Policy control function
PCG	Project Coordination Group (in 3GPP)
PCH	Paging Channel
PCI	Physical Cell Identity
PDC	Personal Digital Cellular
PDCCH	Physical Downlink Control Channel
PDCP	Packet Data Convergence Protocol
PDSCH	Physical Downlink Shared Channel
PDU	Protocol Data Unit
PHS	Personal Handy-phone System
PHY	Physical Layer
PLL	Phase-Locked Loop
PM	Phase Modulation
PMI	Precoding-Matrix Indicator
PN	Phase Noise
PRACH	Physical Random-Access Channel
PRB	Physical Resource Block
P-RNTI	Paging RNTI
PSD	Power Spectral Density
PSS	Primary Synchronization Signal
PUCCH	Physical Uplink Control Channel
PUSCH	Physical Uplink Shared Channel
QAM	Quadrature Amplitude Modulation
QCL	Quasi Co-Location
QoS	Quality-of-Service
QPSK	Quadrature Phase-Shift Keying
RACH	Random Access Channel
RAN	Radio Access Network
RA-RNTI	Random Access RNTI
RAT	Radio Access Technology
RB	Resource Block
RE	Resource Element
RF	Radio Frequency
RFIC	Radio Frequency Integrated Circuit
RI	Rank Indicator
RIB	Radiated Interface Boundary
RIT	Radio Interface Technology
RLC	Radio Link Control
RMSI	Remaining Minimum System Information
RNTI	Radio-Network Temporary Identifier
RoAoA	Range of Angle of Arrival
ROHC	Robust Header Compression
RRC	Radio Resource Control
RRM	Radio Resource Management

RS	Reference Symbol
RSPC	IMT-2000 Radio Interface Specifications
RSRP	Reference Signal Received Power
RV	Redundancy Version
RX	Receiver
SCG	Secondary Cell Group
SCS	Sub-Carrier Spacing
SDL	Supplementary Downlink
SDMA	Spatial Division Multiple Access
SDO	Standards Developing Organization
SDU	Service Data Unit
SEM	Spectrum Emissions Mask
SFI	Slot format indicator
SFI-RNTI	Slot format indicator RNTI
SFN	System Frame Number (in 3GPP).
SI	System Information Message
SIB	System Information Block
SIB1	System Information Block 1
SiGe	Silicon Germanium
SINR	Signal-to-Interference-and-Noise Ratio
SIR	Signal-to-Interference Ratio
SiP	System-in-Package
SI-RNTI	System Information RNTI
SMF	Session management function
SNDR	Signal to Noise-and-Distortion Ratio
SNR	Signal-to-Noise Ratio
SoC	System-on-Chip
SR	Scheduling Request
SRI	SRS resource indicator
SRIT	Set of Radio Interface Technologies
SRS	Sounding Reference Signal
SS	Synchronization Signal
SSB	Synchronization Signal Block
SSS	Secondary Synchronization Signal
SMT	Surface-Mount assembly
SUL	Supplementary Uplink
SU-MIMO	Single-User MIMO
TAB	Transceiver-Array Boundary
TACS	Total Access Communication System
TCI	Transmission configuration indication
TCP	Transmission Control Protocol
TC-RNTI	Temporary C-RNTI
TDD	Time-Division Duplex
TDM	Time Division Multiplexing
TDMA	Time-Division Multiple Access
TD-SCDMA	Time-Division-Synchronous Code-Division Multiple Access
TIA	Telecommunication Industry Association
TR	Technical Report

TRP	Total Radiated Power
TS	Technical Specification
TRS	Tracking Reference Signal
TSDSI	Telecommunications Standards Development Society, India
TSG	Technical Specification Group
TTA	Telecommunications Technology Association
TTC	Telecommunications Technology Committee
TTI	Transmission Time Interval
TX	Transmitter
UCI	Uplink Control Information
UDM	Unified data management
UE	User Equipment, the 3GPP name for the mobile terminal
UEM	Unwanted Emissions Mask
UL	Uplink
UMTS	Universal Mobile Telecommunications System
UPF	User plane function
URLLC	Ultra-reliable low-latency communication
UTRA	Universal Terrestrial Radio Access
V2X	Vehicular-to-Anything
V2V	Vehicular-to-Vehicular
VCO	Voltage-Controlled Oscillator
WARC	World Administrative Radio Congress
WCDMA	Wideband Code-Division Multiple Access
WG	Working Group
WiMAX	Worldwide Interoperability for Microwave Access
WP5D	Working Party 5D
WRC	World Radiocommunication Conference
Xn	The interface between gNBs
ZC	Zadoff-Chu
ZP-CSI-RS	Zero-power CSI-RS

What Is 5G?

Over the last 40 years, the world has witnessed four generations of mobile communication (see Fig. 1.1).

The first generation of mobile communication, emerging around 1980, was based on analog transmission with the main technologies being AMPS (Advanced Mobile Phone System) developed within North America, NMT (Nordic Mobile Telephony) jointly developed by the, at that time, government-controlled public-telephone-network operators of the Nordic countries, and TACS (Total Access Communication System) used in, for example, the United Kingdom. The mobile-communication systems based on first-generation technology were limited to voice services and, for the first time, made mobile telephony accessible to ordinary people.

The second generation of mobile communication, emerging in the early 1990s, saw the introduction of digital transmission on the radio link. Although the target service was still voice, the use of digital transmission allowed for second-generation mobile-communication systems to also provide limited data services. There were initially several different second-generation technologies, including GSM (Global System for Mobile communication) jointly developed by a large number of European countries, D-AMPS (Digital AMPS), PDC (Personal Digital Cellular) developed and solely used in Japan, and, developed at a somewhat later stage, the CDMA-based IS-95 technology. As time went by, GSM spread from Europe to other parts of the world and eventually came to completely dominate among the second-generation technologies. Primarily due to the success of GSM, the second-generation systems also turned mobile telephony from something still being used by only a relatively small fraction of people to a communication tool being a necessary part of life for a large majority of the world's population. Even today there are many places in the world where GSM is the dominating, and in some cases even the only available, technology for mobile communication, despite the later introduction of both third- and fourth-generation technologies.

The third generation of mobile communication, often just referred to as 3G, was introduced in the early 2000. With 3G the true step to high-quality mobile broadband was taken, enabling fast wireless internet access. This was especially enabled by the 3G evolution known as HSPA (High Speed Packet Access) [21]. In addition, while earlier mobile-communication technologies had all been designed for operation in paired spectrum (separate spectrum for network-to-device and device-to-network links) based on the Frequency-Division Duplex

5G NR: The Next Generation Wireless Access Technology. DOI: https://doi.org/10.1016/B978-0-12-814323-0.00001-6

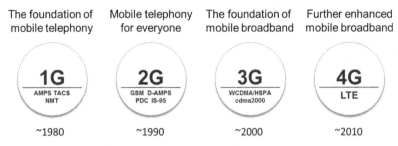

FIGURE 1.1

The different generations of mobile communication.

(FDD), see Chapter 7, 3G also saw the first introduction of mobile communication in unpaired spectrum based on the china-developed TD-SCDMA technology based on Time Division Duplex (TDD).

We are now, and have been for several years, in the fourth-generation (4G) era of mobile communication, represented by the LTE technology [28] LTE has followed in the steps of HSPA, providing higher efficiency and further enhanced mobile-broadband experience in terms of higher achievable end-user data rates. This is provided by means of OFDM-based transmission enabling wider transmission bandwidths and more advanced multi-antenna technologies. Furthermore, while 3G allowed for mobile communication in unpaired spectrum by means of a specific radio-access technology (TD-SCDMA), LTE supports both FDD and TDD operation, that is, operation in both paired and unpaired spectra, within one common radio-access technology. By means of LTE the world has thus converged into a single global technology for mobile communication, used by essentially all mobile-network operators and applicable to both paired and unpaired spectra. As discussed in somewhat more detail in Chapter 4, the later evolution of LTE has also extended the operation of mobile-communication networks into unlicensed spectra.

1.1 3GPP AND THE STANDARDIZATION OF MOBILE COMMUNICATION

Agreeing on multi-national technology specifications and standards has been key to the success of mobile communication. This has allowed for the deployment and interoperability of devices and infrastructure of different vendors and enabled devices and subscriptions to operate on a global basis.

As already mentioned, already the first-generation NMT technology was created on a multinational basis, allowing for devices and subscription to operate over the national borders between the Nordic countries. The next step in multi-national specification/standardization of mobile-communication technology took place when GSM was jointly developed between a large number of European countries within CEPT, later renamed ETSI (European Telecommunications Standards Institute). As a consequence of this, GSM devices and subscriptions were already from the beginning able to operate over a large number of countries, covering a very large number of potential users. This large common market had a profound impact on device availability, leading to an unprecedented number of different device types and substantial reduction in device cost.

However, the final step to true global standardization of mobile communication came with the specification of the 3G technologies, especially WCDMA. Work on 3G technology was initially also carried out on a regional basis, that is, separately within Europe (ETSI), North America (TIA, T1P1), Japan (ARIB), etc. However, the success of GSM had shown the importance of a large technology footprint, especially in terms of device availability and cost. It also become clear that although work was carried out separately within the different regional standard organizations, there were many similarities in the underlying technology being pursued. This was especially true for Europe and Japan which were both developing different but very similar flavors of *wideband CDMA* (WCDMA) technology.

As a consequence, in 1998, the different regional standardization organizations came together and jointly created the *Third-Generation Partnership Project* (3GPP) with the task of finalizing the development of 3G technology based on WCDMA. A parallel organization (3GPP2) was somewhat later created with the task of developing an alternative 3G technology, cdma2000, as an evolution of second-generation IS-95. For a number of years, the two organizations (3GPP and 3GPP2) with their respective 3G technologies (WCDMA and cdma2000) existed in parallel. However, over time 3GPP came to completely dominate and has, despite its name, continued into the development of 4G (LTE, and 5G) technologies. Today, 3GPP is the only significant organization developing technical specifications for mobile communication.

1.2 THE NEXT GENERATION—5G/NR

Discussions on *fifth*-generation (5G) mobile communication began around 2012. In many discussions, the term 5G is used to refer to specific new 5G radio-access technology. However, 5G is also often used in a much wider context, not just referring to a specific radio-access technology but rather to a wide range of new services envisioned to be enabled by future mobile communication.

1.2.1 THE 5G USE CASES

In the context of 5G, one is often talking about three distinctive classes of use cases: *enhanced mobile broadband* (eMBB), *massive machine-type communication* (mMTC), and *ultra-reliable and low-latency communication* (URLLC) (see also Fig. 1.2).

- eMBB corresponds to a more or less straightforward evolution of the mobile-broadband services of today, enabling even larger data volumes and further enhanced user experience, for example, by supporting even higher end-user data rates.
- mMTC corresponds to services that are characterized by a massive number of devices, for example, remote sensors, actuators, and monitoring of various equipment. Key requirements for such services include very low device cost and very low device energy consumption, allowing for very long device battery life of up to at least several years. Typically, each device consumes and generates only a relatively small amount of data, that is, support for high data rates is of less importance.
- URLLC type-of-services are envisioned to require very low latency and extremely high reliability. Examples hereof are traffic safety, automatic control, and factory automation.

It is important to understand that the classification of 5G use cases into these three distinctive classes is somewhat artificial, primarily aiming to simplify the definition of requirements for the technology specification. There will be many use cases that do not fit exactly into one of these classes. Just as an example, there may be services that require very high reliability but for which the latency requirements are not that critical. Similarly, there may be use cases requiring devices of very low cost but where the possibility for very long device battery life may be less important.

eMBB

High data rates, high traffic volumes

mMTC

Massive number of devices,
low cost, low energy consumption

URLLC

Very low latency,
very high reliability and availability

FIGURE 1.2

High-level 5G use-case classification.

1.2.2 **EVOLVING LTE TO 5G CAPABILITY**

The first release of the LTE technical specifications was introduced in 2009. Since then, LTE has gone through several steps of evolution providing enhanced performance and extended capabilities. This has included features for enhanced mobile broadband, including means for higher achievable end-user data rates as well as higher spectrum efficiency. However, it has also included important steps to extend the set of use cases to which LTE can be applied. Especially, there have been important steps to enable truly low-cost devices with very long battery life, in line with the characteristics of massive MTC applications. There have recently also been some significant steps taken to reduce the LTE air-interface latency.

With these finalized, ongoing, and future evolution steps, the evolution of LTE will be able to support a wide range of the use cases envisioned for 5G. Taking into account the more general view that 5G is not a specific radio-access technology but rather defined by the use cases to be supported, the evolution of LTE should thus be seen as an important part of the overall 5G radio-access solution, see Fig. 1.3. Although not being the main aim of this book, an overview of the current state of the LTE evolution is provided in Chapter 4.

1.2.3 **NR—THE NEW 5G RADIO-ACCESS TECHNOLOGY**

Despite LTE being a very capable technology, there are requirements not possible to meet with LTE or its evolution. Furthermore, technology development over the more than 10 years that have passed since the work on LTE was initiated allows for more advanced technical solutions. To meet these requirements and to exploit the potential of new technologies, 3GPP initiated the development of a new radio-access technology known as NR (New Radio). A workshop setting the scope was held in the fall of 2015 and technical work began in the spring of 2016. The first version of the NR specifications was available by the end of 2017 to meet commercial requirements on early 5G deployments already in 2018.

NR reuses many of the structures and features of LTE. However, being a new radio-access technology means that NR, unlike the LTE evolution, is not

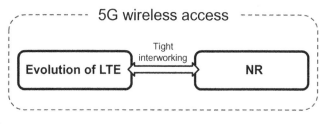

FIGURE 1.3

Evolution of LTE and NR jointly providing the overall 5G radio-access solution.

restricted by a need to retain backwards compatibility. The requirements on NR are also broader than what was the case for LTE, motivating a partly different set of technical solutions.

Chapter 2 discusses the standardization activities related to NR, followed by a spectrum overview in Chapter 3 and a brief summary of LTE and its evolution in Chapter 4. The main part of this book (Chapters 5−19) then provides an in-depth description of the current stage of the NR technical specifications, finishing with an outlook of the future development of NR in Chapter 20.

1.2.4 5GCN—THE NEW 5G CORE NETWORK

In parallel to NR, that is, the new 5G radio-access technology, 3GPP is also developing a new 5G core network referred to as 5GCN. The new 5G radio-access technology will connect to the 5GCN. However, 5GCN will also be able to provide connectivity for the evolution of LTE. At the same time, NR may also connect via the legacy core network EPC when operating in so-called *non-standalone mode* together will LTE, as will be further discussed in Chapter 6.

5G Standardization

The research, development, implementation, and deployment of mobile-communication systems is performed by the wireless industry in a coordinated international effort by which common industry specifications that define the complete mobile-communication system are agreed. The work depends heavily on global and regional regulation, in particular for the spectrum use that is an essential component for all radio technologies. This chapter describes the regulatory and standardization environment that has been, and continues to be, essential for defining the mobile-communication systems.

2.1 OVERVIEW OF STANDARDIZATION AND REGULATION

There are a number of organizations involved in creating technical specifications and standards as well as regulation in the mobile-communications area. These can loosely be divided into three groups: Standards Developing Organizations, regulatory bodies and administrations, and industry forums.

Standards Developing Organizations (SDOs) develop and agree on technical standards for mobile communications systems, in order to make it possible for the industry to produce and deploy standardized products and provide interoperability between those products. Most components of mobile-communication systems, including base stations and mobile devices, are standardized to some extent. There is also a certain degree of freedom to provide proprietary solutions in products, but the communications protocols rely on detailed standards for obvious reasons. SDOs are usually nonprofit industry organizations and not government controlled. They often write standards within a certain area under mandate from governments(s) however, giving the standards a higher status.

There are nationals SDOs, but due to the global spread of communications products, most SDOs are regional and also cooperate on a global level. As an example, the technical specifications of GSM, WCDMA/HSPA, LTE, and NR are all created by 3GPP (Third Generation Partnership Project) which is a global organization from seven regional and national SDOs in Europe (ETSI), Japan (ARIB and TTC), the United States (ATIS), China (CCSA), Korea (TTA), and India (TSDSI). SDOs tend to have a varying degree of transparency, but 3GPP is fully transparent with all technical specifications, meeting documents, reports, and e-mail reflectors publicly available without charge even for nonmembers.

Regulatory bodies and administrations are government-led organizations that set regulatory and legal requirements for selling, deploying, and operating mobile systems and other telecommunication products. One of their most important tasks is to control spectrum use and to set licensing conditions for the mobile operators that are awarded licenses to use parts of the *Radio Frequency* (RF) spectrum for mobile operations. Another task is to regulate "placing on the market" of products through regulatory certification, by ensuring that devices, base stations, and other equipment is type-approved and shown to meet the relevant regulation.

Spectrum regulation is handled both on a national level by national administrations, but also through regional bodies in Europe (CEPT/ECC), the Americas (CITEL), and Asia (APT). On a global level, the spectrum regulation is handled by the *International Telecommunications Union* (ITU). The regulatory bodies regulate what services the spectrum is to be used for and in addition set more detailed requirements such as limits on unwanted emissions from transmitters. They are also indirectly involved in setting requirements on the product standards through regulation. The involvement of ITU in setting requirements on the technologies for mobile communication is explained further in Section 2.2.

Industry forums are industry-led groups promoting and lobbying for specific technologies or other interests. In the mobile industry, these are often led by operators, but there are also vendors creating industry forums. An example of such a group is GSMA (GSM Association) which is promoting mobile-communication technologies based on GSM, WCDMA, LTE, and NR. Other examples of industry forums are *Next Generation Mobile Networks* (NGMN), which is an operator group defining requirements on the evolution of mobile systems, and *5G Americas*, which is a regional industry forum that has evolved from its predecessor 4G Americas.

Fig. 2.1 illustrates the relationship between different organizations involved in setting regulatory and technical conditions for mobile systems. The figure also shows the mobile industry view, where vendors develop products, place them on the market and negotiate with operators who procure and deploy mobile systems. This process relies heavily on the technical standards published by the SDOs, while placing products on the market relies on certification of products on a regional or national level. Note that, in Europe, the regional SDO (ETSI) is producing the so-called *harmonized standards* used for product certification (through the "CE"-mark), based on a mandate from the regulators, in this case the European Commission. These standards are also used for certification in many countries outside of Europe. In Fig. 2.1, full arrows indicate formal documentation such as technical standards, recommendations, and regulatory mandates that define the technologies and regulation. Dashed arrows show more indirect involvement through, for example, liaison statements and white papers.

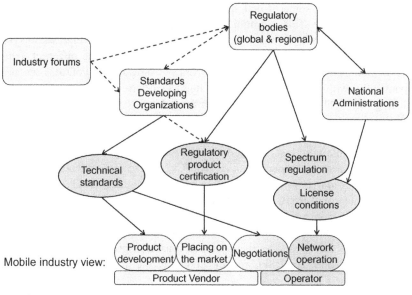

FIGURE 2.1

Simplified view of the relationship between regulatory bodies, standards developing organizations, industry forums, and the mobile industry.

2.2 ITU-R ACTIVITIES FROM 3G TO 5G

2.2.1 THE ROLE OF ITU-R

ITU-R is the radio communications sector of the International Telecommunications Union. ITU-R is responsible for ensuring efficient and economical use of the RF spectrum by all radio communication services. The different subgroups and working parties produce reports and recommendations that analyze and define the conditions for using the RF spectrum. The quite ambitious goal of ITU-R is to "ensure interference-free operations of radio communication systems," by implementing the *Radio Regulations* and regional agreements. The Radio Regulations is an international binding treaty for how RF spectrum is used. A *World Radio-communication Conference* (WRC) is held every 3−4 years. At WRC the Radio Regulations are revised and updated, resulting in revised and updated use of the RF spectrum across the world.

While the technical specification of mobile-communication technologies, such as NR, LTE, and WCDMA/HSPA is done within 3GPP, there is a responsibility for ITU-R in the process of turning the technologies into global standards, in particular for countries that are not covered by the SDOs that are partners in 3GPP. ITU-R defines the spectrum for different services in the RF spectrum, including mobile services, and some of that spectrum is particularly identified for so-called International Mobile Telecommunications (IMT) systems. Within ITU-R, it is *Working Party 5D* (WP5D) that has the responsibility for the overall radio system aspects of IMT systems, which, in practice, corresponds to the different generations of mobile-communication systems from 3G onwards. WP5D has the prime responsibility within ITU-R for issues related to the terrestrial component of IMT, including technical, operational, and spectrum-related issues.

WP5D does not create the actual technical specifications for IMT, but has kept the roles of defining IMT in cooperation with the regional standardization bodies and maintaining a set of recommendations and reports for IMT, including a set of *Radio Interface Specifications* (RSPCs). These recommendations contain "families" of *Radio Interface Technologies* (RITs) for each IMT generation, all included on an equal basis. For each radio interface, the RSPC contains an overview of that radio interface, followed by a list of references to the detailed specifications. The actual specifications are maintained by the individual SDO and the RSPC provides references to the specifications transposed and maintained by each SDO. The following RSPC recommendations are in existence or planned:

- For IMT-2000: ITU-R Recommendation M.1457 [49] containing six different RITs including the 3G technologies such as WCDMA/HSPA.
- For IMT-Advanced: ITU-R Recommendation M.2012 [45] containing two different RITs where the most important is 4G/LTE.
- For IMT-2020: A new ITU-R Recommendation, containing the RITs for 5G technologies, planned to be developed in 2019–20.

Each RSPC is continuously updated to reflect new developments in the referenced detailed specifications, such as the 3GPP specifications for WCDMA and LTE. Input to the updates is provided by the SDOs and the Partnership Projects, nowadays primarily 3GPP.

2.2.2 IMT-2000 AND IMT-ADVANCED

Work on what corresponds to third generation of mobile communication started in the ITU-R in the 1980s. First referred to as *Future Public Land Mobile Systems* (FPLMTS) it was later renamed IMT-2000. In the late 1990s, the work in ITU-R coincided with the work in different SDOs across the world to develop a new generation of mobile systems. An RSPC for IMT-2000 was first published in 2000 and included WCDMA from 3GPP as one of the RITs.

The next step for ITU-R was to initiate work on IMT-Advanced, the term used for systems that include new radio interfaces supporting new capabilities of systems beyond IMT-2000. The new capabilities were defined in a framework recommendation published by the ITU-R [41] and were demonstrated with the "van diagram" shown in Fig. 2.2. The step into IMT-Advanced capabilities by ITU-R coincided with the step into 4G, the next generation of mobile technologies after 3G.

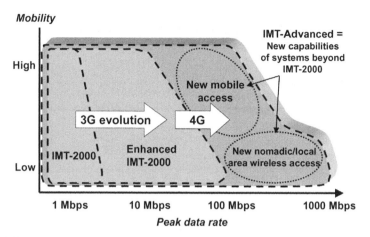

FIGURE 2.2

Illustration of capabilities of IMT-2000 and IMT-Advanced, based on the framework described in ITU-R Recommendation M.1645 [41].

An evolution of LTE as developed by 3GPP was submitted as one candidate technology for IMT-Advanced. While actually being a new release (Release 10) of the LTE specifications and thus an integral part of the continuous evolution of LTE, the candidate was named LTE-Advanced for the purpose of ITU-R submission and this name is also used in the LTE specifications from Release 10. In parallel with the ITU-R work, 3GPP set up its own set of technical requirements for LTE-Advanced, with the ITU-R requirements as a basis [10].

The target of the ITU-R process is always harmonization of the candidates through consensus building. ITU-R determined that two technologies would be included in the first release of IMT-Advanced, those two being LTE-Advanced and WirelessMAN-Advanced [37] based on the IEEE 802.16m specification. The two can be viewed as the "family" of IMT-Advanced technologies as shown in Fig. 2.3. Note that, of these two technologies, LTE has emerged as the dominating 4G technology by far.

2.2.3 IMT-2020 PROCESS IN ITU-R WP5D

Starting in 2012, ITU-R WP5D set the stage for the next generation of IMT systems, named IMT-2020. It is a further development of the terrestrial component of IMT beyond the year 2020 and, in practice, corresponds to what is more commonly referred to as "5G," the fifth generation of mobile systems. The framework and objective for IMT-2020 is outlined in ITU-R Recommendation M.2083 [47], often referred to as the "Vision" recommendation. The recommendation provides the first step for defining the new developments of IMT, looking at the future roles of IMT and how it can serve society, looking at market, user and technology trends,

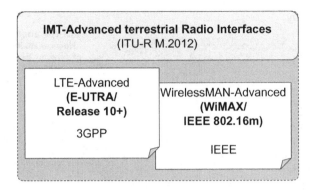

FIGURE 2.3

Radio Interface Technologies IMT-Advanced.

and spectrum implications. The user trends for IMT together with the future role and market lead to a set of *usage scenarios* envisioned for both human-centric and machine-centric communication. The usage scenarios identified are *Enhanced Mobile Broadband* (eMBB), *Ultra-Reliable and Low Latency Communications* (URLLC), and *Massive Machine-Type Communications* (mMTC).

The need for an enhanced mobile broadband experience, together with the new and broadened usage scenarios, leads to an extended set of capabilities for IMT-2020. The Vision recommendation [47] gives a first high-level guidance for IMT-2020 requirements by introducing a set of key capabilities, with indicative target numbers. The key capabilities and the related usage scenarios are discussed further in Section 2.3.

As a parallel activity, ITU-R WP5D produced a report on "Future technology trends of terrestrial IMT systems" [43], with a focus on the time period 2015–20. It covers trends of future IMT technology aspects by looking at the technical and operational characteristics of IMT systems and how they are improved with the evolution of IMT technologies. In this way, the report on technology trends relates to LTE in 3GPP Release 13 and beyond, while the Vision recommendation looks further ahead and beyond 2020. A new aspect on IMT-2020 is that it will be capable of operating in potential new IMT bands above 6 GHz, including mm-wave bands. With this in mind, WP5D produced a separate report studying radio wave propagation, IMT characteristics, enabling technologies, and deployment in frequencies above 6 GHz [44].

At WRC-15, potential new bands for IMT were discussed and an agenda item 1.13 was set up for WRC-19, covering possible additional allocations to the mobile services and for future IMT development. These allocations are identified in a number of frequency bands in the range between 24.25 and 86 GHz. The specific bands and their possible use globally are further discussed in Chapter 3.

After WRC-15, ITU-R WP5D continued the process of setting requirements and defining evaluation methodologies for IMT-2020 systems, based in the

Vision recommendation [47] and the other previous study outcomes. This step of the process was completed in mid-2017, as shown in the IMT-2020 work plan in Fig. 2.4. The result was three documents published late in 2017 that further define the performance and characteristics that are expected from IMT-2020 and that will be applied in the evaluation phase:

- *Technical requirements:* Report ITU-R M.2410 [51] defines 13 minimum requirements related to the technical performance of the IMT-2020 radio interface(s). The requirements are to a large extent based on the key capabilities set out in the Vision recommendation (ITU-R, 2015c). This is further described in Section 2.3.
- *Evaluation guideline:* Report ITU-R M.2412 [50] defines the detailed methodology to use for evaluating the minimum requirements, including test environments, evaluation configurations, and channel models. More details are given in Section 2.3.
- *Submission template:* Report ITU-R M.2411 [52] provides a detailed template to use for submitting a candidate technology for evaluation. It also details the evaluation criteria and requirements on service, spectrum, and technical performance, based on the two previously mentioned ITU-R reports M.2410 and M.2412.

External organizations are being informed of the IMT-2020 process through a circular letter. After a workshop on IMT-2020 was held in October 2017, the IMT-2020 process is open for receiving candidate proposals.

The plan, as shown in Fig. 2.4, is to start the evaluation of proposals in 2018, aiming at an outcome with the RSPC for IMT-2020 being published early in 2020.

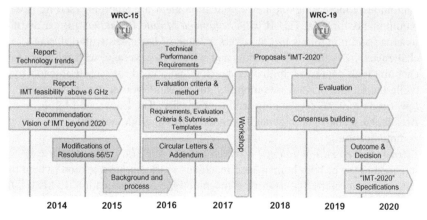

FIGURE 2.4

Work plan for IMT-2020 in ITU-R WP5D [40].

2.3 5G AND IMT-2020

The detailed ITU-R time plan for IMT-2020 was presented above with the most important steps summarized in Fig. 2.4. The ITU-R activities on IMT-2020 started with development of the "vision" recommendation ITU-R M.2083 [47], outlining the expected use scenarios and corresponding required capabilities of IMT-2020. This was followed by definition of more detailed requirements for IMT-2020, requirements that candidate technologies are then to be evaluated against, as documented in the evaluation guidelines. The requirements and evaluation guidelines were finalized mid-2017.

With the requirements finalized, candidate technologies can be submitted to ITU-R. The proposed candidate technology/technologies will be evaluated against the IMT-2020 requirements and the technology/technologies that fulfill the requirements will be approved and published as part of the IMT-2020 specifications in the second half of 2020. Further details on the ITU-R process can be found in Section 2.2.3.

2.3.1 USAGE SCENARIOS FOR IMT-2020

With a wide range of new use cases being one principal driver for 5G, ITU-R has defined three usage scenarios that form a part of the IMT Vision recommendation [47]. Inputs from the mobile industry and different regional and operator organizations were taken into the IMT-2020 process in ITU-R WP5D, and were synthesized into the three scenarios:

- *Enhanced Mobile Broadband (eMBB):* With mobile broadband today being the main driver for use of 3G and 4G mobile systems, this scenario points at its continued role as the most important usage scenario. The demand is continuously increasing and new application areas are emerging, setting new requirements for what ITU-R calls *Enhanced Mobile Broadband*. Because of its broad and ubiquitous use, it covers a range of use cases with different challenges, including both hotspots and wide-area coverage, with the first one enabling high data rates, high user density, and a need for very high capacity, while the second one stresses mobility and a seamless user experience, with lower requirements on data rate and user density. The Enhanced Mobile Broadband scenario is in general seen as addressing human-centric communication.
- *Ultra-reliable and low-latency communications (URLLC):* This scenario is intended to cover both human- and machine-centric communication, where the latter is often referred to as critical machine type communication (C-MTC). It is characterized by use cases with stringent requirements for latency, reliability, and high availability. Examples include vehicle-to-vehicle communication involving safety, wireless control of industrial equipment, remote medical surgery, and distribution automation in a smart grid. An

example of a human-centric use case is 3D gaming and "tactile internet," where the low-latency requirement is also combined with very high data rates.

- *Massive machine type communications (mMTC):* This is a pure machine-centric use case, where the main characteristic is a very large number of connected devices that typically have very sparse transmissions of small data volumes that are not delay-sensitive. The large number of devices can give a very high connection density locally, but it is the total number of devices in a system that can be the real challenge and stresses the need for low cost. Due to the possibility of remote deployment of mMTC devices, they are also required to have a very long battery life time.

The usage scenarios are illustrated in Fig. 2.5, together with some example use cases. The three scenarios above are not claimed to cover all possible use cases, but they provide a relevant grouping of a majority of the presently foreseen use cases and can thus be used to identify the key capabilities needed for the next-generation radio interface technology for IMT-2020. There will most certainly be new use cases emerging, which we cannot foresee today or describe in any detail. This also means that the new radio interface must have a high flexibility to adapt to new use cases and the "space" spanned by the range of the key capabilities supported should support the related requirements emerging from evolving use cases.

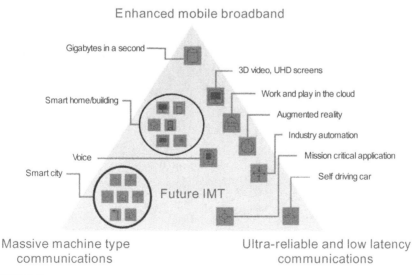

FIGURE 2.5

IMT-2020 use cases and mapping to usage scenarios.

From ITU-R, Recommendation ITU-R M.2083 [47], used with permission from the ITU.

2.3.2 CAPABILITIES OF IMT-2020

As part of developing the framework for the IMT-2020 as documented in the IMT Vision recommendation [47], ITU-R defined a set of capabilities needed for an IMT-2020 technology to support the 5G use cases and usage scenarios identified through the inputs from regional bodies, research projects, operators, administrations, and other organizations. There are a total of 13 capabilities defined in ITU-R [47], where eight were selected as *key capabilities*. Those eight key capabilities are illustrated through two "spider web" diagrams (see Figs. 2.6 and 2.7).

Fig. 2.6 illustrates the key capabilities together with indicative target numbers intended to give a first high-level guidance for the more detailed IMT-2020 requirements that are now under development. As can be seen the target values are partly absolute and partly relative to the corresponding capabilities of IMT-Advanced. The target values for the different key capabilities do not have to be reached simultaneously and some targets are to a certain extent even mutually exclusive. For this reason, there is a second diagram shown in Fig. 2.7 which illustrates the "importance" of each key capability for realizing the three high-level usage scenarios envisioned by ITU-R.

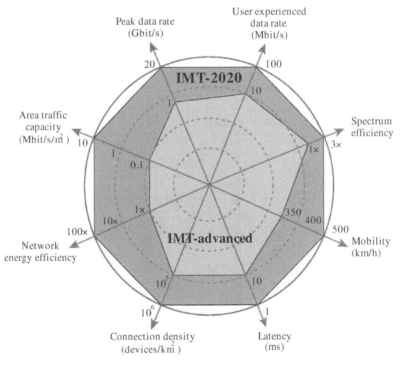

FIGURE 2.6

Key capabilities of IMT-2020.

From ITU-R, Recommendation ITU-R M.2083 [47], used with permission from the ITU.

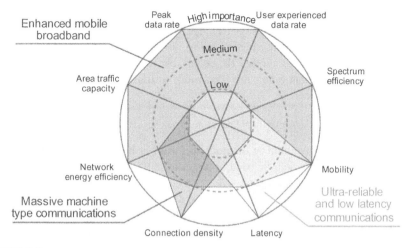

FIGURE 2.7

Relation between key capabilities and the three usage scenarios of ITU-R.

From ITU-R, Recommendation ITU-R M.2083 [47], used with permission from the ITU.

Peak data rate is a number on which there is always a lot of focus, but it is in fact quite an academic exercise. ITU-R defines peak data rates as the maximum achievable data rate under ideal conditions, which means that the impairments in an implementation or the actual impact from a deployment in terms of propagation, etc. does not come into play. It is a dependent *key performance indicator* (KPI) in that it is heavily dependent on the amount of spectrum available for an operator deployment. Apart from that, the peak data rate depends on the peak spectral efficiency, which is the peak data rate normalized by the bandwidth:

$$\text{Peak data rate} = \text{System bandwidth} \times \text{Peak spectral efficiency}$$

Since large bandwidths are really not available in any of the existing IMT bands below 6 GHz, it is expected that really high data rates will be more easily achieved at higher frequencies. This leads to the conclusion that the highest data rates can be achieved in indoor and hotspot environments, where the less favorable propagation properties at higher frequencies are of less importance.

The *user experienced data rate* is the data rate that can be achieved over a large coverage area for a majority of the users. This can be evaluated as the 95th percentile from the distribution of data rates between users. It is also a dependent capability, not only on the available spectrum but also on how the system is deployed. While a target of 100 Mbit/s is set for wide area coverage in urban and suburban areas, it is expected that 5G systems could give 1 Gbit/s data rate ubiquitously in indoor and hotspot environments.

Spectrum efficiency gives the average data throughput per Hz of spectrum and per "cell," or rather per unit of radio equipment (also referred to as *Transmission Reception Point*, TRP). It is an essential parameter for dimensioning networks, but the levels achieved with 4G systems are already very high. The target was set to three times the spectrum efficiency target of 4G, but the achievable increase strongly depends on the deployment scenario.

Area traffic capacity is another dependent capability, which depends not only on the spectrum efficiency and the bandwidth available, but also on how dense the network is deployed:

$$\text{Area Traffic Capacity} = \text{Spectrum efficiency} \cdot \text{BW} \cdot \text{TRP density}$$

By assuming the availability of more spectrum at higher frequencies and that very dense deployments can be used, a target of a 100-fold increase over 4G was set for IMT-2020.

Network energy efficiency is, as already described, becoming an increasingly important capability. The overall target stated by ITU-R is that the energy consumption of the radio access network of IMT-2020 should not be greater than IMT networks deployed today, while still delivering the enhanced capabilities. The target means that the network energy efficiency in terms of energy consumed per bit of data therefore needs to be reduced with a factor at least as great as the envisaged traffic increase of IMT-2020 relative to IMT-Advanced.

These first five key capabilities are of highest importance for the Enhanced Mobile Broadband usage scenario, although mobility and the data rate capabilities would not have equal importance simultaneously. For example, in hotspots, a very high user-experienced and peak data rate, but a lower mobility, would be required than in wide area coverage case.

Latency is defined as the contribution by the radio network to the time from when the source sends a packet to when the destination receives. It will be an essential capability for the URLLC usage scenario and ITU-R envisions that a 10-fold reduction in latency from IMT-Advanced is required.

Mobility is in the context of key capabilities only defined as mobile speed and the target of 500 km/h is envisioned in particular for high-speed trains and is only a moderate increase from IMT-Advanced. As a key capability, it will, however, also be essential for the URLLC usage scenario in the case of critical vehicle communication at high speed and will then be of high importance simultaneously with low latency. Note that mobility and high user-experienced data rates are not targeted simultaneously in the usage scenarios.

Connection density is defined as the total number of connected and/or accessible devices per unit area. The target is relevant for the mMTC usage scenario with a high density of connected devices, but an eMBB dense indoor office can also give a high connection density.

In addition to the eight capabilities given in Fig. 2.6 there are five additional capabilities defined in [47]:

- *Spectrum and bandwidth flexibility*
    Spectrum and bandwidth flexibility refers to the flexibility of the system design to handle different scenarios, and in particular to the capability to

operate at different frequency ranges, including higher frequencies and wider channel bandwidths than today.

- *Reliability*

    Reliability relates to the capability to provide a given service with a very high level of availability.

- *Resilience*

    Resilience is the ability of the network to continue operating correctly during and after a natural or man-made disturbance, such as the loss of mains power.

- *Security and privacy*

    Security and privacy refers to several areas such as encryption and integrity protection of user data and signaling, as well as end-user privacy, preventing unauthorized user tracking, and protection of network against hacking, fraud, denial of service, man in the middle attacks, etc.

- *Operational lifetime*

    Operational life time refers to operation time per stored energy capacity. This is particularly important for machine-type devices requiring a very long battery life (for example more than 10 years), whose regular maintenance is difficult due to physical or economic reasons.

Note that these capabilities are not necessarily less important than the capabilities of Fig. 2.6, despite the fact that the latter are referred to as "key capabilities." The main difference is that the "key capabilities" are more easily quantifiable, while the remaining five capabilities are more of qualitative capabilities that cannot easily be quantified.

2.3.3 IMT-2020 PERFORMANCE REQUIREMENTS AND EVALUATION

Based on the usage scenarios and capabilities described in the Vision recommendation (ITU-R, 2015c), ITU-R developed a set of minimum technical performance requirements for IMT-2020. These are documented in ITU-R report M.2410 [51] and will serve as the baseline for the evaluation of IMT-2020 candidate technologies (see Fig. 2.4). The report describes 14 technical parameters and the corresponding minimum requirements. These are summarized in Table 2.1.

The evaluation guideline of candidate radio interface technologies for IMT-2020 is documented in ITU-R report M.2412 [50] and follows the same structure as the previous evaluation done for IMT-Advanced. It describes the evaluation methodology for the 14 minimum technical performance requirements, plus two additional requirements: support of a wide range of services and support of spectrum bands.

The evaluation is done with reference to five *test environments* that are based on the usage scenarios from the Vision recommendation [47]. Each test environment has a number of *evaluation configurations* that describe the detailed

Table 2.1 Overview of Minimum Technical Performance Requirements for IMT-2020

Parameter	Minimum Technical Performance Requirement
Peak data rate	Downlink: 20 Gbit/s
	Uplink: 10 Gbit/s
Peak spectral efficiency	Downlink: 30 bit/s/Hz
	Uplink: 10 bit/s/Hz
User-experienced data rate	Downlink: 100 Mbit/s
	Uplink: 50 Mbit/s
Fifth percentile user spectral efficiency	$3 \times$ IMT-Advanced
Average spectral efficiency	$3 \times$ IMT-Advanced
Area traffic capacity	10 Mbit/s/m^2 (indoor hotspot for eMBB)
User plane latency	4 ms for eMBB
	1 ms for URLLC
Control plane latency	20 ms
Connection density	1,000,000 devices per km^2
Energy efficiency	Related to two aspects for eMBB:
	a. Efficient data transmission in a loaded case
	b. Low energy consumption when there is no data
	The technology shall have the capability to support a high sleep ratio and long sleep duration
Reliability	$1-10^{-5}$ success probability of transmitting a layer 2 PDU (Protocol Data Unit) of 32 bytes within 1 ms, at coverage edge in Urban Macro for URLLC
Mobility	Normalized traffic channel data rates defined for 10, 30, and 120 km/h at $\sim 1.5 \times$ IMT-Advanced numbers
	Requirement for high-speed vehicular defined for 500 km/h (compared to 350 km/h for IMT-Advanced)
Mobility interruption time	0 ms
Bandwidth	At least 100 MHz and up to 1 GHz in higher-frequency bands. Scalable bandwidth shall be supported

parameters that are to be used in simulations and analysis for the evaluation. The five test environments are:

- *Indoor Hotspot-eMBB:* An indoor isolated environment at offices and/or in shopping malls based on stationary and pedestrian users with very high user density.
- *Dense Urban-eMBB:* An urban environment with high user density and traffic loads focusing on pedestrian and vehicular users.
- *Rural-eMBB:* A rural environment with larger and continuous wide area coverage, supporting pedestrian, vehicular, and high-speed vehicular users.
- *Urban Macro-mMTC:* An urban macro-environment targeting continuous coverage focusing on a high number of connected machine type devices.

- *Urban Macro-URLLC:* An urban macro-environment targeting ultra-reliable and low-latency communications.

There are three fundamental ways that requirements will be evaluated for a candidate technology:

- *Simulation:* This is the most elaborate way to evaluate a requirement and it involves system- or link-level simulations, or both, of the radio interface technology. For system-level simulations, deployment scenarios are defined that correspond to a set of test environments, such as indoor, dense urban, etc. Requirements that will be evaluated through simulation are average and fifth percentile spectrum efficiency, connection density, mobility and reliability.
- *Analysis:* Some requirements can be evaluated through a calculation based on radio interface parameters or be derived from other performance values. Requirements that will be evaluated through analysis are peak spectral efficiency, peak data rate, user-experienced data rate, area traffic capacity, control and user plane latency, and mobility interruption time.
- *Inspection:* Some requirements can be evaluated by reviewing and assessing the functionality of the radio interface technology. Requirements that will be evaluated through simulation are bandwidth, energy efficiency, support of a wide range of services, and support of spectrum bands.

Once candidate technologies are submitted to ITU-R and have entered the process, the evaluation phase will start. Evaluation can be done by the proponent ("self-evaluation") or by an external evaluation group, doing partial or complete evaluation of one or more candidate proposals.

2.4 3GPP STANDARDIZATION

With a framework for IMT systems set up by the ITU-R, with spectrum made available by the WRC and with an ever-increasing demand for better performance, the task of specifying the actual mobile-communication technologies falls on organizations like 3GPP. More specifically, 3GPP writes the technical specifications for 2G GSM, 3G WCDMA/HSPA, 4G LTE, and 5G NR. 3GPP technologies are the most widely deployed in the world, with more than 95% of the world's 7.8 billion mobile subscriptions in Q4 2017 [30]. In order to understand how 3GPP works, it is important to also understand the process of writing specifications.

2.4.1 THE 3GPP PROCESS

Developing technical specifications for mobile communication is not a one-time job; it is an ongoing process. The specifications are constantly evolving, trying to meet new demands for services and features. The process is different in the different fora, but typically includes the four phases illustrated in Fig. 2.8:

1. *Requirements*, where it is decided what is to be achieved by the specification.
2. *Architecture*, where the main building blocks and interfaces are decided.
3. *Detailed specifications*, where every interface is specified in detail.
4. *Testing and verification*, where the interface specifications are proven to work with real-life equipment.

These phases are overlapping and iterative. As an example, requirements can be added, changed, or dropped during the later phases if the technical solutions call for it. Likewise, the technical solution in the detailed specifications can change due to problems found in the testing and verification phase.

The specification starts with the *requirements* phase, where it is decided what should be achieved with the specification. This phase is usually relatively short.

In the *architecture* phase, the architecture is decided—that is, the principles of how to meet the requirements. The architecture phase includes decisions about reference points and interfaces to be standardized. This phase is usually quite long and may change the requirements.

After the architecture phase, the *detailed specification* phase starts. It is in this phase that the details for each of the identified interfaces are specified. During the detailed specification of the interfaces, the standards body may find that previous decisions in the architecture or even in the requirements phases need to be revisited.

Finally, the *testing and verification* phase starts. It is usually not a part of the actual specification, but takes place in parallel through testing by vendors and interoperability testing between vendors. This phase is the final proof of the specification. During the testing and verification phase, errors in the specification may still be found and those errors may change decisions in the detailed specification. Albeit not common, changes may also need to be made to the architecture or the requirements. To verify the specification, products are needed. Hence, the implementation of the products starts after (or during) the detailed specification phase. The testing and verification phase ends when there are stable test specifications that can be used to verify that the equipment is fulfilling the technical specification.

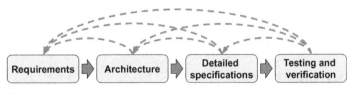

FIGURE 2.8

The standardization phases and iterative process.

Normally, it takes approximately one year from the time when the specification is completed until commercial products are out on the market.

3GPP consists of three *Technical Specifications Groups* (TSGs) (see Fig. 2.9) where TSG RAN (*Radio Access Network*) is responsible for the definition of functions, requirements, and interfaces of the Radio Access. TSG RAN consists of six working groups (WGs):

1. RAN WG1, dealing with the physical layer specifications.
2. RAN WG2, dealing with the layer 2 and layer 3 radio interface specifications.
3. RAN WG3, dealing with the fixed RAN interfaces—for example, interfaces between nodes in the RAN—but also the interface between the RAN and the core network.

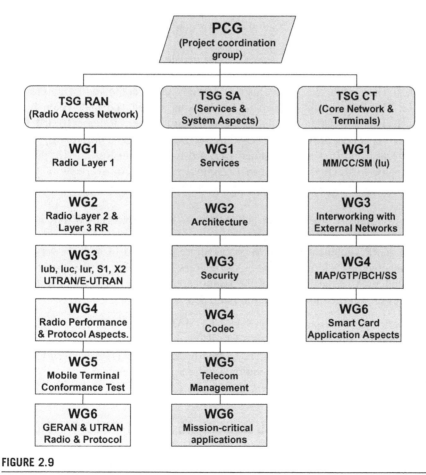

FIGURE 2.9

3GPP organization.

4. RAN WG4, dealing with the *radio frequency* (RF) and *radio resource management* (RRM) performance requirements.
5. RAN WG 5, dealing with the device conformance testing.
6. RAN WG6, dealing with standardization of GSM/EDGE (previously in a separate TSG called GERAN) and HSPA (UTRAN).

The work in 3GPP is carried out with relevant ITU-R recommendations in mind and the result of the work is also submitted to ITU-R as being part of IMT-2000, IMT-Advanced, and now also as a candidate for IMT-2020 in the form of NR. The organizational partners are obliged to identify regional requirements that may lead to options in the standard. Examples are regional frequency bands and special protection requirements local to a region. The specifications are developed with global roaming and circulation of devices in mind. This implies that many regional requirements in essence will be global requirements for all devices, since a roaming device has to meet the strictest of all regional requirements. Regional options in the specifications are thus more common for base stations than for devices.

The specifications of all releases can be updated after each set of TSG meetings, which occur four times a year. The 3GPP documents are divided into releases, where each release has a set of features added compared to the previous release. The features are defined in Work Items agreed and undertaken by the TSGs. LTE is defined from Release 8 and onwards, where Release 10 of LTE is the first version approved by ITU-R as an IMT-Advanced technology and is therefore also the first release named *LTE-Advanced*. From Release 13, the marketing name for LTE is changed to *LTE-Advanced Pro*. An overview of LTE is given in Chapter 4. Further details on the LTE radio interface can be found in [28].

The first release for NR is in 3GPP Release 15. An overview of NR is given in Chapter 5 with further details throughout this book.

The 3GPP Technical Specifications (TS) are organized in multiple series and are numbered TS XX.YYY, where XX denotes the number of the specification series and YYY is the number of the specification within the series. The following series of specifications define the radio access technologies in 3GPP:

- 25-series: Radio aspects for UTRA (WCDMA/HSPA);
- 45-series: Radio aspects for GSM/EDGE;
- 36-series: Radio aspects for LTE, LTE-Advanced and LTE-Advanced Pro;
- 37-series: Aspects relating to multiple radio access technologies;
- 38-series: Radio aspects for NR.

2.4.2 SPECIFICATION OF 5G IN 3GPP AS AN IMT-2020 CANDIDATE

In parallel with the definition and evaluation of the next-generation access initiated in ITU-R, 3GPP started to define the next-generation 3GPP radio access. A workshop on 5G radio access was held in 2014 and a process to define the evaluation criteria for 5G was initiated with a second workshop in early 2015. The evaluation will follow the same process that was used when LTE-Advanced was evaluated and submitted to ITU-R and approved as a 4G technology as part of IMT-advanced. The evaluation and submission of NR follows the ITU-R timeline described in Section 2.2.3.

3GPP TSG RAN documented scenarios, requirements, and evaluation criteria for the new 5G radio access in report TR 38.913 [10] which is in general aligned with the corresponding ITU-R reports [50, 51]. As for the case of the IMT-Advanced evaluation, the corresponding 3GPP evaluation of the next-generation radio access could have a larger scope and may have stricter requirements than the ITU-R evaluation of candidate IMT-2020 radio interface technologies that is defined by ITU-R WP5D.

The standardization work for NR started with a study item phase in Release 14 and continued with development of a first set of specifications through a work item in Release 15. A first set of the Release 15 NR specifications was published in December 2017 and the full specifications are due to be available in mid-2018. Further details on the time plan and the content of the NR releases is given in Chapter 5.

3GPP made a first submission of NR as an IMT-2020 candidate to the ITU-R WP5D meeting in February 2018. NR was submitted both as an RIT by itself and as an SRIT (set of component RITs) together with LTE. The following three candidates were submitted, all including NR as developed by 3GPP:

- 3GPP submitted a candidate named "5G," containing two submissions: the first submission was an SRIT containing two component RITs, these being NR and LTE. The second submission was a separate RIT being NR.
- Korea submitted NR as a RIT, with reference to 3GPP.
- China submitted NR as a RIT, with reference to 3GPP.

Further submissions to ITU-R will be made by 3GPP, giving more details of NR as an IMT-2020 candidate, according to the process described in Fig. 2.4. Simulations for the self-evaluations have also started in 3GPP, targeting the evaluation phase in 2019.

Spectrum for 5G

3

3.1 SPECTRUM FOR MOBILE SYSTEMS

Historically, the bands for the first and second generation of mobile services were assigned at frequencies around 800−900 MHz, but also in a few lower and higher bands. When 3G (IMT-2000) was rolled out, focus was on the 2 GHz band and with the continued expansion of IMT services with 3G and 4G, new bands were added at both lower and higher frequencies, presently spanning from 450 MHz to around 6 GHz. While new, previously unexploited, frequency bands are continuously defined for new mobile generations, the bands used for previous generations are used for the new generation as well. This was the case when 3G and 4G were introduced and it will also be the case for 5G.

Bands at different frequencies have different characteristics. Due to the propagation properties, bands at lower frequencies are good for wide-area coverage deployments, in urban, suburban, and rural environments. Propagation properties of higher frequencies make them more difficult to use for wide-area coverage and, for this reason, higher-frequency bands have to a larger extent been used for boosting capacity in dense deployments.

With the introduction of 5G, the demanding eMBB usage scenario and related new services will require even higher data rates and high capacity in dense deployments. While many early 5G deployments will be in bands already used for previous mobile generations, frequency bands above 24 GHz are being looked at as a complement to the frequency bands below 6 GHz. With the 5G requirements for extreme data rates and localized areas with very high area traffic capacity demands, deployment using even higher frequencies, even above 60 GHz, are considered. Referring to the wavelength, these bands are often called mm-wave bands.

New bands are defined continuously by 3GPP, mainly for the LTE specification, but now also for the new NR specifications. Many new bands are defined for NR operation only. Both paired bands, where separated frequency ranges are assigned for uplink and downlink, and unpaired bands with a single shared frequency range for uplink and downlink, are included in the NR specifications. Paired bands are used for Frequency Division Duplex (FDD) operation, while unpaired bands are used for Time Division Duplex (TDD) operation. The duplex modes of NR are described further in Chapter 7. Note that some unpaired bands are defined as *Supplementary Downlink* (SDL) or *Supplementary Uplink* (SDL) bands. These bands are paired with the uplink or downlink of other bands through *carrier aggregation*, as described in Section 7.6.

5G NR: The Next Generation Wireless Access Technology. DOI: https://doi.org/10.1016/B978-0-12-814323-0.00003-X

3.1.1 SPECTRUM DEFINED FOR IMT SYSTEMS BY THE ITU-R

The ITU-R identifies frequency bands to use for mobile service and specifically for IMT. Many of these were originally identified for IMT-2000 (3G) and new ones came with the introduction of IMT-Advanced (4G). The identification is however technology and generation "neutral," since the identification is for IMT in general, regardless of generation or Radio Interface Technology. The global designations of spectrum for different services and applications are done within the ITU-R and are documented in the *ITU Radio Regulations* [48] and the use of IMT bands globally is described in ITU-R Recommendation M.1036 [46].

The frequency listings in the ITU Radio Regulations [48] do not directly list a band for IMT, but rather allocate a band for the mobile service with a footnote stating that the band is identified for use by administrations wishing to implement IMT. The identification is mostly by region, but is in some cases also specified on a per-country level. All footnotes mention "IMT" only, so there is no specific mentioning of the different generations of IMT. Once a band is assigned, it is therefore up to the regional and local administrations to define a band for IMT use in general or for specific generations. In many cases, regional and local assignments are "technology neutral" and allow for any kind of IMT technology. This means that all existing IMT bands are potential bands for IMT-2020 (5G) deployment in the same way as they have been used for previous IMT generations.

The *World Administrative Radio Congress* WARC-92 identified the bands 1885−2025 and 2110−2200 MHz as intended for implementation of IMT-2000. Out of these 230 MHz of 3G spectrum, 2 × 30 MHz were intended for the satellite component of IMT-2000 and the rest for the terrestrial component. Parts of the bands were used during the 1990s for deployment of 2G cellular systems, especially in the Americas. The first deployments of 3G in 2001−2 by Japan and Europe were done in this band allocation, and for that reason it is often referred to as the IMT-2000 "core band."

Additional spectrum for IMT-2000 was identified at the World Radiocommunication Conference[1] WRC-2000, where it was considered that an additional need for 160 MHz of spectrum for IMT-2000 was forecasted by the ITU-R. The identification includes the bands used for 2G mobile systems at 806−960 and 1710−1885 MHz, and "new" 3G spectrum in the bands at 2500−2690 MHz. The identification of bands previously assigned for 2G was also a recognition of the evolution of existing 2G mobile systems into 3G. Additional spectrum was identified at WRC'07 for IMT, encompassing both IMT-2000 and IMT-Advanced. The bands added were 450−470, 698−806, 2300−2400, and 3400−3600 MHz, but the applicability of the bands varies on a regional and national basis. At WRC'12 there were no additional spectrum allocations

---

[1]The World Administrative Radio Conference (WARC) was reorganized in 1992 and became the World Radio-communication Conference (WRC).

identified for IMT, but the issue was put on the agenda for WRC'15. It was also determined to study the use of the band 694−790 MHz for mobile services in Region 1 (Europe, Middle East, and Africa).

WRC'15 was an important milestone setting the stage for 5G. First a new set of bands were identified for IMT, where many were identified for IMT on a global, or close to global, basis:

- 470−694/698 MHz (600 MHz band): Identified for some countries in Americas and the Asia-Pacific. For Region 1, it is considered for a new agenda item for IMT at WRC-23.
- 694−790 MHz (700 MHz band): This band is now also identified fully for Region 1 and is thereby a global IMT band.
- 1427−1518 MHz (L-band): A new global band identified in all countries.
- 3300−3400 MHz: Global band identified in many countries, but not in Europe or North America.
- 3400−3600 MHz (C-band): Now a global band identified for all countries. The band was already allocated in Europe.
- 3600−3700 MHz (C-band): Global band identified in many countries, but not in Africa and some counties in Asia-Pacific. In Europe, the band has been available since WRC'07.
- 4800−4990 MHz: New band identified for a few countries in Asia-Pacific.

Especially the frequency range from 3300 to 4990 MHz is of interest for 5G, since it is new spectrum in higher frequency bands. This implies that it fits well with the new usage scenarios requiring high data rates and is also suitable for massive MIMO implementation, where arrays with many elements can be implemented with reasonable size. Since it is new spectrum with no widespread use for mobile systems today, it will be easier to assign this spectrum in larger spectrum blocks, thereby enabling wider RF carriers and ultimately higher end-user data rates.

The second major outcome from WRC'15 concerning IMT was the new agenda item (1.13) appointed for the next WRC, to identify high-frequency bands above 24 GHz for 5G mobile services. These bands will be studied by ITU-R until 2019 and be considered for IMT identification at WRC'19. The primary target for the bands is deployment of IMT-2020. A majority of the bands to be studied are already today assigned to the mobile service on a primary basis, in most bands together with fixed and satellite services. They consist of the following band ranges:

- 24.25−27.5 GHz;
- 37−40.5 GHz;
- 42.5−43.5 GHz;
- 45.5−47 GHz;
- 47.2−50.2 GHz;
- 50.4−52.6 GHz;
- 66−76 GHz;
- 81−86 GHz.

There are also bands to be studied for IMT that are presently not allocated to the mobile service on a primary basis and where it will be investigated whether the allocation can be changed to include mobile:

- 31.8–33.4 GHz;
- 40.5–42.5 GHz;
- 47–47.2 GHz.

The complete set of bands is illustrated in Fig. 3.1.

ITU-R has formed a special task group TG 5/1, which will conduct sharing and compatibility studies for the new bands and prepare input for WRC'19 agenda item 1.13. The task group will document spectrum needs, technical and operational characteristics including protection criteria for existing services allocated in or adjacent to the bands studied, based on the studies. As an input to the studies, technical and operational characteristics of IMT-2020 were needed. These characteristics were provided from 3GPP as characteristics of NR, given at an early stage of standardization in January 2017.

It should be noted that there are also a large number of other frequency bands identified for *mobile services*, but not specifically for IMT. These bands are often used also for IMT on a regional or national basis. At WRC'15, there was some interest to also study 27.5–29.5 GHz for IMT, but it was not included in studies of 5G/IMT-2020 bands. Still, the band is planned for 5G mobiles services in at least the US and Korea. There was also support for studies of 5G/IMT-2020 in the frequency bands below 20 GHz, but those bands were ultimately not included. It is expected that several bands in the range 6–20 GHz will be considered for mobile services including IMT, in addition to the bands studied within ITU-R. One example is an FCC inquiry into new use, including next-generation wireless broadband services, in the frequency range 5925–7125 MHz.

The somewhat diverging arrangement between regions of the frequency bands assigned to IMT means that there is not one single band that can be used for roaming worldwide. Large efforts have, however, been put into defining a minimum set of bands that can be used to provide truly global roaming. In this way, multiband devices can provide efficient worldwide roaming for devices. With many of the new bands identified at WRC'15 being global or close to global, global roaming is made possible for devices using fewer bands and it also facilitates economy of scale for equipment and deployment.

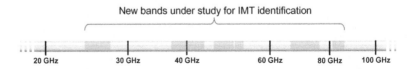

FIGURE 3.1

New IMT bands under study in ITU-R TG 5/1.

3.1.2 GLOBAL SPECTRUM SITUATION FOR 5G

There is a considerable interest globally to make spectrum available for 5G deployments. This is driven by operators and industry organizations such as the Global mobile Suppliers Association [35] and DIGITALEUROPE [29], but is also supported by regulatory bodies in different countries and regions. An overview of the spectrum situation for 5G is given in [56]. In standardization, 3GPP has focused its activities on bands where a high interest is evident (the full list of bands is in Section 3.2). The spectrum of interest can be divided into bands at low, medium, and high frequencies:

Low-frequency bands correspond to existing LTE bands below 2 GHz, which are suitable as a coverage layer, providing wide and deep coverage, including indoor. The bands with highest interest here are the 600 and 700 MHz bands, which correspond to 3GPP NR bands n71 and n28 (see Section 3.2 for further details). Since the bands are not very wide, a maximum of 20 MHz channel bandwidth is expected in the low-frequency bands.

For early deployment, the 600 MHz band is considered for NR in the US, while the 700 MHz band is defined as one of the so-called pioneer bands for Europe. In addition, a number of additional LTE bands in the below 3 GHz range are identified for possible "re-farming" and have been assigned NR band numbers. Since the bands are in general already deployed with LTE, NR is expected to be deployed gradually at a later stage.

Medium-frequency bands are in the range 3−6 GHz and can provide coverage, capacity, as well as high data rates through the wider channel bandwidth possible. The highest interest globally is in the range 3300−4200 MHz, where 3GPP has designated NR bands n77 and n78. Due to the wider bands, channel bandwidths up to 100 MHz are possible. Up to 200 MHz per operator may be assigned in this frequency range in the longer term, where carrier aggregation could then be used to deploy the full bandwidth.

The range 3300−4200 MHz is of global interest, with some variations seen regionally; and 3400−3800 MHz is a pioneer band in Europe, while China and India are planning for 3300−600 MHz and in Japan 3600−4200 MHz is being considered. Similar frequency ranges are considered in North America (3550−3700 MHz and initial discussions about 3700−4200 MHz), Latin America, the Middle East, Africa, India, Australia, etc. A total of 45 countries signed up to the IMT identification of the 3300−3400 MHz band in WRC-15. There is also a large amount of interest for a higher band in China (primarily 4800−5000 MHz) and Japan (4400−4900 MHz). In addition, there are a number of potential LTE re-farming bands in the 2−6 GHz range that have been identified as NR bands.

High-frequency bands are in the mm-Wave range above 24 GHz. They will be best suited for hotspot coverage with locally very high capacity and can provide very high data rates. The highest interest is in the range 24.25−29.5 GHz, with 3GPP NR bands n257 and n258 assigned. Channel bandwidths up to 400 MHz are defined for these bands, with even higher bandwidths possible through carrier aggregation.

The mmWave frequency range is new for IMT deployment, as discussed above. The band 27.5–28.35 was identified at an early stage in the US, while 24.25–27.5 GHz, also called the "26 GHz band," is a pioneer band for Europe, noting that not all of it may be made available for 5G. Different parts of the larger range 24.25–29.5 GHz are being considered globally. The range 27.5–29.5 GHz is the first range planned for Japan and 26.5–29.5 GHz in Korea. Overall, this band can be seen as global with regional variations. The range 37–40 GHz is also planned for the US and similar ranges around 40 GHz are considered in many other regions too, including China.

3.2 FREQUENCY BANDS FOR NR

NR can be deployed both in existing IMT bands and in future bands that may be identified at WRC, or in regional bodies. The possibility of operating a radio-access technology in different frequency bands is a fundamental aspect of global mobile services. Most 2G, 3G, and 4G devices are multiband capable, covering bands used in the different regions of the world to provide global roaming. From a radio-access functionality perspective, this has limited impact and the physical-layer specifications such as those for NR do not assume any specific frequency band. Since NR however spans such a vast range of frequencies, there are certain provisions that are intended only for certain frequency ranges. This includes how the different NR numerologies can be applied (see Chapter 7).

Many RF requirements are specified with different requirements across bands. This is certainly the case for NR, but also for previous generations. Examples of band-specific RF requirements are the allowed maximum transmit power, requirements/limits on out-of-band (OOB) emission and receiver blocking levels. Reasons for such differences are varying external constraints, often imposed by regulatory bodies, in other cases differences in the operational environment that are considered during standardization.

The differences between bands are more pronounced for NR due to the very wide range of frequency bands. For NR operation in the new mm-Wave bands above 24 GHz, both devices and base stations will be implemented with partly novel technology and there will be a more widespread use of massive MIMO, beam forming, and highly integrated advanced antenna systems. This creates differences in how RF requirements are defined, how they are measured for performance assessment and ultimately also what the limits for the requirements are set. Frequency bands within the scope of the present Release 15 work in 3GPP are for this reason divided into two frequency ranges:

- Frequency range 1 (FR1) includes all existing and new bands below 6 GHz.
- Frequency range 2 (FR2) includes new bands in the range 24.25–52.6 GHz.

These frequency ranges may be extended or complemented with new ranges in future 3GPP releases. The impact of the frequency ranges on the RF requirements is further discussed in Chapter 18.

The frequency bands where NR will operate are in both paired and unpaired spectra, requiring flexibility in the duplex arrangement. For this reason, NR supports both FDD and TDD operation. Some ranges are also defined for SDL or SUL. These features are further described in Section 7.7.

3GPP defines *operating bands*, where each operating band is a frequency range for uplink and/or downlink that is specified with a certain set of RF requirements. The operating bands each have a number, where NR bands are numbered n1, n2, n3, etc. When the same frequency range is defined as an operating band for different radio access technologies, the same number is used, but written in a different way. 4G LTE bands are written with Arabic numerals (1, 2, 3, etc.), while 3G UTRA bands are written with Roman numerals (I, II, II, etc.). LTE operating bands that are used with the same arrangement for NR are often referred to as "LTE re-farming bands."

Release 15 of the 3GPP specifications for NR includes 26 operating bands in frequency range 1 and three in frequency range 2. Bands for NR have a numbering scheme with assigned numbers from n1 to n512 using the following rules:

1. For NR in LTE re-farming bands, the LTE band numbers are reused for NR, just adding an "n."
2. New bands for NR are assigned the following numbers:
   - The range n65 to n256 is reserved for NR bands in frequency range 1 (some of these bands can be used for LTE in addition).
   - The range n257 to n512 is reserved for new NR bands in frequency range 2.

The scheme "conserves" band numbers and is backwards compatible with LTE (and UTRA) and does not lead to any new LTE numbers above 256, which is the present maximum possible. Any new LTE-only bands can also be assigned unused numbers below 65. In release 15, the operating bands in frequency range 1 are in the range n1 to n84 as shown in Table 3.1. The bands in frequency range 2 are in the range from n257 to n260, as shown in Table 3.2. All bands for NR are summarized in Figs. 3.2, 3.3, and 3.4, which also show the corresponding frequency allocation defined by the ITU-R.

Table 3.1 Operating Bands Defined by 3GPP for NR in Frequency Range 1

NR Band	Uplink Range (MHz)	Downlink Range (MHz)	Duplex Mode	Main Region(s)
n1	1920–1980	2110–2170	FDD	Europe, Asia
n2	1850–1910	1930–1990	FDD	Americas (Asia)
n3	1710–1785	1805–1880	FDD	Europe, Asia (Americas)
n5	824–849	869–894	FDD	Americas, Asia
n7	2500–2570	2620–2690	FDD	Europe, Asia
n8	880–915	925–960	FDD	Europe, Asia
n20	832–862	791–821	FDD	Europe
n28	703–748	758–803	FDD	Asia/Pacific
n38	2570–2620	2570–2620	TDD	Europe

(Continued)

Table 3.1 Operating Bands Defined by 3GPP for NR in Frequency Range 1
Continued

NR Band	Uplink Range (MHz)	Downlink Range (MHz)	Duplex Mode	Main Region(s)
n41	2496–2690	2496–2690	TDD	US, China
n50	1432–1517	1432–1517	TDD	
n51	1427–1432	1427–1432	TDD	
n66	1710–1780	2110–2200	FDD	Americas
n70	1695–1710	1995–2020	FDD	
n71	663–698	617–652	FDD	Americas
n74	1427–1470	1475–1518	FDD	Japan
n75	N/A	1432–1517	SDL	Europe
n76	N/A	1427–1432	SDL	Europe
n77	3300–4200	3300–4200	TDD	Europe, Asia
n78	3300–3800	3300–3800	TDD	Europe, Asia
n79	4400–5500	4400–5500	TDD	Asia
n80	1710–1785	N/A	SUL	
n81	880–915	N/A	SUL	
n82	832–862	N/A	SUL	
n83	703–748	N/A	SUL	
n84	1920–1980	N/A	SUL	

Table 3.2 Operating Bands Defined by 3GPP for NR in Frequency Range 2

NR Band	Uplink and Downlink Range (MHz)	Duplex Mode	Main Region(s)
n257	26,500–29,500	TDD	Asia, Americas (global)
n258	24,250–27,500	TDD	Europe, Asia (global)
n259	37,000–40,000	TDD	US (global)

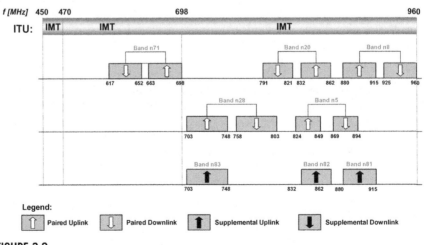

FIGURE 3.2

Operating bands specified in 3GPP release 15 for NR below 1 GHz (in FR1), shown with the corresponding ITU-R allocation. Not fully drawn to scale.

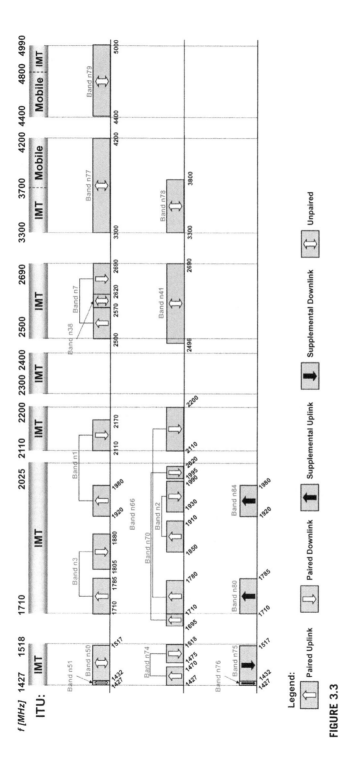

FIGURE 3.3

Operating bands specified in 3GPP release 15 for NR between 1 GHz and 6 GHz (in FR1), shown with the corresponding ITU-R allocation. Not fully drawn to scale.

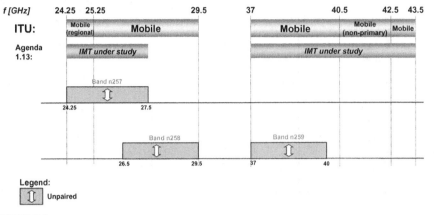

FIGURE 3.4

Operating bands specified in 3GPP release 15 for NR above 24 GHz (in FR2), shown with the corresponding ITU-R allocation, also indicating which parts are for study for IMT under agenda item 1.13. Not fully drawn to scale.

Some of the frequency bands are partly or fully overlapping. In most cases this is explained by regional differences in how the bands defined by the ITU-R are implemented. At the same time, a high degree of commonality between the bands is desired to enable global roaming. Originating in global, regional, and local spectrum developments, a first set of bands was specified as bands for UTRA. The complete set of UTRA bands later transferred to the LTE specifications in 3GPP Release 8. Additional bands have been added in later releases. In release 15, many of the LTE bands are now transferred to the NR specifications.

3.3 RF EXPOSURE ABOVE 6 GHZ

With the expansion of the frequency ranges for 5G mobile communications to bands above 6 GHz, existing regulations on human exposure to RF *electromagnetic fields* (EMFs) may restrict the maximum output power of user devices to levels significantly lower than what are allowed for lower frequencies.

International RF EMF exposure limits, for example those recommended by the *International Commission on Non-Ionizing Radiation* (ICNIRP) and those specified by the *Federal Communications Commission* (FCC) in the US, have been set with wide safety margins to protect against excessive heating of tissue due to energy absorption. In the frequency range of 6−10 GHz, the basic limits change from being specified as specific absorption rate (W/kg) to incident power density (W/m^2). This is mainly because the energy absorption in tissue becomes increasingly superficial with increasing frequency, and thereby more difficult to measure.

It has been shown that for products intended to be used in close proximity to the body, there will be a discontinuity in maximum allowed output power as the transition is made from specific absorption rate to power density-based limits [27]. To be compliant with ICNIRP exposure limits at the higher frequencies, the transmit power might have to be up to 10 dB below the power levels used for current cellular technologies. The exposure limits above 6 GHz appear to have been set with safety margins even larger than those used at lower frequencies, and without any obvious scientific justification.

For the lower-frequency bands, large efforts have been spent over the years to characterize the exposure and to set relevant limits. With a growing interest for utilizing frequency bands above 6 GHz for mobile communications, research efforts are likely to increase which eventually may lead to revised exposure limits. In the most recent RF exposure standards published by IEEE (C95.1-2005, C95.1-2010a), the inconsistency at the transition frequency is less evident. However, these limits have not yet been adopted in any national regulation and it is important also that other standardization organizations and regulators work to address this issue. If not, this might have a large negative impact on coverage at higher frequencies, in particular for user equipment intended to be used near the body, such as wearables, tablets, and mobile phones, for which the maximum transmit power might be heavily limited by the current RF exposure regulations.

LTE—An Overview

The focus of this book is NR, the new 5G radio access. Nevertheless, a brief overview of LTE as background to the coming chapters is relevant. One reason is that both LTE and NR have been developed by 3GPP and hence have a common background and share several technology components. Many of the design choices in NR are also based on experience from LTE. Furthermore, LTE continues to evolve in parallel with NR and is an important component in 5G radio access. For a detailed description of LTE see [28].

The work on LTE was initiated in late 2004 with the overall aim of providing a new radio-access technology focusing on packet-switched data only. The first release of the LTE specifications, release 8, was completed in 2008 and commercial network operation began in late 2009. Release 8 has been followed by subsequent LTE releases, introducing additional functionality and capabilities in different areas, as illustrated in Fig. 4.1. Releases 10 and 13 are particularly interesting. Release 10 is the first release of LTE-Advanced, and release 13, finalized in late 2015, is the first release of LTE-Advanced Pro. Currently, as of this writing, 3GPP is working on release 15 which, in addition to NR, also contains a further evolution of LTE.

4.1 LTE RELEASE 8—BASIC RADIO ACCESS

Release 8 is the first LTE release and forms the basis for all the following LTE releases. In parallel with the LTE radio access scheme, a new core network, the *Evolved Packet Core* (EPC) was developed [63].

One important requirement imposed on the LTE development was spectrum flexibility. A range of carrier bandwidths up to and including 20 MHz is supported for carrier frequencies from below 1 GHz up to around 3 GHz. One aspect of spectrum flexibility is the support of both paired *and* unpaired spectrum using *Frequency-Division Duplex* (FDD) and *Time-Division Duplex* (TDD), respectively, with a common design, albeit two different frame structures. The focus of the development work was primarily macronetworks with above-rooftop antennas and relatively large cells. For TDD, the uplink−downlink allocation is therefore in essence static with the same uplink−downlink allocation across all cells.

The basic transmission scheme in LTE is *orthogonal frequency-division multiplexing* (OFDM). This is an attractive choice due to its robustness to time

5G NR: The Next Generation Wireless Access Technology. DOI: https://doi.org/10.1016/B978-0-12-814323-0.00004-1

FIGURE 4.1

LTE and its evolution.

dispersion and ease of exploiting both the time and frequency domain. Furthermore, it also allows for reasonable receiver complexity also in combination with spatial multiplexing (MIMO) which is an inherent part of LTE. Since LTE was primarily designed with macronetworks in mind with carrier frequencies up to a few GHz, a single subcarrier spacing of 15 kHz and a cyclic prefix of approximately 4.7 μs[1] was found to be a good choice. In total 1200 subcarriers are used in a 20 MHz spectrum allocation.

For the uplink, where the available transmission power is significantly lower than for the downlink, the LTE design settled for a scheme with a low peak-to-average ratio to provide a high power-amplifier efficiency. DFT-precoded OFDM, with the same numerology as in the downlink, was chosen to achieve this. A drawback with DFT-precoded OFDM is the larger complexity on the receiver side, but given that LTE release 8 does not support spatial multiplexing in the uplink this was not seen as a major problem.

In the time domain, LTE organizes transmissions into 10-ms frames, each consisting of ten 1-ms subframes. The subframe duration of 1 ms, which corresponds to 14 OFDM symbols, is the smallest schedulable unit in LTE.

Cell-specific reference signals is a cornerstone in LTE. The base station continuously transmits one or more reference signals (one per layer), regardless of whether there are downlink data to transmit or not. This is a reasonable design for the scenarios which LTE was designed for—relatively large cells with many users per cell. The cell-specific reference signals are used for many functions in LTE: downlink channel estimation for coherent demodulation, channel-state reporting for scheduling purposes, correction of device-side frequency errors, initial access, and mobility measurements to mention just a few. The reference signal density depends on the number of transmission layers set up in a cell, but for the common case of 2×2 MIMO, every third subcarrier in four out of 14 OFDM symbols in a subframe are used for reference signals. Thus, in the time domain there are around 200 μs between reference signal occasions, which limits the possibilities to switch off the transmitter to reduce power consumption.

---

[1]There is also a possibility for 16.7 μs extended cyclic prefix but that option is rarely used in practice.

Data transmission in LTE is primarily scheduled on a dynamic basis in both uplink and downlink. To exploit the typically rapidly varying radio conditions, channel-dependent scheduling can be used. For each 1-ms subframe, the scheduler controls which devices are to transmit or receive and in what frequency resources. Different data rates can be selected by adjusting the code rate of the Turbo code as well as varying the modulation scheme from QPSK up to 64-QAM. To handle transmission errors, *fast hybrid ARQ with soft combining* is used in LTE. Upon downlink reception the device indicates the outcome of the decoding operation to the base station, which can retransmit erroneously received data blocks.

The scheduling decisions are provided to the device through the *Physical Downlink Control Channel* (PDCCH). If there are multiple devices scheduled in the same subframe, which is a common scenario, there are multiple PDCCHs, one per scheduled device. The first up to three OFDM symbols of the subframe are used for transmission of downlink control channels. Each control channel spans the full carrier bandwidth, thereby maximizing the frequency diversity. This also implies that all devices must support the full carrier bandwidth up to the maximum value of 20 MHz. Uplink control signaling from the devices, for example hybrid-ARQ acknowledgments and channel state information for downlink scheduling, is carried on the *Physical Uplink Control Channel* (PUCCH), which has a basic duration of 1 ms.

Multiantenna schemes, and in particular single-user MIMO, are an integral part of LTE. A number of transmission layers are mapped to up to four antennas by means of a precoder matrix of size $N_A \times N_L$, where the number of layers N_L, also known as the transmission rank, is less than or equal to the number of antennas N_A. The transmission rank, as well as the exact precoder matrix, can be selected by the network based on channel-status measurements carried out and reported by the terminal, also known as *closed-loop spatial multiplexing*. There is also a possibility to operate without closed-loop feedback for precoder selection. Up to four layers is possible in the downlink although commercial deployments typically use only two layers. In the uplink only single-layer transmission is possible.

In case of spatial multiplexing, by selecting rank-1 transmission, the precoder matrix, which then becomes an $N_A \times 1$ *precoder vector*, performs a (single-layer) *beam-forming* function. This type of beam-forming can more specifically be referred to as *codebook-based* beam-forming as the beam-forming can only be done according to a limited set of predefined beam-forming (precoder) vectors.

Using the basic features discussed above, LTE release 8 is in theory capable of providing peak data rates up to 150 Mbit/s in the downlink using two-layer transmission in 20 MHz and 75 Mbit/s in the uplink. Latency-wise LTE provides 8 ms roundtrip time in the hybrid-ARQ protocol and (theoretically) less than 5 ms one-way delay in the LTE RAN. In practical deployments, including transport and core network processing, an overall end-to-end latency of some 10 ms is not uncommon in well-deployed networks.

4.2 LTE EVOLUTION

Releases 8 and 9 form the foundation of LTE, providing a highly capable mobile-broadband standard. However, to meet new requirements and expectations, the releases following the basic ones provide additional enhancements and features in different areas. Fig. 4.2 illustrates some of the major areas in which LTE has evolved over the 10 years since its introduction with details provided in the following.

Release 10 marks the start of the LTE evolution. One of the main targets of LTE release 10 was to ensure that the LTE radio-access technology would be fully compliant with the IMT-Advanced requirements, thus the name *LTE-Advanced* is often used for LTE release 10 and later. However, in addition to the ITU requirements, 3GPP also defined its own targets and requirements for LTE-Advanced [10]. These targets/requirements extended the ITU requirements both in terms of being more aggressive as well as including additional requirements. One important requirement was *backwards compatibility*. Essentially this means that an earlier-release LTE device should be able to access a carrier supporting LTE release-10 functionality, although obviously not being able to utilize all the release-10 features of that carrier. The principle of backwards compatibility is important and has been kept for all LTE releases, but also imposes some restrictions on the enhancements possible; restrictions that are not present when defining a new standard such as NR.

LTE release 10 was completed in late 2010 and introduced enhanced LTE spectrum flexibility through carrier aggregation, further extended multiantenna transmission, support for relaying, and improvements around intercell interference coordination in heterogeneous network deployments.

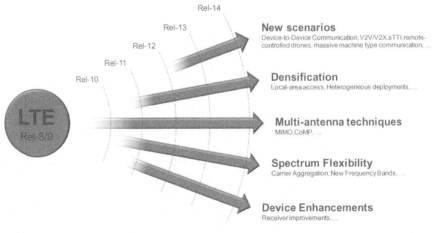

FIGURE 4.2

LTE evolution.

Release 11 further extended the performance and capabilities of LTE. One of the most notable features of LTE release 11, finalized in late 2012, was radio-interface functionality for *coordinated multipoint* (CoMP) transmission and reception. Other examples of improvements in release 11 were carrier-aggregation enhancements, a new control-channel structure (EPDCCH), and performance requirements for more advanced device receivers.

Release 12 was completed in 2014 and focused on small cells with features such as dual connectivity, small-cell on/off, and (semi-)dynamic TDD, as well as on new scenarios with introduction of direct device-to-device communication and provisioning of complexity-reduced machine-type communication.

Release 13, finalized at the end of 2015, marks the start of *LTE Advanced Pro*. It is sometimes in marketing dubbed 4.5G and seen as an intermediate technology step between 4G defined by the first releases of LTE and the 5G NR air interface. License-assisted access to support unlicensed spectra as a complement to licensed spectra, improved support for machine-type communication, and various enhancements in carrier aggregation, multi-antenna transmission, and device-to-device communication are some of the highlights from release 13.

Release 14 was completed in the spring of 2017. Apart from enhancements to some of the features introduced in earlier releases, for example enhancements to operation in unlicensed spectra, it introduced support for vehicle-to-vehicle (V2V) and vehicle-to-everything (V2X) communication, as well as wide-area broadcast support with a reduced subcarrier spacing.

Release 15 will be completed in the middle of 2018. Significantly reduced latency through the so-called sTTI feature, as well as communication using aerials are two examples of enhancements in this release.

In general, expanding LTE to new use cases beyond traditional mobile broadband has been in focus for the later releases and the evolution will continue also in the future. This is also an important part of 5G overall and exemplifies that LTE remains important and a vital part of the overall 5G radio access.

4.3 SPECTRUM FLEXIBILITY

Already the first release of LTE provides a certain degree of spectrum flexibility in terms of multibandwidth support and a joint FDD/TDD design. In later releases this flexibility was considerably enhanced to support higher bandwidths and fragmented spectra using carrier aggregation and access to unlicensed spectra as a complement using license-assisted access (LAA).

4.3.1 CARRIER AGGREGATION

As mentioned earlier, the first release of LTE already provided extensive support for deployment in spectrum allocations of various characteristics, with bandwidths ranging from roughly 1 MHz up to 20 MHz in both paired and unpaired bands. With LTE release 10 the transmission bandwidth can be further extended by means of *carrier aggregation* (CA), where multiple *component carriers* are aggregated and jointly used for transmission to/from a single device. Up to five component carriers, possibly each of different bandwidth, can be aggregated in release 10, allowing for transmission bandwidths up to 100 MHz. All component carriers need to have the same duplex scheme and, in the case of TDD, uplink—downlink configuration. In later releases, this requirement was relaxed. The number of component carriers possible to aggregate was increased to 32, resulting in a total bandwidth of 640 MHz. Backwards compatibility was ensured as each component carrier uses the release-8 structure. Hence, to a release-8/9 device each component carrier will appear as an LTE release-8 carrier, while a carrier-aggregation-capable device can exploit the total aggregated bandwidth, enabling higher data rates. In the general case, a different number of component carriers can be aggregated for the downlink and uplink. This is an important property from a device complexity point of view where aggregation can be supported in the downlink where very high data rates are needed without increasing the uplink complexity.

Component carriers do not have to be contiguous in frequency, which enables exploitation of *fragmented spectra*; operators with a fragmented spectrum can provide high-data-rate services based on the availability of a wide overall bandwidth even though they do not possess a single wideband spectrum allocation.

From a baseband perspective, there is no difference between the cases in Fig. 4.3 and they are all supported by LTE release 10. However, the RF-implementation complexity is vastly different, with the first case being the least complex. Thus, although carrier aggregation is supported by the basic specifications, not all devices will support it. Furthermore, release 10 has some restrictions on carrier aggregation in the RF specifications, compared to what has been specified for physical layer and related signaling, while in later releases there is support for carrier-aggregation within and between a much larger number of bands.

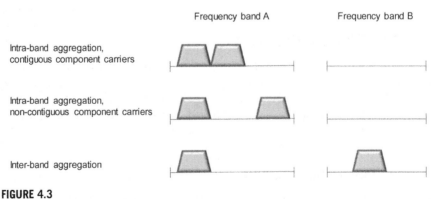

FIGURE 4.3

Carrier aggregation.

Release 11 provided additional flexibility for aggregation of TDD carriers. Prior to release 11, the same downlink—uplink allocation was required for all the aggregated carriers. This can be unnecessarily restrictive in the case of aggregation of different bands as the configuration in each band may be given by coexistence with other radio access technologies in that particular band. An interesting aspect of aggregating different downlink—uplink allocations is that the device may need to receive and transmit simultaneously in order to fully utilize both carriers. Thus, unlike previous releases, a TDD-capable device may, similarly to a FDD-capable device, need a duplex filter. Release 11 also saw the introduction of RF requirements for interband and noncontiguous intraband aggregation, as well as support for an even larger set of interband aggregation scenarios.

Release 12 defined aggregations between FDD and TDD carriers, unlike earlier releases that only supported aggregation within one duplex type. FDD—TDD aggregation allows for efficient utilization of an operator's spectrum assets. It can also be used to improve the uplink coverage of TDD by relying on the possibility for continuous uplink transmission on the FDD carrier.

Release 13 increased the number of carriers possible to aggregate from 5 to 32, resulting in a maximum bandwidth of 640 MHz and a theoretical peak data rate around 25 Gbit/s in the downlink. The main motivation for increasing the number of subcarriers is to allow for very large bandwidths in unlicensed spectra as will be further discussed in conjunction with license-assisted access below.

Carrier aggregation is one of the most successful enhancements of LTE to date with new combinations of frequency band added in every release.

4.3.2 LICENSE-ASSISTED ACCESS

Originally, LTE was designed for licensed spectra where an operator has an exclusive license for a certain frequency range. A licensed spectrum offers many benefits since the operator can plan the network and control the interference situation, but there is typically a cost associated with obtaining the spectrum license and the amount of licensed spectra is limited. Therefore, using unlicensed spectra as a *complement* to offer higher data rates and higher capacity in local areas is of interest. One possibility is to complement the LTE network with Wi-Fi, but higher performance can be achieved with a tighter coupling between licensed and unlicensed spectra. LTE release 13 therefore introduced *license-assisted access*, where the carrier aggregation framework is used to aggregate downlink carriers in unlicensed frequency bands, primarily in the 5 GHz range, with carriers in licensed frequency bands as illustrated in Fig. 4.4. Mobility, critical control signaling, and services demanding high quality-of-service rely on carriers in the licensed spectra while (parts of) less demanding traffic can be handled by the carriers using unlicensed spectra. Operator-controlled small-cell deployments are the target. Fair sharing of the spectrum resources with other systems, in particular Wi-Fi, is an important characteristic of LAA which therefore incudes a listen-before-talk mechanism. In release 14, license-assisted access was enhanced to

Primary Carrier   **Secondary Carrier**
Licensed Spectrum   Unlicensed Spectrum

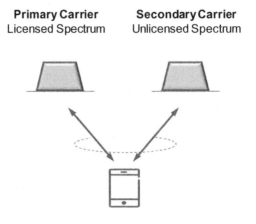

FIGURE 4.4

License-assisted access.

address also uplink transmissions. Although the LTE technology standardized in 3GPP supports license-assisted access only, where a licensed carrier is needed, there has been work outside 3GPP in the MulteFire alliance resulting in a stand-alone mode-of-operation based on the 3GPP standard.

4.4 MULTI-ANTENNA ENHANCEMENTS

Multi-antenna support has been enhanced over the different releases, increasing the number of transmission layers in the downlink to eight and introducing uplink spatial multiplexing of up to four layers. Full-dimension MIMO and two-dimensional beamforming are other enhancements, as is the introduction of coordinated multipoint transmission.

4.4.1 EXTENDED MULTI-ANTENNA TRANSMISSION

In release 10, downlink spatial multiplexing was expanded to support up to eight transmission layers. This can be seen as an extension of the release-9 dual-layer beam-forming to support up to eight antenna ports and eight corresponding layers. Together with the support for carrier aggregation this enables downlink data rates up to 3 Gbit/s in 100 MHz of spectra in release 10, increased to 25 Gbit/s in release 13 using 32 carriers, eight layers spatial multiplexing, and 256QAM.

Uplink spatial multiplexing of up to four layers was also introduced as part of LTE release 10. Together with the possibility for uplink carrier aggregations this allows for uplink data rates up to 1.5 Gbit/s in 100 MHz of spectrum. Uplink spatial multiplexing consists of a codebook-based scheme under the control of the base station, which means that the structure can also be used for uplink transmitter-side beam-forming.

An important consequence of the multiantenna extensions in LTE release 10 was the introduction of an enhanced downlink *reference-signal structure* that more extensively separated the function of channel estimation and the function of acquiring channel-state information. The aim of this was to better enable novel antenna arrangements and new features such as more elaborate multipoint coordination/transmission in a flexible way.

In release 13, and continued in release 14, improved support for massive antenna arrays was introduced, primarily in terms of more extensive feedback of channel-state information. The larger degrees of freedom can be used for, for example, beamforming in both elevation and azimuth and massive multiuser MIMO where several spatially separated devices are simultaneously served using the same time-frequency resource. These enhancements are sometimes termed full-dimension MIMO and form a step into massive MIMO with a very large number of steerable antenna elements.

4.4.2 MULTIPOINT COORDINATION AND TRANSMISSION

The first release of LTE included specific support for coordination between transmission points, referred to as *Inter-Cell Interference Coordination* (ICIC), to control the interference between cells. However, the support for such coordination was significantly expanded as part of LTE release 11, including the possibility for much more dynamic coordination between transmission points.

In contrast to release 8 ICIC, which was limited to the definition of certain messages between base stations to assist coordination between cells, the release 11 activities focused on radio-interface features and device functionality to assist different coordination means, including the support for channel-state feedback for multiple transmission points. Jointly these features and functionality go under the name *Coordinated Multi-Point* (CoMP) transmission/reception. Refinement to the reference-signal structure was also an important part of the CoMP support, as was the enhanced control-channel structure introduced as part of release 11, see below.

Support for CoMP includes *multipoint coordination*—that is, when transmission to a device is carried out from one specific transmission point but where scheduling and link adaptation are coordinated between the transmission points, as well as *multipoint transmission* in which case transmission to a device can be carried out from multiple transmission points either in such a way that that transmission can switch dynamically between different transmission points (*Dynamic Point Selection*) or be carried out jointly from multiple transmission points (*Joint Transmission*) (see Fig. 4.5).

A similar distinction can be made for uplink where one can distinguish between (uplink) multipoint coordination and multipoint *reception*. In general,

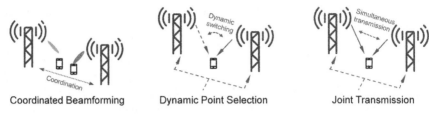

Coordinated Beamforming   Dynamic Point Selection   Joint Transmission

FIGURE 4.5

Different types of CoMP.

uplink CoMP is mainly a network implementation issue and has very little impact on the device and very little visibility in the radio-interface specifications.

The CoMP work in release 11 assumed "ideal" backhaul, in practice implying centralized baseband processing connected to the antenna sites using low-latency fiber connections. Extensions to relaxed backhaul scenarios with non-centralized baseband processing were introduced in release 12. These enhancements mainly consisted of defining new X2 messages between base stations for exchanging information about so-called CoMP hypotheses, essentially a potential resource allocation, and the associated gain/cost.

4.4.3 ENHANCED CONTROL CHANNEL STRUCTURE

In release 11, a new complementary control channel structure was introduced to support intercell interference coordination and to exploit the additional flexibility of the new reference-signal structure not only for data transmission, which was the case in release 10, but also for control signaling. The new control-channel structure can thus be seen as a prerequisite for many CoMP schemes, although it is also beneficial for beamforming and frequency-domain interference coordination as well. It is also used to support narrow-band operation for MTC enhancements in releases 12 and 13.

4.5 DENSIFICATION, SMALL CELLS, AND HETEROGENEOUS DEPLOYMENTS

Small cells and dense deployment has been in focus for several releases as means to provide very high capacity and data rates. Relaying, small-cell on/off, dynamic TDD, and heterogeneous deployments are some examples of enhancements over the releases. License-assisted access, discussed in Section 4.3.2, is another feature primarily targeting small cells.

4.5.1 RELAYING

In the context of LTE, *relaying* implies that the device communicates with the network via a *relay node* that is *wirelessly connected* to a *donor cell* using the

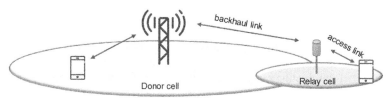

FIGURE 4.6

Example of relaying.

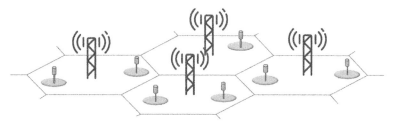

FIGURE 4.7

Example of heterogeneous deployment with low-power nodes inside macro cells.

LTE radio-interface technology (see Fig. 4.6). From a device point of view, the relay node will appear as an ordinary cell. This has the important advantage of simplifying the device implementation and making the relay node backwards compatible—that is, LTE release-8/9 devices can also access the network via the relay node. In essence, the relay is a low-power base station wirelessly connected to the remaining part of the network.

4.5.2 HETEROGENEOUS DEPLOYMENTS

Heterogeneous deployments refer to deployments with a mixture of network nodes with different transmit power and overlapping geographical coverage (Fig. 4.7). A typical example is a pico node placed within the coverage area of a macrocell. Although such deployments were already supported in release 8, release 10 introduced new means to handle the interlayer interference that may occur between, for example, a pico layer and the overlaid macro. The multipoint-coordination techniques introduced in release 11 further extend the set of tools for supporting heterogeneous deployments. Enhancements to improve mobility between the pico layer and the macro layer were introduced in release 12.

4.5.3 SMALL-CELL ON/OFF

In LTE, cells are continuously transmitting cell-specific reference signals and broadcasting system information, regardless of the traffic activity in the cell. One reason for this is to enable idle-mode devices to detect the presence of a cell; if there are no transmissions from a cell there is nothing for the device to measure upon and the cell would therefore not be detected. Furthermore, in a large macro-cell deployment there is a relatively high likelihood of at least one device being active in a cell motivating continuous transmission of reference signals.

However, in a dense deployment with many relatively small cells, the likelihood of not all cells serving the device at the same time can be relatively high in some scenarios. The downlink interference scenario experienced by a device may also be more severe with devices experiencing very low signal-to-interference ratios due to interference from neighboring, potentially empty, cells, especially if there is a large amount of line-of-sight propagation. To address this, release 12 introduced mechanisms for turning on/off individual cells as a function of the traffic situation to reduce the average intercell interference and reduce power consumption.

4.5.4 DUAL CONNECTIVITY

Dual connectivity implies a device is simultaneously connected to two cells, see Fig. 4.8, as opposed to the baseline case with the device connected to a single device only. User-plane aggregation, where the device is receiving data transmission from multiple sites, separation of control and user planes, and uplink—downlink separation where downlink transmissions originate from a different node than the uplink reception node are some examples of the benefits with dual connectivity. To some extent it can be seen as carrier aggregation extended to the case of nonideal backhaul. The dual connectivity framework has also turned out to be very promising for integrating other radio-access schemes such as WLAN into 3GPP networks. It is also essential for NR when operating in non-standalone mode with LTE providing mobility and initial access.

4.5.5 DYNAMIC TDD

In TDD, the same carrier frequency is shared in the time domain between uplink and downlink. The fundamental approach to this in LTE, as well as in many other TDD systems, is to statically split the resources in to uplink and downlink.

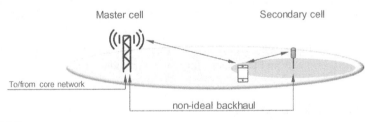

FIGURE 4.8

Example of dual connectivity.

Having a static split is a reasonable assumption in larger macrocells as there are multiple users and the aggregated per-cell load in uplink and downlink is relatively stable. However, with an increased interest in local-area deployments, TDD is expected to become more important compared to the situation for wide-area deployments to date. One reason is unpaired spectrum allocations being more common in higher-frequency bands not suitable for wide-area coverage. Another reason is that many problematic interference scenarios in wide-area TDD networks are not present with below-rooftop deployments of small nodes. An existing wide-area FDD network could be complemented by a local-area layer using TDD, typically with low output power per node.

To better handle the high traffic dynamics in a local-area scenario, where the number of devices transmitting to/receiving from a local-area access node can be very small, dynamic TDD is beneficial. In dynamic TDD, the network can dynamically use resources for either uplink or downlink transmissions to match the instantaneous traffic situation, which leads to an improvement of the end-user performance compared to the conventional static split of resources between uplink and downlink. To exploit these benefits, LTE release 12 includes support for dynamic TDD, or *enhanced Interference Mitigation and Traffic Adaptation* (eIMTA) as is the official name for this feature in 3GPP.

4.5.6 WLAN INTERWORKING

The 3GPP architecture allows for integrating non-3GPP access, for example WLAN, but also cdma2000 [12]. Essentially, these solutions connect the non-3GPP access to the EPC and are thus not visible in the LTE radio-access network. One drawback of this way of WLAN interworking is the lack of network control; the device may select Wi-Fi even if staying on LTE would provide a better user experience. One example of such a situation is when the Wi-Fi network is heavily loaded while the LTE network enjoys a light load. Release 12 therefore introduced means for the network to assist the device in the selection procedure. Basically, the network configures a signal-strength threshold controlling when the device should select LTE or Wi-Fi.

Release 13 provided further enhancements in WLAN interworking with more explicit control from the LTE RAN on when a device should use Wi-Fi and when to use LTE. Furthermore, release 13 also includes LTE–WLAN aggregation where LTE and WLAN are aggregated at the PDCP level using a framework very similar to dual connectivity.

4.6 DEVICE ENHANCEMENTS

Fundamentally, a device vendor is free to design the device receiver in any way as long as it supports the minimum requirements defined in the specifications.

There is an incentive for the vendors to provide significantly better receivers as this could be directly translated into improved end-user data rates. However, the network may not be able to exploit such receiver improvements to their full extent as it might not know which devices have significantly better performance. Network deployments therefore need to be based on the minimum requirements. Defining performance requirements for more advanced receiver types to some extent alleviates this as the minimum performance of a device equipped with an advanced receiver is known. Both releases 11 and 12 saw a lot of focus on receiver improvements with cancellation of some overhead signals in release 11 and more generic schemes in release 12, including network-assisted interference cancellation (NAICS), where the network can provide the devices with information assisting intercell interference cancellation.

4.7 NEW SCENARIOS

LTE was originally designed as a mobile broadband system, aiming at providing high data rates and high capacity over wide areas. The evolution of LTE has added features improving capacity and data rates, but also enhancements making LTE highly relevant also for new use cases. Operation in areas without network coverage, for example in a disaster area, is one example, resulting in support for device-to-device commination being included in the LTE. Massive machine-type communication, where a large number of low-cost devices, for example sensors, are connected to a cellular network is another example. V2V/V2X and remote-controlled drones are yet other examples of new scenarios.

4.7.1 DEVICE-TO-DEVICE COMMUNICATION

Cellular systems, such as LTE, are designed assuming that devices connect to a base station to communicate. In most cases this is an efficient approach as the server with the content of interest is typically not in the vicinity of the device. However, if the device is interested in communicating with a neighboring device, or just detecting whether there is a neighboring device that is of interest, the network-centric communication may not be the best approach. Similarly, for public safety, such as a first responder officer searching for people in need in a disaster situation, there is typically a requirement that communication should also be possible in the absence of network coverage.

To address these situations, release 12 introduced network-assisted device-to-device communication using parts of the uplink spectrum (Fig. 4.9). Two scenarios were considered when developing the device-to-device enhancements, in coverage as well as out-of-coverage communication for public safety, and in coverage discovery of neighboring devices for commercial use cases. In release 13, device-to-device communication was further enhanced with relaying solutions

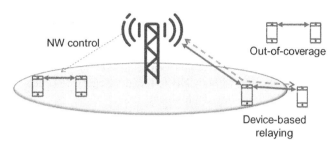

FIGURE 4.9

Device-to-device communciation.

for extended coverage. The device-to-device design also served as the basis for the V2V and V2X work in release 14.

4.7.2 MACHINE-TYPE COMMUNICATION

Machine-type communication (MTC) is a very wide term, basically covering all types of communication between machines. Although spanning a wide range of different applications, many of which are yet unknown, MTC applications can be divided into two main categories, massive MTC and ultrareliable low-latency communication (URLLC).

Examples of massive MTC scenarios are different types of sensors, actuators, and similar devices. These devices typically have to be of very low cost and have very low energy consumption, enabling very long battery life. At the same time, the amount of data generated by each device is normally very small and very low latency is not a critical requirement. URLLC, on the other hand, corresponds to applications such as traffic safety/control or wireless connectivity for industrial processes, and in general scenarios where very high reliability and availability is required, combined with low latency.

To better support massive MTC, several enhancements have been introduced, starting with release 12 and the introduction of a new, low-end device category, category 0, supporting data rates up to 1 Mbit/s. A power-save mode for reduced device power consumption was also defined. Release 13 further improved the MTC support by defining category-M1 with further extended coverage and support for 1.4 MHz device bandwidth, irrespective of the system bandwidth, to further reduce device cost. From a network perspective these devices are normal LTE devices, albeit with limited capabilities, and can be freely mixed with more capable LTE devices on a carrier.

Narrow-band Internet-of-Things (NB-IoT) is a parallel LTE track completed in release 13. It targets even lower cost and data rates than category-M1, 250 kbit/s or less, in a bandwidth of 180 kHz, and even further enhanced coverage. Thanks to the use of OFDM with 15-kHz subcarrier spacing, it can be deployed inband on top of an LTE carrier, outband in a separate spectrum allocation, or in the guard bands of LTE, providing a high degree of flexibility for an operator. In the uplink, transmission on a single tone is supported to obtain very large coverage for the lowest data rates. NB-IoT uses the same family of higher-layer protocols (MAC, RLC, and PDCP) as LTE, with extensions for faster connection setup applicable to both NB-IoT and category-M1, and can therefore easily be integrated into existing deployments.

Both eMTC and NB-IoT will play an important role in 5G networks for massive machine-type communication. Special means for deploying NR on top of an already-existing carrier used for massive machine-type communication has therefore been included (see Chapter 17).

Improved support for URLLC has been added in the later LTE releases. Examples hereof are the sTTI feature in release 15 (see below) and the general work on the reliability part of URLLC in release 15.

4.7.3 LATENCY REDUCTION—STTI

In release 15, work on reducing the overall latency has been carried out, resulting in the so-called *short TTI* (sTTI) feature. The target with this feature is to provide very low latency for use cases where this is important, for example factory automation. It uses similar techniques as used in NR, such as a transmission duration of a few OFDM symbols and reduced device processing delay, but incorporated in LTE in a backwards-compatible manner. This allows for low-latency services to be included in existing networks, but also implies certain limitations compared to a clean-slate design such as NR.

4.7.4 V2V AND V2X

Intelligent transportation systems (ITSs) refer to services to improve traffic safety and increase efficiency. Examples are vehicle-to-vehicle communication for safety, for example to transmit messages to vehicles behind when the car in front breaks. Another example is platooning where several trucks drive very close to each other and follow the first truck in the platoon, thereby saving fuel and reducing CO_2 emissions. Communication between vehicles and infrastructure is also useful, for example to obtain information about the traffic situation, weather updates, and alternative routes in case of congestion (Fig. 4.10).

In release 14, 3GPP specified enhancements in this area, based on the device-to-device technologies introduced in release 12 and quality-of-service enhancements in the network. Using the same technology for communication both between vehicles and between vehicles and infrastructure is attractive, both to improve the performance but also to reduce cost.

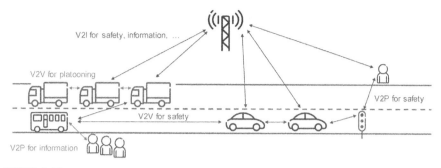

FIGURE 4.10

Illustration of V2V and V2X.

4.7.5 AERIALS

The work on aerials in release 15 covers communication via a drone acting as a relay to provide cellular coverage in an otherwise noncovered area, but also remote control of drones for various industrial and commercial applications. Since the propagation conditions between the ground and an airborne drone are different than in a terrestrial network, new channel models are developed as part of release 15. The interference situation for a drone is different than for a device on the groups due to the larger number of base stations visible to the drone, calling for interference-mitigation techniques such as beamforming, as well as enhancements to the power-control mechanism.

NR Overview

5

Fig. 5.1 outlines the timeline for the NR development within 3GPP. The technical work on NR was initiated in the spring of 2016 as a study item in 3GPP release 14, based on a kick-off workshop in the fall of 2015. During the study item phase, different technical solutions were studied, but given the tight time schedule, some technical decisions were taken already in this phase. The work continued into a work item phase in release 15, resulting in the first version of the NR specifications available by the end of 2017, before the closure of 3GPP release 15 in mid-2018. The reason for the intermediate release of the specifications, before the end of release-15, is to meet commercial requirements on early 5G deployments.

The first specification from December 2017, which is the focus of this book, is limited to non-standalone NR operation (see Chapter 6), implying that NR devices rely on LTE for initial access and mobility. The final release-15 specifications support standalone NR operation as well. The difference between standalone and non-standalone primarily affects higher layers and the interface to the core network; the basic radio technology is the same in both cases.

During the development of release 15, the focus was on eMBB and (to some extent) URLLC type of services. For massive machine-type communication (mMTC), LTE-based technologies such as eMTC and NB-IoT [28,58] can be used with excellent results. The support for LTE-based massive MTC on a carrier overlapping with an NR carrier has been accounted for in the design of NR (see Chapter 17), resulting in an integrated overall system capable of handling a very wide range of services. Native NR support for extended mMTC, as well as special technology features such as direct device-to-device connectivity, in 3GPP referred to as *sidelink* transmission, will be addressed in later releases.

In parallel to the work on the NR radio-access technology in 3GPP, a new 5G core network has been developed, responsible for functions not related to the radio access but needed for providing a complete network. However, it is possible to connect the NR radio-access network also to the legacy LTE core network known as the *Evolved Packet Core* (EPC). In fact, this is the case when operating NR in non-standalone mode where LTE and EPC handle functionality like connection set-up and paging and NR primarily provides a data-rate and capacity booster. Later releases will introduce standalone operation with NR connecting to the 5G core.

The remaining part of this chapter provides an overview of NR radio access including basic design principles and the most important technology components

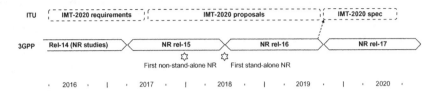

FIGURE 5.1

3GPP timeline.

of NR release 15. The chapter can either be read on its own to get a high-level overview of NR, or as an introduction to the subsequent Chapters 6−19, which provide a detailed description of the NR.

Compared to LTE, NR provides many benefits. Some of the main ones are:

- exploitation of much higher-frequency bands as a mean to obtain additional spectra to support very wide transmission bandwidths and the associated high data rates;
- ultra-lean design to enhance network energy performance and reduce interference;
- forward compatibility to prepare for future, yet unknown, use cases and technologies;
- low latency to improve performance and enable new use cases; and
- a beam-centric design enabling extensive usage of beamforming and a massive number of antenna elements not only for data transmission (which to some extent is possible in LTE) but also for control-plane procedures such as initial access.

The first three can be classified as design principles (or requirements on the design) and will be discussed first, followed by a discussion of the key technology components applied to NR.

5.1 HIGHER-FREQUENCY OPERATION AND SPECTRUM FLEXIBILITY

One key feature of NR is a substantial expansion in terms of the range of spectra in which the radio-access technology can be deployed. Unlike LTE, where support for licensed spectra at 3.5 GHz and unlicensed spectra at 5 GHz are just being introduced, NR supports licensed-spectrum operation from below 1 GHz up to 52.6 GHz[1] already from its first release, with extension to unlicensed spectra also already being planned for.

---

[1]The upper limit of 52.6 GHz is due to some very specific spectrum situations.

Operation at mm-wave frequencies offers the possibility for large amounts of spectrum and associated very wide transmission bandwidths, thereby enabling very high traffic capacity and extreme data rates. However, higher frequencies are also associated with higher radio-channel attenuation, limiting the network coverage. Although this can partly be compensated for by means of advanced multi-antenna transmission/reception, which is one of the motivating factors for the beam-centric design in NR, a substantial coverage disadvantage remains, especially in non-line-of-sight and outdoor-to-indoor propagation conditions. Thus, operation in lower-frequency bands will remain a vital component for wireless communication also in the 5G era. Especially, joint operation in lower *and* higher spectra, for example 2 GHz and 28 GHz, can provide substantial benefits. A higher-frequency layer, with access to a large amount of spectra can provide service to a large fraction of the users despite the more limited coverage. This will reduce the load on the more bandwidth-constrained lower-frequency spectrum, allowing the use of this to focus on the worst-case users [66].

Another challenge with operation in higher-frequency bands is the regulatory aspects. For non-technical reasons, the rules defining the allowed radiation changes at 6 GHz, from a SAR-based limitation to a more EIRP-like limitation. Depending on the device type (handheld, fixed, etc.), this may result in a reduced transmission power, making the link budget more challenging than what propagation conditions alone may indicate and further stressing the benefit of combined low-frequency/high-frequency operation.

5.2 ULTRA-LEAN DESIGN

An issue with current mobile-communication technologies is the amount of transmissions carried by network nodes regardless of the amount of user traffic. Such signals, sometimes referred to as "always-on" signals, include, for example, signals for base-station detection, broadcast of system information, and always-on reference signals for channel estimation. Under the typical traffic conditions for which LTE was designed, such transmissions constitute only a minor part of the overall network transmissions and thus have a relatively small impact on the network performance. However, in very dense networks deployed for high peak data rates, the average traffic load per network node can be expected to be relatively low, making the always-on transmissions a more substantial part of the overall network transmissions.

The always-on transmissions have two negative impacts:

- they impose an upper limit on the achievable network energy performance; and
- they cause interference to other cells, thereby reducing the achievable data rates.

The *ultra-lean design* principle aims at minimizing the always-on transmissions, thereby enabling higher network energy performance and higher achievable data rates.

In comparison, the LTE design is heavily based on cell-specific reference signals, signals that a device can assume are always present and use for channel estimation, tracking, mobility measurements, etc. In NR, many of these procedures have been revisited and modified to account for the ultra-lean design principle. For example, the cell-search procedures have been redesigned in NR compared to LTE to support the ultra-lean paradigm. Another example is the demodulation reference-signal structure where NR relies heavily on reference signals being present only when data are transmitted but not otherwise.

5.3 FORWARD COMPATIBILITY

An important aim in the development of the NR specification was to ensure a high degree of *forward compatibility* in the radio-interface design. In this context, forward compatibility implies a radio-interface design that allows for substantial future evolution, in terms of introducing new technology and enabling new services with yet unknown requirements and characteristics, while still supporting legacy devices on the same carrier.

Forward compatibility is inherently difficult to guarantee. However, based on experience from the evolution of previous generations, 3GPP agreed on some basic design principles related to NR forward compatibility as quoted from [3]:

- Maximizing the amount of time and frequency resources that can be flexibly utilized or that can be left blank without causing backward compatibility issues in the future;
- Minimizing transmission of always-on signals;
- Confining signals and channels for physical layer functionalities within a configurable/allocable time/frequency resource.

According to the third bullet one should, as much as possible, avoid having transmissions on time/frequency resources fixed by the specification. In this way one retains flexibility for the future, allowing for later introduction of new types of transmissions with limited constraints from legacy signals and channels. This differs from the approach taken in LTE where, for example, a synchronous hybrid-ARQ protocol is used, implying that a retransmission in the uplink occurs at a fixed point in time after the initial transmission. The control channels are also vastly more flexible in NR compared to LTE in order not to unnecessarily block resources.

Note that these design principles partly coincide with the aim of ultra-lean design as described above. There is also a possibility in NR to configure *reserved resources*, that is, time-frequency resources that, when configured, are not used for transmission and thus available for future radio-interface extensions. The same mechanism can also be used for LTE-NR coexistence in the case of overlapping LTE and NR carriers.

5.4 TRANSMISSION SCHEME, BANDWIDTH PARTS, AND FRAME STRUCTURE

Similar to LTE [28], OFDM was found to be a suitable waveform for NR due to its robustness to time dispersion and ease of exploiting both the time and frequency domain when defining the structure for different channels and signals. However, unlike LTE where DFT-precoded OFDM is the sole transmission scheme in the uplink, NR uses conventional, that is, non-DFT-precoded OFDM, as the baseline uplink transmission scheme due to the simpler receiver structures in combination with spatial multiplexing and an overall desire to have the same transmission scheme in both uplink and downlink. Nevertheless, DFT-precoding can be used as a complement in the uplink for similar reasons as in LTE, namely to enable high power-amplifier efficiency on the device side by reducing the *cubic metric* [60]. Cubic metric is a measure of the amount of additional power back-off needed for a certain signal waveform.

To support a wide range of deployment scenarios, from large cells with sub-1 GHz carrier frequency up to mm-wave deployments with very wide spectrum allocations, NR supports a flexible OFDM numerology with subcarrier spacings ranging from 15 kHz up to 240 kHz with a proportional change in cyclic prefix duration. A small subcarrier spacing has the benefit of providing a relatively long cyclic prefix in absolute time at a reasonable overhead while higher subcarrier spacings are needed to handle, for example, the increased phase noise at higher carrier frequencies. Up to 3300 subcarriers are used although the maximum total bandwidth is limited to 400 MHz, resulting in the maximum carrier bandwidths of 50/100/200/400 MHz for subcarrier spacings of 15/30/60/120 kHz, respectively. If even larger bandwidths are to be supported, carrier aggregation can be used.

Although the NR physical-layer specification is band-agnostic, not all supported numerologies are relevant for all frequency bands (see Fig. 5.2). For each frequency band, radio requirements are therefore defined for a subset of the supported numerologies as illustrated in Fig. 5.2. The frequency range 0.45−6 GHz is commonly referred to as *frequency range* 1 (FR1) in the specifications, while the range 24.25−52.6 GHz is known as FR2. Currently, there is no NR spectrum identified between 6 GHz and 24.25 GHz. However, the basic NR radio-access technology is spectrum agnostic and the NR specifications can easily be extended to cover additional spectra, for example, spectra from 6 GHz up to 24.25 GHz.

In LTE, all devices support the maximum carrier bandwidth of 20 MHz. However, given the very wide bandwidths possible in NR, it is not reasonable to

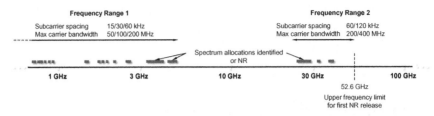

FIGURE 5.2

Spectra identified for NR and corresponding subcarrier spacings.

require all devices to support the maximum carrier bandwidth. This has implications on several areas and requires a design different from LTE, for example the design of control channels as discussed later. Furthermore, NR allows for device-side *receiver-bandwidth adaptation* as a means to reduce the device energy consumption. Bandwidth adaptation refers to the use of a relatively modest bandwidth for monitoring control channels and receiving medium data rates, and dynamically opens up a wideband receiver only when needed to support very high data rates.

To handle these two aspects NR defines *bandwidth parts* that indicate the bandwidth over which a device is currently assumed to receive transmissions of a certain numerology. If a device is capable of simultaneous reception of multiple bandwidths parts, it is in principle possible to, on a single carrier, mix transmissions of different numerologies for a single device, although release 15 only supports a single active bandwidth part at a time.

The NR time-domain structure is illustrated in Fig. 5.3 with a 10-ms radio frame divided into ten 1-ms subframes. A subframe is in turn divided into slots consisting of 14 OFDM symbols each, that is, the duration of a slot in milliseconds depends on the numerology. For the 15-kHz subcarrier spacing, an NR slot has a structure that is identical to the structure of an LTE subframe, which is beneficial from a coexistence perspective. Since a slot is defined as a fixed number of OFDM symbols, a higher subcarrier spacing leads to a shorter slot duration. In principle this could be used to support lower-latency transmission, but as the cyclic prefix also shrinks when increasing the subcarrier spacing, it is not a feasible approach in all deployments. Therefore, NR supports a more efficient approach to low latency by allowing for transmission over a fraction of a slot, sometimes referred to as "mini-slot" transmission. Such transmissions can also preempt an already ongoing slot-based transmission to another device, allowing for immediate transmission of data requiring very low latency.

Having the flexibility of starting a data transmission not only at the slot boundaries is also useful when operating in unlicensed spectra. In unlicensed spectra the transmitter is typically required to ensure that the radio channel is not occupied by other transmissions prior to starting a transmission, a procedure commonly known as "listen-before-talk." Clearly, once the channel is found to be

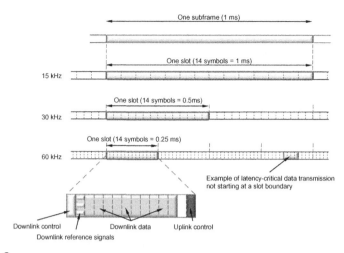

FIGURE 5.3

Frame structure (TDD assumed in this example).

available it is beneficial to start the transmission immediately, rather than wait until the start of the slot, in order to avoid some other transmitter initiating a transmission on the channel.

Operation in the mm-wave domain is another example of the usefulness of "mini-slot" transmissions as the available bandwidth in such deployments is often very large and even a few OFDM symbols can be sufficient to carry the available payload. This is of particular use in conjunction with *analog beamforming*, discussed below, where transmissions to multiple devices in different beams cannot be multiplexed in the frequency domain but only in the time domain.

Unlike LTE, NR does not include cell-specific reference signals but solely relies on user-specific demodulation reference signals for channel estimation. Not only does this enable efficient beamforming and multi-antenna operation as discussed below, it is also in line with the ultra-lean design principle described above. In contrast to cell-specific reference signals, demodulation reference signals are not transmitted unless there are data to transmit, thereby improving network energy performance and reducing interference.

The overall NR time/frequency structure, including bandwidth parts, is the topic of Chapter 7.

5.5 DUPLEX SCHEMES

The duplex scheme to use is typically given by the spectrum allocation at hand. For lower-frequency bands, allocations are often paired, implying frequency-division duplex (FDD) as illustrated in Fig. 5.4. At higher-frequency bands,

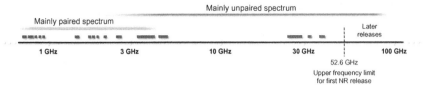

FIGURE 5.4

Spectrum and duplex schemes.

unpaired spectrum allocations are increasingly common, calling for time-division duplex (TDD). Given the significantly higher carrier frequencies supported by NR compared to LTE, efficient support for unpaired spectra is an even more critical component of NR, compared to LTE.

NR can operate in both paired and unpaired spectra using *one* common frame structure, unlike LTE were two different frame structures were used (and later expanded to three when support for unlicensed spectra was introduced in release 13). The basic NR frame structure is designed such that it can support both half-duplex and full-duplex operation. In half duplex, the device cannot transmit and receive at the same time. Examples hereof are TDD and half-duplex FDD. In full-duplex operation, on the other hand, simultaneous transmission and reception is possible with FDD as a typical example.

As already mentioned, TDD increases in importance when moving to higher-frequency bands where unpaired spectrum allocations are more common. These frequency bands are less useful for wide-area coverage with very large cells due to their propagation conditions but are highly relevant for local-area coverage with smaller cell sizes. Furthermore, some of the problematic interference scenarios in wide-area TDD networks are less pronounced in local area deployments with lower transmission power and below-rooftop antenna installations. In such denser deployments with smaller cell sizes, the per-cell traffic variations are more rapid compared to large-cell deployments with a large number of active devices per cell. To address such scenarios, *dynamic TDD*, that is, the possibility for dynamic assignment and reassignment of time-domain resources between the downlink and uplink transmission directions, is a key NR technology component. This is in contrast to LTE where the uplink−downlink allocation does not change over time.[2] Dynamic TDD enables following rapid traffic variations which are particularly pronounced in dense deployments with a relatively small number of users per cell. For example, if a user is (almost) alone in a cell and needs to download a large object, most of the resources should be utilized in the downlink direction and only a small fraction in the uplink direction. At a later point in

---

[2]In later LTE releases, the eIMTA features allows some dynamics in the uplink−downlink allocation.

time, the situation may be different and most of the capacity is needed in the uplink direction.

The basic approach to dynamic TDD is for the device to monitor for downlink control signaling and follow the scheduling decisions. If the device is instructed to transmit, it transmits in the uplink, otherwise it will attempt to receive any downlink transmissions. The uplink−downlink allocation is then completely under the control of the scheduler and any traffic variations can be dynamically tracked. There are deployment scenarios where dynamic TDD may not be useful, but it is much simpler to restrict the dynamics of a dynamic scheme in those scenarios when needed rather than trying to add dynamics to a fundamentally semistatic design as LTE. For example, in a wide-area macro network with above-rooftop antennas, the intercell interference situation requires coordination of the uplink−downlink allocation between the cells. In such situations, a semistatic allocation is appropriate with operation along the lines of LTE. This can be obtained by the appropriate scheduling implementation. There is also the possibility to semistatically configure the transmission direction of some or all of the slots, a feature that can allow for reduced device energy consumption as it is not necessary to monitor for downlink control channels in slots that are a priori known to be reserved for uplink usage.

5.6 LOW-LATENCY SUPPORT

The possibility for very low latency is an important characteristic of NR and has impacted many of the NR design details. One example is the use of "front-loaded" reference signals and control signaling, as illustrated in Fig. 5.3. By locating the reference signals and downlink control signaling carrying scheduling information at the beginning of the transmission and not using time-domain interleaving across OFDM symbols, a device can start processing the received data immediately without prior buffering, thereby minimizing the decoding delay. The possibility for transmission over a fraction of a slot, sometimes referred to as "mini-slot" transmission, is another example.

The requirements on the device (and network) processing times are tightened significantly in NR compared to LTE. As an example, a device has to respond with a hybrid-ARQ acknowledgment of approximately one slot (or even less depending on device capabilities) after receiving a downlink data transmission. Similarly, the time from grant reception to uplink data transfer is in the same range.

The higher-layer protocols MAC and RLC have also been designed with low latency in mind with header structures chosen to enable processing without knowing the amount of data to transmit (see Chapter 6). This is especially important in the uplink direction as the device may only have a few OFDM symbols after receiving the uplink grant until the transmission should take place. In contrast, the LTE protocol design requires the MAC and RLC protocol layers to know the amount of data to transmit before any processing can take place, which makes support for a very low latency more challenging.

5.7 SCHEDULING AND DATA TRANSMISSION

One key characteristic of mobile radio communication is the large and typically rapid variations in the instantaneous channel conditions stemming from frequency-selective fading, distance-dependent path loss, and random interference variations due to transmissions in other cells and by other devices. Instead of trying to combat these variations, they can be exploited through *channel-dependent scheduling* where the time-frequency resources are dynamically shared between users (see Chapter 14 for details). Dynamic scheduling is used in LTE as well and on a high level, the NR scheduling framework is similar to the one in LTE. The scheduler, residing in the base station, takes scheduling decisions based on channel-quality reports obtained from the devices. It also takes different traffic priorities and quality-of-service requirements into account when forming the scheduling decisions sent to the scheduled devices.

Each device monitors several *physical downlink control channels* (PDCCHs), typically once per slot, although it is possible to configure more frequent monitoring to support traffic requiring very low latency. Upon detection of a valid PDCCH, the device follows the scheduling decision and receives (or transmits) one unit of data known as a transport block in NR.

In the case of downlink data transmission, the device attempts to decode the downlink transmission. Given the very high data rates supported by NR, channel-coding data transmission is based on low-density parity-check (LDPC) codes [68]. LDPC codes are attractive from an implementation perspective, especially at higher code rates where they can offer a lower complexity than Turbo codes as used in LTE.

Hybrid automatic repeat-request (ARQ) retransmission using incremental redundancy is used where the device reports the outcome of the decoding operation to the base station (see Chapter 13 for details). In the case of erroneously received data, the network can retransmit the data and the device combines the soft information from multiple transmission attempts. However, retransmitting the whole transport block could in this case become inefficient. NR therefore supports retransmissions on a finer granularity known as *code-block group* (CBG). This can also be useful when handling *preemption*. An urgent transmission to a second device may use only one or a few OFDM symbols and therefore cause high interference to the first device in some OFDM symbols only. In this case it may be sufficient to retransmit the interfered CBGs only and not the whole data block. Handling of preempted transmission can be further assisted by the possibility to indicate to the first device the impacted time-frequency resources such that it can take this information into account in the reception process.

Although dynamic scheduling is the basic operation of NR, operation without a dynamic grant can be configured. In this case, the device is configured in advance with resources that can be used for uplink data transmission (or downlink data reception). Once a device has data available it can immediately commence uplink transmission without going through the scheduling request—grant cycle, thereby enabling lower latency.

5.8 CONTROL CHANNELS

Operation of NR requires a set of physical-layer control channels to carry the scheduling decisions in the downlink and to provide feedback information in the uplink. A detailed description of the structure of these control channels is provided in Chapter 10.

Downlink control channels are known as PDCCHs (*physical downlink control channels*). One major difference compared to LTE is the more flexible time-frequency structure of downlink control channels where PDCCHs are transmitted in one or more *control resource sets* (CORESETs) which, unlike LTE where the full carrier bandwidth is used, can be configured to occupy only part of the carrier bandwidth. This is needed in order to handle devices with different bandwidth capabilities and also in line with the principles for forward compatibility as discussed above. Another major difference compared to LTE is the support for beamforming of the control channels, which has required a different reference signal design with each control channel having its own dedicated reference signal.

Uplink control information, such as hybrid-ARQ acknowledgments, channel-state feedback for multi-antenna operation, and scheduling request for uplink data awaiting transmission, are transmitted using the *physical uplink control channel* (PUCCH). There are several different PUCCH formats, depending on the amount of information and the duration of the PUCCH transmission. The *short PUCCH* is transmitted in the last one or two symbols of a slot and can support very fast feedback of hybrid-ARQ acknowledgments in order to realize so-called self-contained slots where the delay from the end of the data transmission to the reception of the acknowledgment from the device is in the order of an OFDM symbol, corresponding to a few tens of microseconds depending on the numerology used. This can be compared to almost 3 ms in LTE and is yet another example on how the focus on low latency has impacted the NR design. For situations when the duration of the short PUCCH is too short to provide sufficient coverage, there are also possibilities for longer PUCCH durations.

For the physical-layer control channels, for which the information blocks are small compared to data transmission and hybrid-ARQ is not used, polar codes [17] have been selected. For the smallest control payloads, Reed−Muller codes are used.

5.9 BEAM-CENTRIC DESIGN AND MULTI-ANTENNA TRANSMISSION

Support for a for a large number of steerable antenna elements for both transmission and reception is a key feature of NR. At higher-frequency bands, the large number of antenna elements are primarily used for beamforming to extend coverage, while at lower-frequency bands they enable full-dimensional MIMO, sometimes referred to as massive MIMO, and interference avoidance by spatial separation.

NR channels and signals, including those used for control and synchronization, have all been designed to support beamforming (Fig. 5.5). Channel-state information (CSI) for operation of massive multi-antenna schemes can be obtained by feedback of CSI reports based on transmission of CSI reference signals in the downlink, as well as using uplink measurements exploiting channel reciprocity.

To provide implementation flexibility, NR is deliberately supporting functionality to support analog beam-forming a well as digital precoding/beam-forming (see Chapter 11). At high frequencies, analog beamforming, where the beam is shaped after digital-to-analog conversion, may be necessary from an implementation perspective, at least initially. Analog beamforming results in the constraint that a receive or transmit beam can only be formed in one direction at a given time instant and requires beam-sweeping where the same signal is repeated in multiple OFDM symbols but in different transmit beams. By having beam-sweeping possibility, it is ensured that any signal can be transmitted with a high gain, narrow beam to reach the entire intended coverage area.

Signaling to support beam-management procedures is specified, such as an indication to the device to assist selection of a receive beam (in the case of analog receive beamforming) to be used for data and control reception. For a large number of antennas, beams are narrow and beam tracking can fail, therefore beam-recovery procedures have also been defined where a device can trigger a beam-recovery procedure. Moreover, a cell may have multiple transmission points, each with beams and the beam-management procedures allow for device transparent mobility for seamless handover between the beams of different points. Additionally, uplink-centric and reciprocity-based beam management is possible by utilizing uplink signals.

With the use of a massive number of antenna elements for lower-frequency bands, the possibility to separate users spatially increases both in uplink and downlink, but requires that the transmitter has channel knowledge. For NR, extended support for such multi-user spatial multiplexing is introduced, either by using a high-resolution channel-state-information feedback using a linear

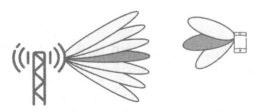

FIGURE 5.5

Beamforming in NR.

combination of DFT vectors, or uplink sounding reference signals targeting the utilization of channel reciprocity.

Twelve orthogonal demodulation reference signals are specified for multi-user MIMO transmission purposes, while an NR device can maximally receive eight MIMO layers in the downlink and up to four layers in the uplink. Moreover, additional configuration of a phase tracking reference signal is introduced in NR since the increased phase noise power at high carrier frequency bands otherwise will degrade demodulation performance for larger modulation constellations, for example 64 QAM.

In addition, NR is prepared to support distributed MIMO, although the support is not complete in release 15. Distributed MIMO implies that the device can receive multiple independent physical data shared channels (PDSCHs) per slot to enable simultaneous data transmission from multiple transmission points to the same user. In essence, some MIMO layers are transmitted from one site while other layers are transmitted from another site.

Multi-antenna transmission in general, as well as a more detailed discussion on NR multi-antenna precoding, is described in Chapter 11 with beam management being the subject of Chapter 12.

5.10 INITIAL ACCESS

Initial access is the procedures allowing a device to find a cell to camp on, receive the necessary system information, and to request a connection through random access. The basic structure of NR initial access, described in Chapter 16, is similar to the corresponding functionality of LTE [28]:

- There is a pair of downlink signals, the *primary synchronization signal* (PSS) and the *secondary synchronization signal* (SSS), that is used by devices to find, synchronize to, and identify a network;
- There is a downlink *physical broadcast channel* (PBCH) transmitted together with the PSS/SSS. The PBCH carries a minimum amount of system information including an indication where the remaining broadcast system information is transmitted. In the context of NR, the PSS, SSS, and PBCH are jointly referred to as a *synchronization signal* (SS) *block*;
- There is a four-stage random-access procedure, commencing with the uplink transmission of a *random-access preamble.*

However, there are some important differences between LTE and NR in terms of initial access. These differences come mainly from the ultra-lean principle and the beam-centric design, both of which impact the initial access procedures and partly lead to different solutions compared to LTE.

In LTE, the PSS, SSS, and PBCH are located at the center of the carrier and are transmitted once every 5 ms. Thus, by dwelling on each possible carrier frequency during at least 5 ms, a device is guaranteed to receive at least one PSS/SSS/PBCH transmission if a carrier exists at the specific frequency. Without any a priori knowledge a device must search all possible carrier frequencies over a carrier raster of 100 kHz.

To enable higher NR network energy performance in line with the ultra-lean principle, the SS block is, by default, transmitted once every 20 ms. Due to the longer period between consecutive SS blocks, compared to the corresponding signals/channels in LTE, a device searching for NR carriers must dwell on each possible frequency for a longer time. To reduce the overall search time while keeping the device complexity comparable to LTE, NR supports a *sparse frequency raster* for SS block. This implies that the possible frequency-domain positions of the SS block could be significantly sparser, compared to the possible positions of an NR carrier (the *carrier raster*). As a consequence, the SS block will typically not be located at the center of the NR carrier, which has impacted the NR design.

The sparse SS-block raster enables a significantly reduced time for initial cell search, at the same time as the network energy performance can be significantly improved due to the longer SS-block period.

Network-side beam-sweeping is supported for both downlink SS-slock transmission and uplink random-access reception as a means to improve coverage, especially in the case of operation at higher frequencies. It is important to realize that beam sweeping is a *possibility* enabled by the NR design. It does not imply that it must be used. Especially at lower carrier frequencies, beam sweeping may not be needed.

5.11 INTERWORKING AND LTE COEXISTENCE

As it is difficult to provide full coverage at higher frequencies, interworking with systems operating at lower frequencies is important. In particular, a coverage imbalance between uplink and downlink is a common scenario, especially if they are in different frequency bands. The higher transmit power for the base station compared to the mobile device results in the downlink achievable data rates often being bandwidth limited, making it more relevant to operate the downlink in a higher spectrum where wider bandwidth may be available. In contrast, the uplink is more often power-limited, reducing the need for wider bandwidth. Instead, higher data rates may be achieved on lower-frequency spectra, despite there being less available bandwidth, due to less radio-channel attenuation.

Through interworking, a high-frequency NR system can complement a low-frequency system (see Chapter 17 for details). The lower-frequency system can be either NR or LTE, and NR supports interworking with either of these. The interworking can be realized at different levels, including intra-NR carrier

Downlink and uplink co-existence   Uplink-only co-existence

FIGURE 5.6

Example of NR−LTE coexistence.

aggregation, dual connectivity[3] with a common packet data convergence protocol (PDCP) layer, and handover.

However, the lower-frequency bands are often already occupied by current technologies, primarily LTE. Furthermore, an additional low-frequency spectrum is planned to be deployed with LTE in the relatively near future. LTE/NR *spectrum coexistence*, that is, the possibility for an operator to deploy NR in the same spectrum as an already existing LTE deployment has therefore been identified as a way to enable early NR deployment in lower-frequency spectra without reducing the amount of spectrum available to LTE.

Two coexistence scenarios were identified in 3GPP and guided the NR design:

- In the first scenario, illustrated in the left part of Fig. 5.6, there is LTE/NR coexistence in both downlink and uplink. Note that this is relevant for both paired and unpaired spectra although a paired spectrum is used in the illustration.
- In the second scenario, illustrated in the right part of Fig. 5.6, there is coexistence only in the uplink transmission direction, typically within the uplink part of a lower-frequency paired spectrum, with NR downlink transmission taking place in the spectrum dedicated to NR, typically at higher frequencies. This scenario attempts to address the uplink−downlink imbalance discussed above. NR supports a *supplementary uplink* (SUL) to specifically handle this scenario.

The possibility for an LTE-compatible NR numerology based on 15-kHz subcarrier spacing, enabling identical time/frequency resource grids for NR and LTE, is one of the fundamental tools for such coexistence. The flexible NR scheduling with a scheduling granularity as small as one symbol can then be used to avoid scheduled NR transmissions to collide with key LTE signals, such as cell-specific reference signals, CSI-RS, and the signals/channels used for LTE initial access. Reserved resources, introduced for forward compatibility (see Section 5.3), can also be used to further enhance NR−LTE coexistence. It is possible to configure reserved resources matching the cell-specific reference signals in LTE, thereby enabling an enhanced NR−LTE overlay in the downlink.

---

[3]In the December version of release 15, dual connectivity is only supported between NR and LTE. Dual connectivity between NR and NR is part of the final June 2018 release 15.

Radio-Interface Architecture

This chapter contains a brief overview of the overall architecture of an NR radio-access network and the associated core network, followed by descriptions of the radio-access network user-plane and control-plane protocols.

6.1 OVERALL SYSTEM ARCHITECTURE

In parallel to the work on the NR (New Radio) radio-access technology in 3GPP, the overall system architectures of both the *Radio-Access Network* (RAN) and the *Core Network* (CN) were revisited, including the split of functionality between the two networks.

The RAN is responsible for all radio-related functionality of the overall network including, for example, scheduling, radio-resource handling, retransmission protocols, coding, and various multi-antenna schemes. These functions will be discussed in detail in the subsequent chapters.

The 5G core network is responsible for functions not related to the radio access but needed for providing a complete network. This includes, for example, authentication, charging functionality, and setup of end-to-end connections. Handling these functions separately, instead of integrating them into the RAN, is beneficial as it allows for several radio-access technologies to be served by the same core network.

However, it is possible to connect the NR radio-access network also to the legacy LTE (Long-Term Evolution) core network known as the *Evolved Packet Core* (EPC). In fact, this is the case when operating NR in non-standalone mode, where LTE and EPC handle functionality like connection set-up and paging. Later releases will introduce standalone operation with NR connecting to the 5G core, as well as LTE connecting to the 5G core. Thus, the LTE and NR radio-access schemes and their corresponding core networks are closely related, unlike the transition from 3G to 4G where the 4G LTE radio-access technology cannot connect to a 3G core network.

Although this book focuses on the NR radio access, a brief overview of the 5G core network, as well as how it connects to the RAN, is useful as a background.

5G NR: The Next Generation Wireless Access Technology. DOI: https://doi.org/10.1016/B978-0-12-814323-0.00006-5

6.1.1 5G CORE NETWORK

The 5G core network builds upon the EPC with three new areas of enhancement compared to EPC: service-based architecture, support for network slicing, and control-plane/user-plane split.

A service-based architecture is the basis for the 5G core. This means that the specification focuses on the services and functionalities provided by the core network, rather than nodes as such. This is natural as the core network today is already often highly virtualized with the core network functionality running on generic computer hardware.

Network slicing is a term commonly seen in the context of 5G. A network slice is a logical network serving a certain business or customer need and consists of the necessary functions from the service-based architecture configured together. For example, one network slice can be set up to support mobile broadband applications with full mobility support, similar to what is provided by LTE, and another slice can be set up to support a specific non-mobile, latency-critical industry-automation application. These slices will all run on the same underlying physical core and radio networks, but, from the end-user application perspective, they appear as independent networks. In many aspects it is similar to configuring multiple virtual computers on the same physical computer. Edge computing, where parts of the end-user application run close to the core network edge to provide low latency, can also be part of such a network slice.

Control-plane/user-plane split is emphasized in the 5G core network architecture, including independent scaling of the capacity of the two. For example, if more control plane capacity is need, it should be straightforward to add it without affecting the user-plane of the network.

On a high level, the 5G core can be illustrated as shown in Fig. 6.1. The figure uses a service-based representation, where the services and functionalities are

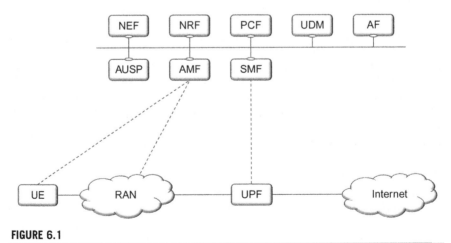

FIGURE 6.1

High-level core network architecture (service-based description).

in focus. In the specifications there is also an alternative, reference-point description, focusing on the point-to-point interaction between the functions, but that description is not captured in the figure.

The user-plane function consists of the *User Plane Function* (UPF) which is a gateway between the RAN and external networks such as the Internet. Its responsibilities include packet routing and forwarding, packet inspection, quality-of-service handling and packet filtering, and traffic measurements. It also serves as an anchor point for (inter-RAT) mobility when necessary.

The control-plane functions consist of several parts. The *Session Management Function* (SMF) handles, among other functions, IP address allocation for the device (also known as *User Equipment*, UE), control of policy enforcement, and general session-management functions. The *Access and Mobility Management Function* (AMF) is in charge of control signaling between the core network and the device, security for user data, idle-state mobility, and authentication. The functionality operating between the core network, more specifically the AMF, and the device is sometimes referred to as the *Non-Access Stratum* (NAS), to separate it from the *Access Stratum* (AS), which handles functionality operating between the device and the radio-access network.

In addition, the core network can also handle other types of functions, for example, the *Policy Control Function* (PCF) responsible for policy rules, the *Unified Data Management* (UDM) responsible for authentication credentials and access authorization, the *Network Exposure Function* (NEF), the *NR Repository Function* (NRF), the *Authentication Server Function* (AUSF) handing authentication functionality, and the *Application Function* (AF). These functions are not discussed further in this book and the reader is referred to [13] for further details.

It should be noted that the core network functions can be implemented in many ways. For example, all the functions can be implemented in a single physical node, distributed across multiple nodes, or executed on a cloud platform.

The description above focused on the new 5G core network, developed in parallel to the NR radio access and capable of handling both NR and LTE radio accesses. However, to allow for an early introduction of NR in existing networks, it is also possible to connect NR to EPC, the LTE core network. This is illustrated as "option 3" in Fig. 6.2 and is also known as "non-standalone operation" as LTE is used for control-plane functionality such as initial access, paging, and mobility. The nodes denoted eNB and gNB will be discussed in more detail in the next

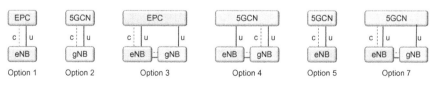

FIGURE 6.2

Different combinations of core networks and radio-access technologies.

section; for the time being eNB and gNB can be thought of as base stations for LTE and NR, respectively.

In option 3, the EPC core network is connected to the eNB. All control-plane functions are handled by LTE, and NR is used only for the user-plane data. The gNB is connected to the eNB and user-plane data from the EPC can be forwarded from the eNB to the gNB. There are also variants of this: option 3a and option 3x. In option 3a, the user-plane parts of both the eNB and gNB are directly connected to the EPC. In option 3x, only the gNB user plane is connected to the EPC and user-plane data to the eNB are routed via the gNB.

For standalone operation, the gNB is connected directly to the 5G core as shown in option 2. Both user-plane and control-plane functions are handled by the gNB. Options 4, 5, and 7 show various possibilities for connecting an LTE eNB to the 5GCN.

6.1.2 RADIO-ACCESS NETWORK

The radio-access network can have two types of nodes connected to the 5G core network:

- A gNB, serving NR devices using the NR user-plane and control-plane protocols; or
- An ng-eNB, serving LTE devices using the LTE user-plane and control-plane protocols.[1]

A radio-access network consisting of both ng-eNBs for LTE radio access and gNBs for NR radio access is known as an NG-RAN, although the term RAN will be used in the following for simplicity. Furthermore, it will be assumed that the RAN is connected to the 5G core and hence 5G terminology, such as gNB, will be used. In other words, the description will assume a 5G core network and an NR-based RAN as shown in option 2 in Fig. 6.2. However, as already mentioned, the first version of NR operates in non-standalone mode where NR is connected to the EPC using option 3. The principles are in this case similar, although the naming of the nodes and interfaces differs slightly.

The gNB (or ng-eNB) is responsible for all radio-related functions in one or several cells, for example, radio resource management, admission control, connection establishment, routing of user-plane data to the UPF and control-plane information to the AMF, and QoS flow management. It is important to note that an gNB is a *logical* node and not a physical implementation. One common implementation of an gNB is a three-sector site, where a base station is handling transmissions in three cells, although other implementations can be found as well, such as one baseband processing unit to which several remote radio heads are connected. Examples of the latter are a large number of indoor cells, or several

---

[1]Fig. 6.2 is simplified as it does not make a distinction between eNB connected to the EPC and ng-eNB connected to the 5GCN.

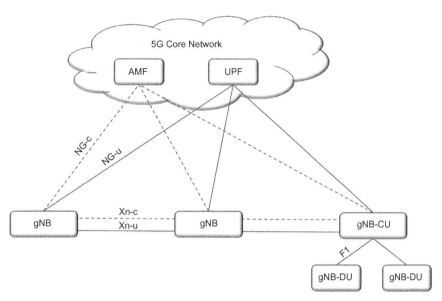

FIGURE 6.3

Radio-access network interfaces.

cells along a highway, belonging to the same gNB. Thus, a base station is a *possible* implementation of, but not *the same* as, a gNB.

As can be seen in Fig. 6.3, the gNB is connected to the 5G core network by means of the *NG interface*, more specifically to the UPF by means of the *NG user-plane part* (NG-u), and to the AMF by means of the *NG control-plane part* (NG-c). One gNB can be connected to multiple UPFs/AMFs for the purpose of load sharing and redundancy.

The *Xn interface*, connecting gNBs to each other, is mainly used to support active-mode mobility and dual connectivity. This interface may also be used for multicell *Radio Resource Management* (RRM) functions. The Xn interface is also used to support lossless mobility between neighboring cells by means of packet forwarding.

There is also a standardized way to split the gNB into two parts, a central unit (gNB-CU) and one or more distributed units (gNB-DU) using the *F1 interface*. In the case of a split gNB, the RRC, PDCP, and SDAP protocols, described in more detail below, reside in the gNB-CU and the remaining protocol entities (RLC, MAC, PHY) in the gNB-DU.

The interface between the gNB (or the gNB-DU) and the device is known as the *Uu interface*.

For a device to communicate, at least one connection between the device and the network is required. As a baseline, the device is connected to one cell handling all the uplink as well as downlink transmissions. All data flows, user data

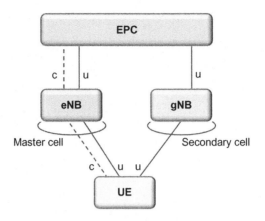

FIGURE 6.4

LTE—NR dual connectivity using option 3.

as well as RRC signaling, are handled by this cell. This is a simple and robust approach, suitable for a wide range of deployments. However, allowing the device to connect to the network through multiple cells can be beneficial in some scenarios. One example is user-plane aggregation, where flows from multiple cells are aggregated in order to increase the data rate. Another example is control-plane/ user-plane separation where the control plane communication is handled by one node and the user plane by another. The scenario of a device connected to two cells[2] is known as *dual connectivity*.

Dual connectivity between LTE and NR is of particular importance as it is the basis for non-standalone operation using option 3 as illustrated in Fig. 6.4. The LTE-based master cell handles control-plane and (potentially) user-plane signaling, and the NR-based secondary cell handles user-plane only, in essence boosting the data rates.

Dual connectivity between NR and NR is not part of the December 2017 version of release 15 but is possible in the final June 2018 version of release 15.

6.2 QUALITY-OF-SERVICE HANDLING

Handling of different quality-of-service (QoS) requirements is possible already in LTE, and NR builds upon and enhances this framework. The key principles of LTE are kept, namely that the network is in charge of the QoS control and that

---

[2]Actually, two cell groups, the master cell group (MCG) and the secondary cell group (SCG) in the case of carrier aggregation as carrier aggregation implies multiple cells in each of the two cell groups.

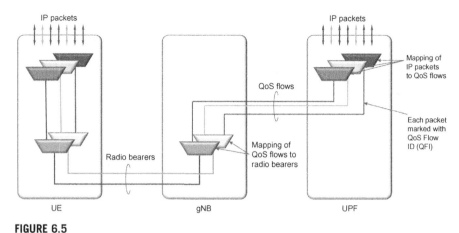

FIGURE 6.5

QoS flows and radio bearers during a PDU session.

the 5G core network but not the radio-access network is aware of the service. QoS handling is essential for the realization of network slicing.

For each connected device, there is one or more *PDU sessions*, each with one or more *QoS flows* and *data radio bearers*. The IP packets are mapped to the QoS flows according to the QoS requirements, for example in terms of delay or required data rate, as part of the UDF functionality in the core network. Each packet can be marked with a *QoS Flow Identifier* (QFI) to assist uplink QoS handling. The second step, mapping of QoS flows to data radio bearers, is done in the radio-access network. Thus, the core network is aware of the service requirements, while the radio-access network only maps the QoS flows to radio bearers. The QoS-flow-to-radio-bearer mapping is not necessarily a one-to-one mapping; multiple QoS flows can be mapped to the same data radio bearer (Fig. 6.5).

There are two ways of controlling the mapping from quality-of-service flows to data radio bearers in the uplink: reflective mapping and explicit configuration.

In the case of reflective mapping, which is a new feature in NR when connected to the 5G core network, the device observes the QFI in the downlink packets for the PDU session. This provides the device with knowledge about which IP flows are mapped to which QoS flow and radio bearer. The device then uses the same mapping for the uplink traffic.

In the case of explicit mapping, the quality-of-service flow to data radio bearer mapping is configured in the device using RRC signaling.

6.3 RADIO PROTOCOL ARCHITECTURE

With the overall network architecture in mind, the RAN protocol architecture for the user and control planes can be discussed. Fig. 6.6 illustrates the RAN protocol

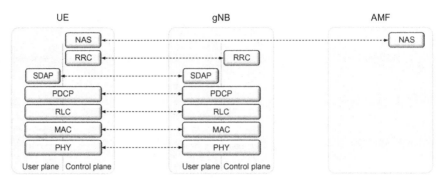

FIGURE 6.6

User-plane and control plane protocol stack.

architecture (the AMF is, as discussed in the previous section, not part of the RAN but is included in the figure for completeness).

In the following, the user-plane protocols will be described in Section 6.4, followed by the control plane protocols in Section 6.5. As seen in Fig. 6.6, many of the protocol entities are common to the user and control planes and hence PDCP, RLC, MAC, and PHY will only be described in the user-plane section.

6.4 USER-PLANE PROTOCOLS

A general overview of the NR user-plane protocol architecture for the downlink is illustrated in Fig. 6.7. Many of the protocol layers are similar to those in LTE, although there are some differences as well. One of the differences is the quality-of-service handling in NR when connected to a 5G core network, where the SDAP protocol layer accepts one or more QoS flows carrying IP packets according to their Quality-of-Service requirements. In the case of the NR user plane connected to the EPC, the SDAP is not used.

As will become clear in the subsequent discussion, not all the entities illustrated in Fig. 6.7 are applicable in all situations. For example, ciphering is not used for broadcasting of the basic system information. The uplink protocol structure is similar to the downlink structure in Fig. 6.7, although there are some differences with respect to, for example, transport-format selection and the control of logical-channel multiplexing.

The different protocol entities of the radio-access network are summarized below and described in more detail in the following sections.

- *Service Data Application Protocol* (SDAP) is responsible for mapping QoS bearers to radio bearers according to their quality-of-service requirements. This protocol layer is not present in LTE but introduced in NR when connecting to the 5G core network due to the new quality-of-service handling.

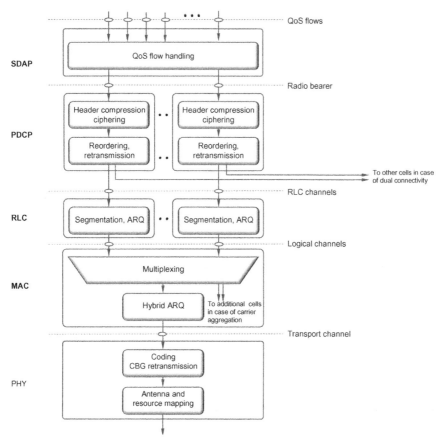

FIGURE 6.7

NR downlink user-plane protocol architecture as seen from the device.

- *Packet Data Convergence Protocol* (PDCP) performs IP header compression, ciphering, and integrity protection. It also handles retransmissions, in-sequence delivery, and duplicate removal[3] in the case of handover. For dual connectivity with split bearers, PDCP can provide routing and duplication. There is one PDCP entity per radio bearer configured for a device.
- *Radio-Link Control* (RLC) is responsible for segmentation and retransmission handling. The RLC provides services to the PDCP in the form of *RLC channels*. There is one RLC entity per RLC channel (and hence per radio bearer) configured for a device. Compared to LTE, the NR RLC does not support in-sequence delivery of data to higher protocol layers, a change motivated by the reduced delays as discussed below.

---

[3]Duplicate detection is part of the June 2018 release and not present in the December 2017 release of NR.

- *Medium-Access Control* (MAC) handles multiplexing of logical channels, hybrid-ARQ retransmissions, and scheduling and scheduling-related functions. The scheduling functionality is located in the gNB for both uplink and downlink. The MAC provides services to the RLC in the form of *logical channels*. The header structure in the MAC layer has been changed in NR to allow for more efficient support of low-latency processing than in LTE.
- *Physical Layer* (PHY) handles coding/decoding, modulation/demodulation, multi-antenna mapping, and other typical physical-layer functions. The physical layer offers services to the MAC layer in the form of *transport channels*.

To summarize the flow of downlink data through all the protocol layers, an example illustration with three IP packets, two on one radio bearer and one on another radio bearer, is given in Fig. 6.8. In this example, there are two radio bearers and one RLC SDU is segmented and transmitted in two different transports. The data flow in the case of uplink transmission is similar.

The SDAP protocol maps the IP packets to the different radio bearers; in this example IP packets *n* and *n* + 1 are mapped to radio bearer *x* and IP packet *m* is mapped to radio bearer *y*. In general, the data entity from/to a higher protocol layer is known as a *Service Data Unit* (SDU) and the corresponding entity to/from a lower protocol layer entity is called a *Protocol Data Unit* (PDU). Hence, the output from the SDAP is an SDAP PDU, which equals an PDCP SDU.

The PDCP protocol performs (optional) IP-header compression, followed by ciphering, for each radio bearer. A PDCP header is added, carrying information required for deciphering in the device as well as a sequence number used for retransmission and in-sequence delivery, if configured. The output from the PDCP is forwarded to the RLC.

The RLC protocol performs segmentation of the PDCP PDUs if necessary and adds an RLC header containing a sequence number used for handing retransmissions. Unlike LTE, the NR RLC is not providing in-sequence delivery of data to higher layers. The reason is additional delay incurred by the reordering

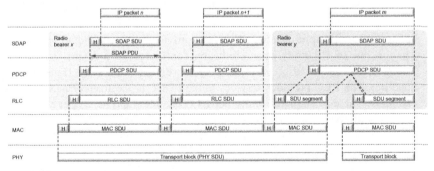

FIGURE 6.8

Example of user-plane data flow.

mechanism, a delay that might be detrimental for services requiring very low latency. If needed, in-sequence delivery can be provided by the PDCP layer instead.

The RLC PDUs are forwarded to the MAC layer, which multiplexes a number of RLC PDUs and attaches a MAC header to form a transport block. Note that the MAC headers are distributed across the MAC PDU, such that the MAC header related to a certain RLC PDU is located immediately prior to the RLC PDU. This is different compared to LTE, which has all the header information at the beginning of the MAC PDU and is motivated by efficient low-latency processing. With the structure in NR, the MAC PDU can be assembled "on the fly" as there is no need to assemble the full MAC PDU before the header fields can be computed. This reduces the processing time and hence the overall latency.

The remainder of this chapter contains an overview of the SDAP, RLC, MAC, and physical layers.

6.4.1 SERVICE DATA ADAPTATION PROTOCOL (SDAP)

The Service Data Adaptation Protocol (SDPA) is responsible for mapping between a quality-of-service flow from the 5G core network and a data radio bearer, as well as marking the quality-of-service flow identifier (QFI) in uplink and downlink packets. The reason for the introduction of SDAP in NR is the new quality-of-service handling compared to LTE when connected to the 5G core. In this case the SDAP is responsible for the mapping between QoS flows and radio bearers as described in Section 6.2. If the gNB is connected to the EPC, as is the case for non-standalone mode, the SDAP is not used.

6.4.2 PACKET-DATA CONVERGENCE PROTOCOL (PDCP)

The PDCP performs IP header compression to reduce the number of bits to transmit over the radio interface. The header-compression mechanism is based on robust header compression (ROHC) framework [38], a set of standardized header-compression algorithms also used for several other mobile-communication technologies. PDCP is also responsible for ciphering to protect against eavesdropping and, for the control plane, integrity protection to ensure that control messages originate from the correct source. At the receiver side, the PDCP performs the corresponding deciphering and decompression operations.

The PDCP is also responsible for duplicate removal and (optional) in-sequence delivery, functions useful, for example, in the case of intra-gNB handover. Upon handover, undelivered downlink data packets will be forwarded by the PDCP from the old gNB to the new gNB. The PDCP entity in the device will also handle retransmission of all uplink packets not yet delivered to the gNB as the hybrid-ARQ buffers are flushed upon handover. In this case, some PDUs may be received in duplicate, both over the connection to the old gNB and the new gNB. The PDCP will in this case remove any duplicates. The PDCP can also be

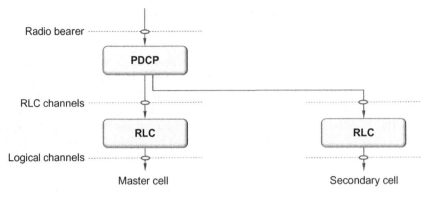

FIGURE 6.9

Dual connectivity with split bearer.

configured to perform reordering to ensure in-sequence delivery of SDUs to higher-layer protocols if desirable.

Duplication in PDCP can also be used for additional diversity. Packets can be duplicated and transmitted on multiple cells, increasing the likelihood of at least one copy being correctly received. This can be useful for services requiring very high reliability. At the receiving end, the PDCP duplicate removal functionality removes any duplicates. In essence, this results in selection diversity.

Dual connectivity is another area where PDCP plays an important role. In dual connectivity, a device is connected to two cells, or in general, two cell groups,[4] the *Master Cell Group* (MCG) and the *Secondary Cell Group* (SCG). The two cell groups can be handled by different gNBs. A radio bearer is typically handled by one of the cell groups, but there is also the possibility for *split bearers*, in which case one radio bearer is handled by both cell groups. In this case the PDCP is in charge of distributing the data between the MCG and the SCG, as illustrated in Fig. 6.9.

The June 2018 version of release 15 supports dual connectivity in general, while the December 2017 version is limited to dual connectivity between LTE and NR. This is of particular importance as it is the basis for non-standalone operation using option 3 as illustrated in Fig. 6.4. The LTE-based master cell handles control-plane and (potentially) user-plane signaling, and the NR-based secondary cell handles user-plane only, in essence boosting the data rates.

6.4.3 RADIO-LINK CONTROL

The RLC protocol is responsible for segmentation of RLC SDUs from the PDCP into suitably sized RLC PDUs. It also handles retransmission of erroneously

---

[4]The reason for the term *cell group* is to cover also the case of carrier aggregation where there are multiple cells, one per aggregated carriers, in each cell group.

received PDUs, as well as removal of duplicate PDUs. Depending on the type of service, the RLC can be configured in one of three modes—transparent mode, unacknowledged mode, and acknowledged mode—to perform some or all of these functions. Transparent mode is, as the name suggests, transparent, and no headers are added. Unacknowledged mode supports segmentation and duplicate detection, while acknowledged mode in addition supports retransmission of erroneous packets.

One major difference compared to LTE is that the RLC does not ensure in-sequence delivery of SDUs to upper layers. Removing in-sequence delivery from the RLC reduces the overall latency as later packets do not have to wait for retransmission of an earlier missing packet before being delivered to higher layers but can be forwarded immediately. Another difference is the removal of concate-nation from the RLC protocol to allow RLC PDUs to be assembled in advance, prior to receiving the uplink scheduling grant. This also helps reduce the overall latency, as discussed in Chapter 13.

Segmentation, one of the main RLC functions, is illustrated in Fig. 6.10. Included in the figure is also the corresponding LTE functionality, which also supports concatenation. Depending on the scheduler decision, a certain amount of data, that is, certain transport-block size, is selected. As part of the overall low-latency design of NR, the scheduling decision in case of an uplink transmission is known to the device just before transmission, in the order of a few OFDM sym-bols before. In the case of concatenation in LTE, the RLC PDU cannot be assem-bled until the scheduling decision is known, which results in an additional delay until the uplink transmission and cannot meet the low-latency requirement of NR. By removing the concatenation from RLC, the RLC PDUs can be assembled in advance and upon receipt of the scheduling decision the device only has to for-ward a suitable number of RLC PDUs to the MAC layer, the number depending on the scheduled transport block size. To completely fill up the transport block

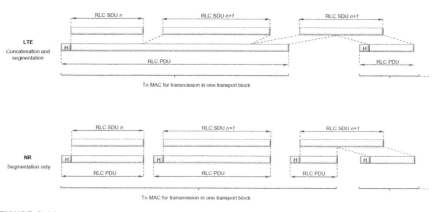

FIGURE 6.10

RLC segmentation.

size, the last RLC PDU may contain a segment of an SDU. The segmentation operation is simple. Upon receiving the scheduling grant, the device includes the amount of data needed to fill up the transport block and updates the header to indicate it is a segmented SDU.

The RLC retransmission mechanism is also responsible for providing error-free delivery of data to higher layers. To accomplish this, a retransmission protocol operates between the RLC entities in the receiver and transmitter. By monitoring the sequence numbers indicated in the headers of the incoming PDUs, the receiving RLC can identify missing PDUs (the RLC sequence number is independent of the PDCP sequence number). Status reports are fed back to the transmitting RLC entity, requesting retransmission of missing PDUs. Based on the received status report, the RLC entity at the transmitter can take the appropriate action and retransmit the missing PDUs if needed.

Although the RLC is capable of handling transmission errors due to noise, unpredictable channel variations, etc., error-free delivery is in most cases handled by the MAC-based hybrid-ARQ protocol. The use of a retransmission mechanism in the RLC may therefore seem superfluous at first. However, as will be discussed in Chapter 13, this is not the case and the use of both RLC- and MAC-based retransmission mechanisms is in fact well motivated by the differences in the feedback signaling.

The details of RLC are further described in Section 13.2.

6.4.4 MEDIUM-ACCESS CONTROL

The MAC layer handles logical-channel multiplexing, hybrid-ARQ retransmissions, and scheduling and scheduling-related functions, including handling of different numerologies. It is also responsible for multiplexing/demultiplexing data across multiple component carriers when carrier aggregation is used.

6.4.4.1 Logical Channels and Transport Channels

The MAC provides services to the RLC in the form of *logical channels*. A logical channel is defined by the *type* of information it carries and is generally classified as a *control channel*, used for transmission of control and configuration information necessary for operating an NR system, or as a *traffic channel*, used for the user data. The set of logical-channel types specified for NR includes:

- The *Broadcast Control Channel* (BCCH), used for transmission of *system information* from the network to all devices in a cell. Prior to accessing the system, a device needs to acquire the system information to find out how the system is configured and, in general, how to behave properly within a cell. Note that, in the case of non-standalone operation, system information is provided by the LTE system and there is no BCCH.
- The *Paging Control Channel* (PCCH), used for paging of devices whose location on a cell level is not known to the network. The paging message

therefore needs to be transmitted in multiple cells. Note that, in the case of non-standalone operation, paging is provided by the LTE system and there is no PCCH.

- The *Common Control Channel* (CCCH), used for transmission of control information in conjunction with random access.
- The *Dedicated Control Channel* (DCCH), used for transmission of control information to/from a device. This channel is used for individual configuration of devices such as setting various parameters in devices.
- The *Dedicated Traffic Channel* (DTCH), used for transmission of user data to/from a device. This is the logical channel type used for transmission of all unicast uplink and downlink user data.

The above logical channels are in general present also in an LTE system and used for similar functionality. However, LTE provides additional logical channels for features not yet supported by NR (but likely to be introduced in upcoming releases).

From the physical layer, the MAC layer uses services in the form of *transport channels*. A transport channel is defined by *how* and *with what characteristics* the information is transmitted over the radio interface. Data on a transport channel are organized into *transport blocks*. In each *Transmission Time Interval* (TTI), at most one transport block of dynamic size is transmitted over the radio interface to/from a device (in the case of spatial multiplexing of more than four layers, there are two transport blocks per TTI).

Associated with each transport block is a *Transport Format* (TF), specifying *how* the transport block is to be transmitted over the radio interface. The transport format includes information about the transport-block size, the modulation-and-coding scheme, and the antenna mapping. By varying the transport format, the MAC layer can thus realize different data rates, a process known as *transport-format selection*.

The following transport-channel types are defined for NR:

- The *Broadcast Channel* (BCH) has a fixed transport format, provided by the specifications. It is used for transmission of parts of the BCCH system information, more specifically the so-called *Master Information Block* (MIB), as described in Chapter 16.
- The *Paging Channel* (PCH) is used for transmission of paging information from the PCCH logical channel. The PCH supports *discontinuous reception* (DRX) to allow the device to save battery power by waking up to receive the PCH only at predefined time instants.
- The *Downlink Shared Channel* (DL-SCH) is the main transport channel used for transmission of downlink data in NR. It supports key NR features such as dynamic rate adaptation and channel-dependent scheduling in the time and frequency domains, hybrid ARQ with soft combining, and spatial multiplexing. It also supports DRX to reduce device power consumption while still providing an always-on experience. The DL-SCH is also used for

transmission of the parts of the BCCH system information not mapped to the BCH. Each device has a DL-SCH per cell it is connected to. In slots where system information is received there is one additional DL-SCH from the device perspective.

- The *Uplink Shared Channel* (UL-SCH) is the uplink counterpart to the DL-SCH—that is, the uplink transport channel used for transmission of uplink data.

In addition, the *Random-Access Channel* (RACH) is also defined as a transport channel, although it does not carry transport blocks.

Part of the MAC functionality is multiplexing of different logical channels and mapping of the logical channels to the appropriate transport channels. The mapping between logical-channel types and transport-channel types is given in Fig. 6.11. This figure clearly indicates how DL-SCH and UL-SCH are the main downlink and uplink transport channels, respectively. In the figures, the corresponding physical channels, described further below, are also included and the mapping between transport channels and physical channels is illustrated.

To support priority handling, multiple logical channels, where each logical channel has its own RLC entity, can be multiplexed into one transport channel by the MAC layer. At the receiver, the MAC layer handles the corresponding demultiplexing and forwards the RLC PDUs to their respective RLC entity. To support the demultiplexing at the receiver, a MAC header is used. The placement of the MAC headers has been improved compared to LTE, again with low-latency operation in mind. Instead of locating all the MAC header information at the beginning of a MAC PDU, which implies that assembly of the MAC PDU cannot start until the scheduling decision is available, the subheader corresponding to a certain MAC SDU is placed immediately before the SDU, as shown in Fig. 6.12. This allows the PDUs to be preprocessed before having received the scheduling decision. If necessary, padding can be appended to align the transport block size with those supported in NR.

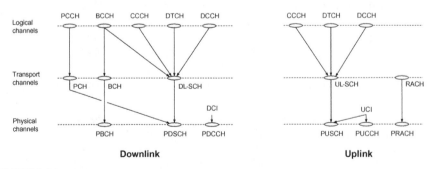

Downlink   **Uplink**

FIGURE 6.11

Mapping between logical, transport, and physical channels.

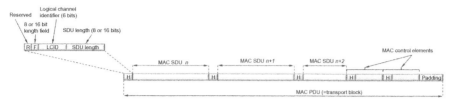

FIGURE 6.12

MAC SDU multiplexing and header insertion (uplink case).

The subheader contains the identity of the logical channel (LCID) from which the RLC PDU originated and the length of the PDU in bytes. There is also a flag indicating the size of the length indicator, as well as a reserved bit for future use.

In addition to multiplexing of different logical channels, the MAC layer can also insert *MAC control elements* into the transport blocks to be transmitted over the transport channels. A MAC control element is used for inband control signaling and identified with reserved values in the LCID field, where the LCID value indicates the type of control information. Both fixed- and variable-length MAC control elements are supported, depending on their usage. For downlink transmissions, MAC control elements are located at the beginning of the MAC PDU, while for uplink transmissions the MAC control elements are located at the end, immediately before the padding (if present). Again, the placement is chosen in order to facilitate low-latency operation in the device.

MAC control elements are, as mentioned above, used for inband control signaling. It provides a faster way to send control signaling than RLC, without having to resort to the restrictions in terms of payload sizes and reliability offered by physical-layer L1/L2 control signaling (PDCCH or PUCCH). There are multiple MAC control elements, used for various purposes, for example:

- Scheduling-related MAC control elements, such as buffer status reports and power headroom reports used to assist uplink scheduling as described in Chapter 14, and the configured grant confirmation MAC control element used when configuring semipersistent scheduling;
- Random-access-related MAC control elements such as the C-RNTI and contention-resolution MAC control elements;
- Timing-advance MAC control elements to handle timing advance as described in Chapter 15;
- Activation and deactivation of previously configured components;
- DRX-related MAC control elements;
- Activation/deactivation of PDCP duplication detection; and
- Activation/deactivation of CSI reporting and SRS transmission (see Chapter 8).

The MAC entity is also responsible for distributing data from each flow across the different component carriers, or cells, in the case of carrier aggregation. The

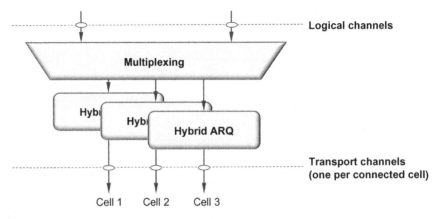

FIGURE 6.13

Carrier aggregation.

basic principle for carrier aggregation is independent processing of the component carriers in the physical layer, including control signaling, scheduling, and hybrid-ARQ retransmissions, while carrier aggregation is invisible above the MAC layer. Carrier aggregation is therefore mainly seen in the MAC layer, as illustrated in Fig. 6.13, where logical channels, including any MAC control elements, are multiplexed to form transport blocks per component carrier with each component carrier having its own hybrid-ARQ entity.

Both carrier aggregation and dual connectivity result in the device being connected to more than one cell. Despite this similarity, there are fundamental differences, primarily related to how tightly the different cells are coordinated and whether they reside in the same or in different gNBs.

Carrier aggregation implies very tight coordination, with all the cells belonging to the same gNB. Scheduling decisions are taken jointly for all the cells the device is connected to by one joint scheduler.

Dual connectivity, on the other hand, allows for a much looser coordination between the cells. The cells can belong to different gNBs, and they may even belong to different radio-access technologies as is the case for NR-LTE dual connectivity in case of non-standalone operation.

Carrier aggregation and dual connectivity can also be combined. This is the reason for the terms master cell group and secondary cell group. Within each of the cell groups, carrier aggregation can be used.

6.4.4.2 Scheduling

One of the basic principles of NR radio access is shared-channel transmission—that is, time—frequency resources are dynamically shared between users. The *scheduler* is part of the MAC layer (although often better viewed as a separate entity) and controls the assignment of uplink and downlink resources in terms of so-called *resource blocks* in the frequency domain and OFDM symbols and slots in the time domain.

The basic operation of the scheduler is *dynamic* scheduling, where the gNB takes a scheduling decision, typically once per slot, and sends scheduling information to the selected set of devices. Although per-slot scheduling is a common case, neither the scheduling decisions, nor the actual data transmission is restricted to start or end at the slot boundaries. This is useful to support low-latency operation as well as future extensions to unlicensed spectrum operation as mentioned in Chapter 6.

Uplink and downlink scheduling are separated in NR, and uplink and downlink scheduling decisions can be taken independently of each other (within the limits set by the duplex scheme in the case of half-duplex operation).

The downlink scheduler is responsible for (dynamically) controlling which device(s) to transmit to and, for each of these devices, the set of resource blocks upon which the device's DL-SCH should be transmitted. Transport-format selection (selection of transport-block size, modulation scheme, and antenna mapping) and logical-channel multiplexing for downlink transmissions are controlled by the gNB, as illustrated in the left part of Fig. 6.14.

The uplink scheduler serves a similar purpose, namely to (dynamically) control which devices are to transmit on their respective UL-SCH and on which uplink time−frequency resources (including component carrier). Despite the fact that the gNB scheduler determines the transport format for the device, it is important to point out that the uplink scheduling decision does not explicitly schedule a certain logical channel but rather the device as such. Thus, although the gNB scheduler controls the payload of a scheduled device, the device is responsible for

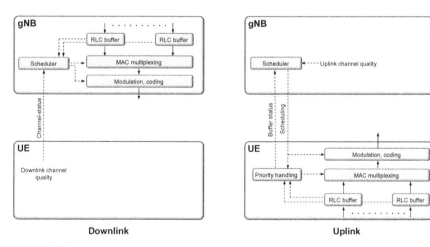

FIGURE 6.14

Transport-format selection in (a) downlink and (b) uplink.

selecting *from which radio bearer(s)* the data are taken according to a set of rules, the parameters of which can be configured by the gNB. This is illustrated in the right part of Fig. 6.14, where the gNB scheduler controls the transport format and the device controls the logical-channel multiplexing.

Although the scheduling strategy is implementation specific and not specified by 3GPP, the overall goal of most schedulers is to take advantage of the channel variations between devices and preferably schedule transmissions to a device on resources with advantageous channel conditions in both the time and frequency domain, often referred to as *channel-dependent scheduling*.

Downlink channel-dependent scheduling is supported through *channel-state information* (CSI), reported by the device to the gNB and reflecting the instantaneous downlink channel quality in the time and frequency domains, as well as information necessary to determine the appropriate antenna processing in the case of spatial multiplexing. In the uplink, the channel-state information necessary for uplink channel-dependent scheduling can be based on a *sounding reference signal* transmitted from each device for which the gNB wants to estimate the uplink channel quality. To aid the uplink scheduler in its decisions, the device can transmit buffer-status and power-headroom information to the gNB using MAC control elements. This information can only be transmitted if the device has been given a valid scheduling grant. For situations when this is not the case, an indicator that the device needs uplink resources is provided as part of the uplink L1/L2 control-signaling structure (see Chapter 10).

Although dynamic scheduling is the baseline mode-of-operation, there is also a possibility for transmission/reception without a dynamic grant to reduce the control-signaling overhead. The details differ between downlink and uplink.

In the downlink, a scheme similar to semipersistent scheduling in LTE is used. A semistatic scheduling pattern is signaled in advance to the device. Upon activation by L1/L2 control signaling, which also includes parameters such as the time−frequency resources and coding-and-modulation scheme to use, the device receives downlink data transmissions according to the preconfigured pattern.

In the uplink, there are two slightly different schemes, type 1 and type 2, differing on how to activate the scheme. In type 1, RRC configures all parameters, including the time−frequency resources and the modulation-and-coding scheme to use, and also activates the uplink transmission according to the parameters. Type 2, on the other hand, is similar to semipersistent scheduling where RRC configures the scheduling pattern in time. Activation is done using L1/L2 signaling, which includes the necessary transmission parameters (except the periodicity which is provides through RRC signaling). In both type 1 and type 2, the device does not transmit in the uplink unless there are data to convey.

6.4.4.3 Hybrid ARQ With Soft Combining

Hybrid ARQ with soft combining provides robustness against transmission errors. As hybrid-ARQ retransmissions are fast, many services allow for one or multiple retransmissions, and the hybrid-ARQ mechanism therefore forms an implicit (closed loop) rate-control mechanism. The hybrid-ARQ protocol is part of the MAC layer, while the physical layer handles the actual soft combining.[5]

Hybrid ARQ is not applicable for all types of traffic. For example, broadcast transmissions, where the same information is intended for multiple devices, typically do not rely on hybrid ARQ. Hence, hybrid ARQ is only supported for the DL-SCH and the UL-SCH, although its usage is up to the gNB implementation.

The hybrid-ARQ protocol uses multiple parallel stop-and-wait processes in a similar way to LTE. Upon receipt of a transport block, the receiver tries to decode the transport block and informs the transmitter about the outcome of the decoding operation through a single acknowledgment bit indicating whether the decoding was successful or if a retransmission of the transport block is required. Clearly, the receiver must know to which hybrid-ARQ process a received acknowledgment is associated. This is solved by using the timing of the acknowledgment for association with a certain hybrid-ARQ process or by using the position of the acknowledgment in the hybrid-ARQ codebook in case of multiple acknowledgments transmitted at the same time (see Section 13.1 for further details).

An asynchronous hybrid-ARQ protocol is used for both downlink and uplink—that is, an explicit hybrid-ARQ process number is used to indicate which process is being addressed. In an asynchronous hybrid-ARQ protocol, the retransmissions are in principle scheduled similarly to the initial transmissions. The use of an asynchronous uplink protocol, instead of a synchronous one as in LTE, is necessary to support dynamic TDD where there is no fixed uplink/downlink allocation. It also offers better flexibility in terms of prioritization between data flows and devices and is beneficial for future extension to unlicensed spectrum operation.[6]

Up to 16 hybrid-ARQ processes are supported. Having a larger maximum number of hybrid-ARQ processes than in LTE[7] is motivated by the possibility for remote radio heads, which incurs a certain front-haul delay, together with the shorter slot durations at high frequencies. It is important though, that the larger number of maximum hybrid-ARQ processes does not imply a longer roundtrip time as not all processes need to be used, it is only an upper limit of the number of processes possible.

The use of multiple parallel hybrid-ARQ processes, illustrated in Fig. 6.15, for a device can result in data being delivered from the hybrid-ARQ mechanism out of sequence. For example, transport block 3 in the figure was successfully decoded before transport block 2, which required retransmissions. For many applications this is acceptable and, if not, in-sequence delivery can be provided

---

[5]The soft combining is done before or as part of the channel decoding, which clearly is a physical-layer functionality. Also, the per-CBG retransmission handling is formally part of the physical layer.

[6]LTE changed to an asynchronous uplink hybrid-ARQ protocol for LAA.

[7]In LTE, eight processes are used for FDD and up to 15 processes for TDD, depending on the uplink−downlink configuration.

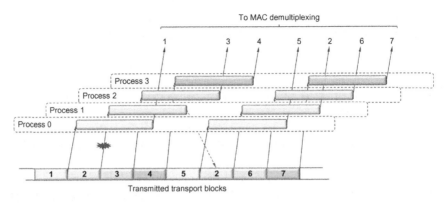

FIGURE 6.15

Multiple parallel hybrid-ARQ processes.

through the PDCP protocol. The reason for not providing in-sequence delivery in the RLC protocol is to reduce latency. If in-sequence delivery would be enforced in Fig. 6.15, packet numbers 3, 4, and 5 would have to be delayed until packet number 2 is correctly received before delivering them to higher layers, while without in-sequence delivery each packet can be forwarded as soon as it is correctly received.

One additional feature of the hybrid-ARQ mechanism in NR compared to LTE is the possibility for retransmission of *codeblock groups*, a feature that can be beneficial for very large transport blocks or when a transport block is partially interfered by another preempting transmission. As part of the channel-coding operation in the physical layer, a transport block is split into one or more codeblocks with error-correcting coding applied to each of the codeblocks of at most 8448 bits[8] in order to keep the channel-coding complexity reasonable. Thus, even for modest data rates there can be multiple code blocks per transport block and at Gbps data rates there can be hundreds of code blocks per transport block. In many cases, especially if the interference is bursty and hits a small number of OFDM symbols in the slot, only a few of these code blocks in the transport block may be corrupted, while the majority of code blocks are correctly received. To correctly receive the transport block, it is sufficient to retransmit the erroneous code blocks. At the same time, the control signaling overhead would be too large if individual code blocks can be addressed by the hybrid-ARQ mechanism. Therefore, *codeblock groups* (CBGs) are defined. If per-CBG retransmission is configured, feedback is provided per CBG and only the erroneously received code block groups are retransmitted (Fig. 6.16). This can consume less resource than retransmitting the whole transport block. CBG retransmissions are invisible to the MAC layer and are handled in the physical layer, despite being part of the

---

[8]For code rates below ¼, the code block size is 3840.

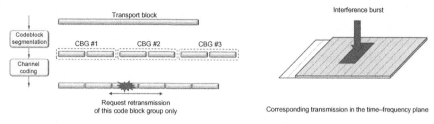

FIGURE 6.16

Codeblock group retransmission.

hybrid-ARQ mechanism. The reason for this is not technical but purely related to the specification structure. From a MAC perspective, the transport block is not correctly received until all the CBGs are correctly received. It is not possible, in the same hybrid-ARQ process, to mix transmission of new CBGs belonging to another transport block with retransmissions of CBGs belonging to the incorrectly received transport block.

The hybrid-ARQ mechanism will rapidly correct transmission errors due to noise or unpredictable channel variations. As discussed above, the RLC is also capable of requesting retransmissions, which at first sight may seem unnecessary. However, the reason for having two retransmission mechanisms on top of each other can be seen in the feedback signaling—hybrid ARQ provides fast retransmissions but due to errors in the feedback the residual error rate is typically too high for, for example, good TCP performance, while RLC ensures (almost) error-free data delivery but slower retransmissions than the hybrid-ARQ protocol. Hence, the combination of hybrid ARQ and RLC provides an attractive combination of small round-trip time and reliable data delivery.

6.4.5 PHYSICAL LAYER

The physical layer is responsible for coding, physical-layer hybrid-ARQ processing, modulation, multi-antenna processing, and mapping of the signal to the appropriate physical time—frequency resources. It also handles mapping of transport channels to physical channels, as shown in Fig. 6.11.

As mentioned in the introduction, the physical layer provides services to the MAC layer in the form of transport channels. Data transmissions in downlink and uplink use the DL-SCH and UL-SCH transport-channel types, respectively. There is at most one transport block (two transport blocks in the case of spatial multiplexing of more than four layers in the downlink) to a single device per TTI on a DL-SCH or UL-SCH. In the case of carrier aggregation, there is one DL-SCH (or UL-SCH) per component carrier seen by the device.

A *physical channel* corresponds to the set of time—frequency resources used for transmission of a particular transport channel and each transport channel is mapped to a corresponding physical channel, as shown in Fig. 6.11. In addition to the physical channels with a corresponding transport channel, there are also physical channels without a corresponding transport channel. These channels, known as L1/L2 control channels, are used for *downlink control information* (DCI), providing the device with the necessary information for proper reception and decoding of the downlink data transmission, and *uplink control information* (UCI) used for providing the scheduler and the hybrid-ARQ protocol with information about the situation at the device.

The following physical-channel types are defined for NR:

- The *Physical Downlink Shared Channel* (PDSCH) is the main physical channel used for unicast data transmission, but also for transmission of, for example, paging information, random-access response messages, and delivery of parts of the system information.
- The *Physical Broadcast Channel* (PBCH) carries part of the system information, required by the device to access the network.
- The *Physical Downlink Control Channel* (PDCCH) is used for downlink control information, mainly scheduling decisions, required for reception of PDSCH, and for scheduling grants enabling transmission on the PUSCH.
- The *Physical Uplink Shared Channel* (PUSCH) is the uplink counterpart to the PDSCH. There is at most one PUSCH per uplink component carrier per device.
- The *Physical Uplink Control Channel* (PUCCH) is used by the device to send hybrid-ARQ acknowledgments, indicating to the gNB whether the downlink transport block(s) was successfully received or not, to send channel-state reports aiding downlink channel-dependent scheduling, and for requesting resources to transmit uplink data upon.
- The *Physical Random-Access Channel* (PRACH) is used for random access.

Note that some of the physical channels, more specifically the channels used for downlink and uplink control information (PDCCH and PUCCH) do not have a corresponding transport channel mapped to them.

6.5 CONTROL-PLANE PROTOCOLS

The control-plane protocols are, among other things, responsible for connection setup, mobility, and security.

The NAS control-plane functionality operates between the AMF in the core network and the device. It includes authentication, security, and different idle-mode procedures such as paging (described below). It is also responsible for assigning an IP address to a device.

The *Radio Resource Control* (RRC) control-plane functionality operates between the RRC located in the gNB. RRC is responsible for handling the RAN-related control-plane procedures, including:

- Broadcast of system information necessary for the device to be able to communicate with a cell. Acquisition of system information is described in Chapter 16.
- Transmission of paging messages originating from the MME to notify the device about incoming connection requests. Paging is used in the RRC_IDLE state (described further below) when the device is not connected to a cell. Indication of system-information updates is another use of the paging mechanism, as is public warning systems.
- Connection management, including setting up bearers and mobility. This includes establishing an RRC context—that is, configuring the parameters necessary for communication between the device and the radio-access network.
- Mobility functions such as cell (re)selection.
- Measurement configuration and reporting.
- Handling of device capabilities; when connection is established the device will announce its capabilities as not all devices are capable of supporting all the functionality described in the specifications.

RRC messages are transmitted to the device using *signaling radio bearers* (SRBs), using the same set of protocol layers (PDCP, RLC, MAC, and PHY) as described in Section 6.4. The SRB is mapped to the common control channel (CCCH) during establishment of connection and, once a connection is established, to the dedicated control channel (DCCH). Control-plane and user-plane data can be multiplexed in the MAC layer and transmitted to the device in the same TTI. The aforementioned MAC control elements can also be used for control of radio resources in some specific cases where low latency is more important than ciphering, integrity protection, and reliable transfer.

6.5.1 RRC STATE MACHINE

In most wireless communication systems, the device can be in different states depending on the traffic activity. This is true also for NR and an NR device can be in one of three RRC states, RRC_IDLE, RRC_ACTIVE, and RRC_INACTIVE (see Fig. 6.17). The first two RRC states, RRC_IDLE and RRC_CONNECTED, are similar to the counterparts in LTE, while RRC_INACTIVE is a new state

FIGURE 6.17

RRC states.

introduced in NR and not present in the original LTE design. There are also core network states not discussed further herein, CN_IDLE and CN_CONNECTED, depending on whether the device has established a connection with the core network or not.

In RRC_IDLE, there is no RRC context—that is, the parameters necessary for communication between the device and the network—in the radio-access network and the device does not belong to a specific cell. From a core network perspective, the device is in the CN_IDLE state. No data transfer may take place as the device sleeps most of the time to reduce battery consumption. In the downlink, devices in idle state periodically wake up to receive paging messages, if any, from the network. Mobility is handled by the device through cell reselection (see Section 6.5.2). Uplink synchronization is not maintained and hence the only uplink transmission activity that may take place is random access, discussed in Chapter 16, to move to a connected state. As part of moving to a connected state, the RRC context is established in both the device and the network.

In RRC_CONNECTED, the RRC context is established and all parameters necessary for communication between the device and the radio-access network are known to both entities. From a core network perspective, the device is in the CN_CONNECTED state. The cell to which the device belongs is known and an identity of the device, the *Cell Radio-Network Temporary Identifier* (C-RNTI), used for signaling purposes between the device and the network, has been configured. The connected state is intended for data transfer to/from the device, but *discontinuous reception* (DRX) can be configured to reduce device power consumption (DRX is described in further detail in Section 14.5). Since there is an RRC context established in the gNB in the connected state, leaving DRX and starting to receive/transmit data is relatively fast as no connection setup with its associated signaling is needed. Mobility is managed by the radio-access network, that is, the device provides neighboring-cell measurements to the network which commands the device to perform a handover when relevant. Uplink time alignment may or may not exist but need to be established using random access and maintained as described in Section 16.2 for data transmission to take place.

In LTE, only idle and connected states are supported. A common case in practice is to use the idle state as the primary sleep state to reduce the device power consumption. However, as frequent transmission of small packets is common for many smartphone applications, the result is a significant amount of idle-to-active transitions in the core network. These transitions come at a cost in terms of signaling load and associated delays. Therefore, to reduce the signaling load and in general reduce the latency, a third state is defined in NR, the RRC_INACTIVE state.

In RRC_INACTIVE, the RRC context is kept in both the device and the gNB. The core network connection is also kept, that is, the device is in CN_CONNECTED from a core network perspective. Hence, transition to connected state for data transfer is fast. No core network signaling is needed. The RRC context is already in place in the network and idle-to-active transitions can be handled in the radio-access network. At the same time, the device is allowed to sleep in a similar way as in the idle state and mobility is handled through cell reselection, that is, without involvement of the network. Thus, RRC_INACTIVE can be seen as a mix of the idle and connected states.[9]

As seen from the discussion above, one important difference between the different states is the mobility mechanisms involved. Efficient mobility handling is a key part of any mobile communication system. For the idle and inactive states, mobility is handled by the device through cell reselection, while for the connected mode, mobility is handled by the radio-access network based on measurements. The different mobility mechanisms are described below, starting with idle- and inactive-mode mobility.

6.5.2 IDLE-STATE AND INACTIVE-STATE MOBILITY

The purpose of the mobility mechanism in idle and inactive states is to ensure that a device is reachable by the network. The network does this by notifying the device by means of a paging message. The area over which such a paging message is transmitted is a key aspect of the paging mechanism and in idle and inactive modes, the device is in control on when to update this information. This is sometimes referred to as cell reselection. In essence, the device searches for and measures on candidate cells similar to the initial cell search as described in Chapter 16. Once the device discovers a cell with a received power sufficiently higher than its current one, it considered this as the best cell and, if necessary, contacts the network through random access.

6.5.2.1 Tracking the Device

In principle, the network could transmit the page to the device over the entire coverage of the network, by broadcasting the paging message from every cell. However, that would obviously imply a very high overhead in terms of paging-message transmissions as the vast majority of the paging transmissions would take place in cells where the target device is not located. On the other hand, if the paging message is only to be transmitted in the cell in which the device is located, there is a need to track the device on a cell level. This would imply that the device would have to inform the network every time it moves out of the coverage of one cell and into the coverage of another cell. This would also lead to very high overhead, in this case in terms of the signaling needed to inform the network about the updated device location. For this reason, a compromise between these two extremes is typically used, where devices are only tracked on a cell-group level:

---

[9]In LTE release 13, the RRC suspend/resume mechanism was introduced to provide similar functionality as RRC_INACTIVE in NR. However, the connection to the core network is not maintained in RRC suspend/resume.

- The network only receives new information about the device location if the device moves into a cell outside of the current cell group;
- When paging the device, the paging message is broadcast over all cells within the cell group.

For NR, the basic principle for such tracking is the same for idle state and inactive state, although the grouping is somewhat different in the two cases.

As illustrated in Fig. 6.18, NR cells are grouped into *RAN Areas*, where each RAN Area is identified by an *RAN Area Identifier* (RAI). The RAN Areas, in turn, are grouped into even larger *Tracking Areas*, with each Tracking Area being identified by a *Tracking Area Identifier* (TAI). Thus, each cell belongs to one RAN Area and one Tracking Area, the identities of which are provided as part of the cell system information.

The Tracking Areas are the basis for device tracking on core-network level. Each device is assigned a *UE Registration Area* by the core network, consisting of a list of tracking area identifiers. When a device enters a cell that belongs to a Tracking Area not included in the assigned UE Registration Area it accesses the network, including the core network, and performs a *NAS Registration Update*. The core network registers the device location and updates the device UE Registration Area, in practice providing the device with a new TAI list that includes the new TAI.

The reason the device is assigned a set of TAIs, that is, a set of Tracking Areas, is to avoid repeated NAS Registration Updates if a device moves back and forth over the border of two neighbor Tracking Areas. By keeping the old TAI within the updated UE Registration Area no new update is needed if the device moves back into the old TAI.

The RAN Area is the basis for device tracking on radio-access-network level. UEs in inactive state can be assigned a *RAN Notification Area* that consists of either of the following:

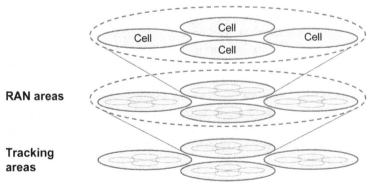

FIGURE 6.18

RAN Areas and Tracking Areas.

- A list of cell identities;
- A list of RAIs, in practice a list of RAN Areas; or
- A list of TAIs, in practice a list of Tracking Areas.

Note the first case is essentially the same as having each RAN Area consist of a single cell, while the last case is essentially the same as having the RAN Areas coincide with the Tracking Areas.

The procedure for RAN Notification Area updates is similar to updates of the UE Registration Area. When a device enters a cell that is not directly or indirectly (via a RAN/Tracking Area) included in the RAN Notification Area, the device accesses the network and makes an *RRC RAN Notification Area Update*. The radio network registers the device location and updates the device RAN Notification Area. As a change of Tracking Area always implies a change also of the device RAN Area, an RRC RAN Notification Area update is done implicitly every time a device makes a UE Registration update.

In order to track its movement within the network, the device searches for and measures on SS blocks similar to the initial cell search as described in Chapter 16. Once the device discovers an SS block with a received power that exceeds the received power of its current SS block by a certain threshold it reads the system information (SIB1) of the new cell in order to acquire information about the Tracking and RAN Areas.

6.5.2.2 Paging Message Transmission

Similar to the delivery of system information, paging messages are provided by means of ordinary scheduled PDSCH transmissions. In order to allow for low device energy consumption, a device is only supposed to wake up at specific time instances, for example, once every 100 ms or even less often, to monitor for paging messages. Paging messages are indicated by a specific PI-RNTI carried within the DCI. Once detecting such a DCI, the device demodulates and decodes the corresponding PDSCH to extract the paging message(s). Note that there can be multiple paging messages, corresponding to different devices, within the same paging transmission. The PI-RNTI is thus a shared identity.

6.5.3 CONNECTED-STATE MOBILITY

In a connected state the device has a connection established to the network. The aim of connected-state mobility is to ensure that this connectivity is retained without any interruption or noticeable degradation as the device moves within the network.

To ensure this, the device continuously searches for new cells both on the current carrier frequency (intra-frequency measurements) and on different carrier frequencies (inter-frequency measurements) that the device has been informed about. Such measurements can be done on an SS block in essentially the same way as for initial access and cell search in idle and inactive mode (see above). However, measurements can also be done on configured CSI-RS.

In a connected state, the device does not make any decisions of its own when it comes to handover to a different cell. Rather, based on different triggering conditions, for example, the relative power of a measured SS block compared to the current cell, the device reports the result of the measurements to the network. Based on this reporting the network makes a decision as to whether or not the device is to handover to a new cell. It should be pointed out that this reporting is done using RRC signaling, that is, it is not covered by the Layer-1 measurement and reporting framework (Chapter 8) used, for example, for beam management.

Except for very small cells that are tightly synchronized to each other, the current uplink transmission timing of a device will typically not match the new cell to which a device is assumed to handover. To establish synchronization to a new cell a device thus has to carry out a procedure similar to the random-access procedure of Chapter 16. However, this may then be a contention-free random access using resources specifically assigned to the device with no risk for collision but only aiming at establishing synchronization to the new cell. Thus, only the two first steps of the random-access procedure are needed, that is, the preamble transmission and corresponding random-access response providing the device with updated transmission timing.

Overall Transmission Structure

7

Prior to discussing the detailed NR downlink and uplink transmission schemes, a description of the basic time−frequency transmission resource of NR will be provided in this chapter, including bandwidth parts, supplementary uplink, carrier aggregation, duplex schemes, antenna ports, and quasi-colocation.

7.1 TRANSMISSION SCHEME

OFDM was found to be a suitable waveform for NR due to its robustness to time dispersion and ease of exploiting both the time and frequency domains when defining the structure for different channels and signals. It is therefore the basic transmission scheme for both the downlink and uplink transmission directions in NR. However, unlike LTE where DFT-precoded OFDM is the sole transmission scheme in the uplink, NR uses OFDM as the baseline uplink transmission scheme with the possibility for complementary DFT-precoded OFDM. The reasons for DFT-precoded OFDM in the uplink are the same as in LTE, namely to reduce the cubic metric and obtain a higher power-amplifier efficiency, but the use of DFT-precoding also has several drawbacks including:

- Spatial multiplexing ("MIMO") receivers become more complex. This was not an issue when DFT-precoding was agreed in the first LTE release as it did not support uplink spatial multiplexing but becomes important when supporting uplink spatial multiplexing.
- Maintaining symmetry between uplink and downlink transmission schemes is in many cases beneficial, something which is lost with an DFT-precoded uplink. One example of the benefits with symmetric schemes is sidelink transmission, that is, direct transmissions between devices. When sidelinks were introduced in LTE, it was agreed to keep the uplink transmission scheme which requires the devices to implement a receiver for DFT-precoded OFDM in addition to the OFDM receiver being already equipped for downlink transmissions. Introducing sidelink support in NR in the future is thus simpler as the device already has support for OFDM transmission and reception.

- DFT-precoded OFDM implies scheduling restrictions as only contiguous allocations in the frequency domain are possible. In many cases this is an acceptable restriction, but there are also situations when it is desirable to use a noncontiguous allocation, for example, to obtain frequency diversity.

Hence, NR has adopted OFDM in the uplink with *complementary* support for DFT-precoding for data transmission. When DFT-precoding is used, uplink transmissions are restricted to a single layer only, while uplink transmissions of up to four layers are possible with OFDM. Support for DFT-precoding is mandatory in the device and the network can therefore configure DFT-precoding if/when needed. The waveform to use for the uplink random-access messages is configured as part of the system information.

One important aspect of OFDM is the selection of the numerology, in particular the subcarrier spacing and the cyclic prefix length. A large subcarrier spacing is beneficial from a frequency-error perspective as it reduces the impact from frequency errors and phase noise. However, for a certain cyclic prefix length in microseconds, the relative overhead increases the larger the subcarrier spacing and from this perspective a smaller cyclic prefix would be preferable. The selection of the subcarrier spacing therefore needs to carefully balance overhead from the cyclic prefix against sensitivity to Doppler spread/shift and phase noise.

For LTE, a choice of 15 kHz subcarrier spacing and a cyclic prefix of approximately 4.7 μs was found to offer a good balance between these different constraints for scenarios for which LTE was originally designed—outdoor cellular deployments up to approximately 3 GHz carrier frequency.

NR, on the other hand, is designed to support a wide range of deployment scenarios, from large cells with sub-1 GHz carrier frequency up to mm-wave deployments with very wide spectrum allocations. Having a single numerology for all these scenarios is not efficient or even possible. For the lower range of carrier frequencies, from below 1 GHz up to a few GHz, the cell sizes can be relatively large and a cyclic prefix capable of handling the delay spread expected in these type of deployments, a couple of microseconds, is necessary. Consequently, a subcarrier spacing in the LTE range or somewhat higher, in the range of 15−30 kHz, is needed. For higher carrier frequencies approaching the mm-wave range, implementation limitations such as phase noise become more critical, calling for higher subcarrier spacings. At the same time, the expected cell sizes are smaller at higher frequencies as a consequence of the more challenging propagation conditions. The extensive use of beamforming at high frequencies also helps reduce the expected delay spread. Hence, for these types of deployments a higher subcarrier spacing and a shorter cyclic prefix are suitable.

From the discussion above it is seen that a scalable numerology is needed. NR therefore supports a flexible numerology with a range of subcarrier spacings, based on scaling a baseline subcarrier spacing of 15 kHz. The reason for the choice of 15 kHz is coexistence with LTE and the LTE-based NB-IoT on the same carrier. This is an important requirement, for example, for an operator which has deployed NB-IoT or eMTC to support machine-type communication. Unlike smartphones, such MTC devices can have a relatively long replacement cycle, 10 years or longer. Without provisioning for coexistence, the operator would not be able to migrate the carrier to NR until all the MTC devices have been replaced. Another example is gradual migration where the limited spectrum availability may force an operator to share a single carrier between LTE and NR in the time domain. LTE coexistence is further discussed in Chapter 17.

Consequently, 15 kHz subcarrier spacing was selected as the baseline for NR. From the baseline subcarrier spacing, subcarrier spacings ranging from 15 kHz up to 240 kHz with a proportional change in cyclic prefix duration as shown in Table 7.1 are derived. Note that 240 kHz is supported for the SS block only (see Section 16.1), and not for regular data transmission. Although the NR physical-layer specification is band-agnostic, not all supported numerologies are relevant for all frequency bands. For each frequency band, radio requirements are therefore defined for a subset of the supported numerologies as discussed in Chapter 18, RF Characteristics and Requirements.

To provide consistent and exact timing definitions, different time intervals within the NR specifications are defined as multiples of a basic time unit $T_c = 1/(480000 \cdot 4096)$. The basic time unit T_c can thus be seen as the sampling time of an FFT-based transmitter/receiver implementation for a subcarrier spacing of 480 kHz with an FFT size equal to 4096. This is similar to the approach taken in LTE, which uses a basic time unit $T_s = 64T_c$.

As discussed above, the choice of 15 kHz subcarrier spacing is motivated by coexistence with LTE. Efficient coexistence also requires alignment in the time domain and for this reason the NR slot structure for 15 kHz is identical to the LTE subframe structure. This means that the cyclic prefix for the first and eighth symbols are somewhat larger than for the other symbols. The slot structure for higher subcarrier spacings in NR is then derived by scaling this baseline structure by powers of two. In essence, an OFDM symbol is split into two OFDM symbols of the next higher numerology (see Fig. 7.1). Scaling by powers of two is beneficial as it maintains the symbol boundaries across numerologies, which simplifies mixing different numerologies on the same carrier. For the OFDM symbols with

Table 7.1 Subcarrier Spacings Supported by NR

Subcarrier Spacing (kHz)	Useful Symbol Time, T_u (μs)	Cyclic Prefix, T_{CP} (μs)
15	66.7	4.7
30	33.3	2.3
60	16.7	1.2
120	8.33	0.59
240	4.17	0.29

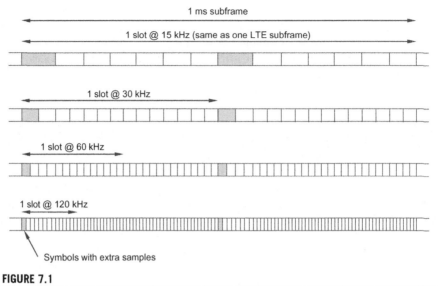

FIGURE 7.1

Symbol alignment.

a somewhat larger cyclic prefix, the excess samples are allocated to the first of the two symbols obtained when splitting one symbol.

The useful symbol time T_u depends on the subcarrier spacing as shown in Table 7.1, with the overall OFDM symbol time being the sum of the useful symbol time and the cyclic-prefix length T_{CP}. In LTE, two different cyclic prefixes are defined, normal cyclic prefix and extended cyclic prefix. The extended cyclic prefix, although less efficient from a cyclic-prefix-overhead point of view, was intended for specific environments with excessive delay spread where performance was limited by time dispersion. However, extended cyclic prefix was not used in practical deployments (except for MBSFN transmission), rendering it an unnecessary feature in LTE for unicast transmission. With this in mind, NR defines a normal cyclic prefix only, with the exception of 60 kHz subcarrier spacing, where both normal and extended cyclic prefix are defined for reasons discussed below.

7.2 TIME-DOMAIN STRUCTURE

In the time domain, NR transmissions are organized into *frames* of length 10 ms, each of which is divided into 10 equally sized *subframes* of length 1 ms. A subframe is in turn divided into slots consisting of 14 OFDM symbols each, that is, the duration of a slot in milliseconds depends on the numerology as illustrated in Fig. 7.2. On a higher level, each frame is identified by a *system frame number*

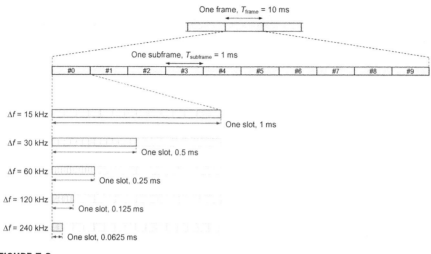

FIGURE 7.2

Frames, subframes, and slots in NR.

(SFN). The SFN is used to define different transmission cycles that have a period longer than one frame, for example, paging sleep-mode cycles. The SFN period equals 1024, thus the SFN repeats itself after 1024 frames or 10.24 s.

For the 15 kHz subcarrier spacing, an NR slot thus has the same structure as an LTE subframe with normal cyclic prefix, which is beneficial from a coexistence perspective as discussed above. Note that a subframe in NR serves as a numerology-independent time reference, which is useful, especially in the case of multiple numerologies being mixed on the same carrier, while a slot is the typical dynamic scheduling unit. In contrast, LTE with its single subcarrier spacing uses the term subframe for both these purposes.

Since a slot is defined as a fixed number of OFDM symbols, a higher subcarrier spacing leads to a shorter slot duration. In principle this can be used to support lower-latency transmission, but as the cyclic prefix also shrinks when increasing the subcarrier spacing, it is not a feasible approach in all deployments. Therefore, to facilitate a fourfold reduction in the slot duration and the associated delay while maintaining a cyclic prefix similar to the 15 kHz case, an extended cyclic prefix is defined for 60 kHz subcarrier spacing. However, it comes at the cost of increased overhead in terms of cyclic prefix and is a less efficient way of providing low latency. The subcarrier spacing is therefore primarily selected to meet the deployment scenario in terms of, for example, carrier frequency, expected delay spread in the radio channel, and any coexistence requirements with LTE-based systems on the same carrier.

An alternative and more efficient way to support low latency is to decouple the transmission duration from the slot duration. Instead of changing subcarrier spacing and/or slot duration, the latency-critical transmission uses whatever number of OFDM symbols necessary to deliver the payload. NR therefore supports occupying only part of a slot for the transmission, sometimes referred to as "mini-slot transmission." In other words, the term slot is primarily a numerology-dependent time reference and only loosely coupled with the actual transmission duration.

There are multiple reasons why it is beneficial to allow transmission to occupy only a part of a slot as illustrated in Fig. 7.3. One reason is, as already discussed, support of very low latency. Such transmissions can also preempt an already ongoing, longer transmission to another device as discussed in Section 14.1.2, allowing for immediate transmission of data requiring very low latency.

Another reason is support for analog beamforming as discussed in Chapters 11 and 12 where at most one beam at a time can be used for transmission. Different devices therefore need to be time-multiplexed and with the very large bandwidths available in the mm-wave range, a few OFDM symbols can be sufficient even for relatively large payloads.

A third reason is operation in unlicensed spectra. Unlicensed operation is not part of release 15 but will be introduced in a later release. In unlicensed spectra, listen-before-talk is typically used to ensure the radio channel is available for transmission. Once the listen-before-talk operation has declared the channel available, it is beneficial to start transmission immediately to avoid another device occupying the channel. If data transmission would have to wait until the start of a slot boundary, some form of dummy data or reservation signal needs to be transmitted from the successful listen-before-talk operation until the start of the slot, which would degrade the efficiency of the system.

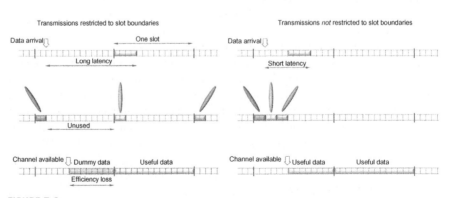

FIGURE 7.3

Decoupling transmissions from slot boundaries to achieve low latency (top), more efficient beam sweeping (middle), and better support for unlicensed spectra (bottom).

7.3 FREQUENCY-DOMAIN STRUCTURE

When the first release of LTE was designed, it was decided that all devices should be capable of the maximum carrier bandwidth of 20 MHz, which was a reasonable assumption at the time given the relatively modest bandwidth, compared to NR. On the other hand, NR is designed to support very wide bandwidths, up to 400 MHz for a single carrier. Mandating all devices to handle such wide carriers is not reasonable from a cost perspective. Hence, an NR device may see only a part of the carrier and, for efficient utilization of the carrier, the part of the carrier received by the device may not be centered around the carrier frequency. This has implications for, among other things, the handling of the DC subcarrier.

In LTE, the DC subcarrier is not used as it may be subject to disproportionally high interference due to, for example, local-oscillator leakage. Since all LTE devices can receive the full carrier bandwidth and are centered around the carrier frequency, this was straightforward.[1] NR devices, on the other hand, may not be centered around the carrier frequency and each NR device may have its DC located at different locations in the carrier, unlike LTE where all devices typically have the DC coinciding with the center of the carrier. Therefore, having special handling of the DC subcarrier would be cumbersome in NR and instead it was decided to exploit also the DC subcarrier for data as illustrated in Fig. 7.4, accepting that the quality of this subcarrier may be degraded in some situations.

A *resource element*, consisting of one subcarrier during one OFDM symbol, is the smallest physical resource in NR. Furthermore, as illustrated in Fig. 7.5, 12 consecutive subcarriers in the frequency domain are called a *resource block*.

Note that the NR definition of a resource block differs from the LTE definition. An NR resource block is a one-dimensional measure spanning the frequency domain only, while LTE uses two-dimensional resource blocks of 12 subcarriers in the frequency domain and one slot in the time domain. One reason for defining resource blocks in the frequency domain only in NR is the flexibility in time duration for different transmissions whereas, in LTE, at least in the original release, transmissions occupied a complete slot.[2]

NR supports multiple numerologies on the same carrier and, consequently, there are multiple resource sets of resource grids, one for each numerology (Fig. 7.6). Since a resource block is 12 subcarriers, the frequency span measured in Hz is different. The resource block boundaries are aligned across numerologies such that two resource blocks at a subcarrier spacing of Δf occupy the same frequency range as one resource block at a subcarrier spacing of $2\Delta f$. In the NR specifications, the alignment across numerologies in terms of resource block

---

[1] In the case of carrier aggregation, multiple carriers may use the same power amplifier, in which case the DC subcarrier of the transmission does not necessarily coincide with the unused DC subcarrier in the LTE grid.
[2] There are some situations in LTE, for example, the DwPTS in LTE/TDD, where a transmission does not occupy a full slot.

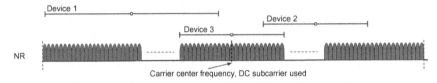

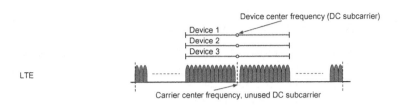

FIGURE 7.4

Handling of the DC subcarrier in LTE and NR.

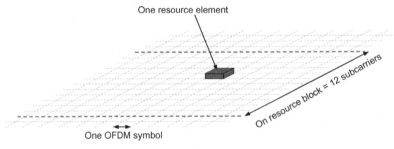

FIGURE 7.5

Resource element and resource block.

boundaries, as well as symbol boundaries, is described through multiple *resource grids* where there is one resource grid per subcarrier spacing and antenna port (see Section 7.9 for a discussion of antenna ports), covering the full carrier bandwidth in the frequency domain and one subframe in the time domain.

The resource grid models the transmitted signal as seen by the device for a given subcarrier spacing. However, the device needs to know where in the carrier the resource blocks are located. In LTE, where there is a single numerology and all devices support the full carrier bandwidth, this is straightforward. NR, on the other hand, supports multiple numerologies and, as discussed further below in conjunction with bandwidth parts, not all devices may support the full carrier bandwidth. Therefore, a common reference point, known as *point A*, together with the notion of two types of resource blocks, *common resource blocks* and

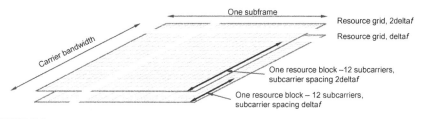

FIGURE 7.6

Resource grids for two different subcarrier spacings.

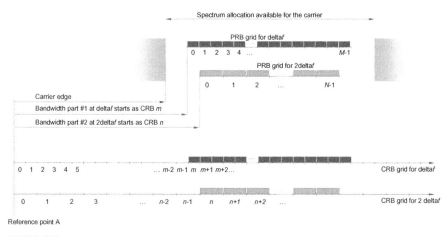

FIGURE 7.7

Common and physical resource blocks.

physical resource blocks, are used.[3] Reference point A coincides with subcarrier 0 of common resource block 0 for all subcarrier spacings. This point serves as a reference from which the frequency structure can be described and point A may be located outside the actual carrier. Upon detecting an SS block as part of the initial access (see Section 16.1), the device is signalled the location of point A as part of the broadcast system information (SIB1).

The physical resource blocks, which are used to describe the actual transmitted signal, are then located relative to this reference point, as illustrated in Fig. 7.7. For example, physical resource block 0 for subcarrier spacing Δf is located m resource blocks from reference point A or, expressed differently, corresponds to common resource block m. Similarly, physical resource block 0 for

---

[3]There is a third type of resource block, *virtual resource blocks*, which are mapped to physical resource blocks when describing the mapping of the PDSCH/PUSCH (see Chapter 9: Transport-Channel Processing).

subcarrier spacing $2\Delta f$ corresponds to common resource block n. The starting points for the physical resource blocks are signaled independently for each numerology (m and n in the example in Fig. 7.7), a feature that is useful for implementing the filters necessary to meet the out-of-band emission requirements (see Chapter 18). The guard in Hz needed between the edge of the carrier and the first used subcarrier is larger, the larger the subcarrier spacing, which can be accounted for by independently setting the offset between the first used resource block and reference point A. In the example in Fig. 7.7, the first used resource block for subcarrier spacing $2\Delta f$ is located further from the carrier edge than for subcarrier spacing Δf to avoid excessively steep filtering requirements for the higher numerology or, expressed differently, to allow a larger fraction of the spectrum to be used for the lower subcarrier spacing.

The location of the first usable resource block, which is the same as the start of the resource grid in the frequency domain, is signaled to the device. Note that this may or may not be the same as the first resource block of a bandwidth part (bandwidth parts are described in Section 7.4).

An NR carrier should at most be 275 resource blocks wide, which corresponds to $275 \cdot 12 = 3300$ used subcarriers. This also defines the largest possible carrier bandwidth in NR for each numerology. However, there is also an agreement to limit the per-carrier bandwidth to 400 MHz, resulting in the maximum carrier bandwidths of 50/100/200/400 MHz for subcarrier spacings of 15/30/60/120 kHz, respectively, as mentioned in Chapter 5. The smallest possible carrier bandwidth of 11 resource blocks is given by the RF requirements on spectrum utilization (see Chapter 18). However, for the numerology used for the SS block (see Chapter 16) at least 20 resource blocks are required in order for the device to be able to find and synchronize to the carrier.

7.4 BANDWIDTH PARTS

As discussed above, LTE is designed under the assumption that all devices are capable of the maximum carrier bandwidth of 20 MHz. This avoided several complications, for example, around the handling of the DC subcarrier as already discussed, while having a negligible impact on the device cost. It also allowed control channels to span the full carrier bandwidth to maximize frequency diversity.

The same assumption—all devices being able to receive the full carrier bandwidth—is not reasonable for NR, given the very wide carrier bandwidth supported. Consequently, means for handling different device capabilities in terms of bandwidth support must be included in the design. Furthermore, reception of a very wide bandwidth can be costly in terms of device energy consumption compared to receiving a narrower bandwidth. Using the same approach as in LTE where the downlink control channels would occupy the full carrier bandwidth would therefore significantly increase the power consumption of the device. A better approach is, as done in NR, to use *receiver-bandwidth adaptation* such that the device can use a narrower bandwidth for monitoring control channels and to receive small-to-medium-sized data transmissions and to open the full bandwidth when a large amount of data is scheduled.

To handle these two aspects—support for devices not capable of receiving the full carrier bandwidth and receiver-side bandwidth adaptation—NR defines *bandwidth parts* (BWPs) (see Fig. 7.8). A bandwidth part is characterized by a numerology (subcarrier spacing and cyclic prefix) and a set of consecutive resource blocks in the numerology of the BWP, starting at a certain common resource block.

When a device enters the connected state it has obtained information from the PBCH about the *control resource set* (CORESET; see Section 10.1.2) where it can find the control channel used to schedule the remaining system information (see Chapter 16 for details). The CORESET configuration obtained from the PBCH also defines and activates the *initial* bandwidth part in the downlink. The initial active uplink bandwidth part is obtained from the system information scheduled using the downlink PDCCH.

Once connected, a device can be configured with up to four downlink bandwidth parts and up to four uplink bandwidth parts for each serving cell. In the case of SUL operation (see Section 7.7), there can be up to four additional uplink bandwidth parts on the supplementary uplink carrier.

On each serving cell, at a given time instant one of the configured downlink bandwidth parts is referred to as the *active downlink bandwidth part* for the serving cell and one of the configured uplink bandwidth parts is referred to as the *active uplink bandwidth part* for the serving cell. For unpaired spectra a device may assume that the active downlink bandwidth part and the active uplink bandwidth part of a serving cell have the same center frequency. This simplifies the implementation as a single oscillator can be used for both directions. The gNB

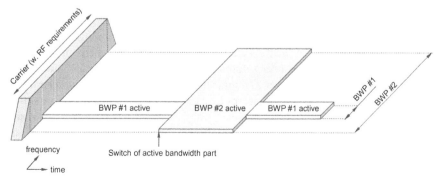

FIGURE 7.8

Example of bandwidth adaptation using bandwidth parts.

can activate and deactivate bandwidth parts using the same downlink control signaling as for scheduling information (see Chapter 10), thereby achieving rapid switching between different bandwidth parts.

In the downlink, a device is not assumed to be able to receive downlink data transmissions, more specifically the PDCCH or PDSCH, outside the active bandwidth part. Furthermore, the numerology of the PDCCH and PDSCH are restricted to the numerology configured for the bandwidth part. Thus, in release 15, a device can only receive one numerology at a time as multiple bandwidth parts cannot be simultaneously active. Mobility measurements can still be done outside an active bandwidth part but require a measurement gap similarly to inter-cell measurements. Hence, a device is not expected to monitor downlink control channels while doing measurements outside the active bandwidth part.

In the uplink, a device transmits PUSCH and PUCCH in the active uplink bandwidth part only.

Given the above discussion, a relevant question is why two mechanisms, carrier aggregation and bandwidth parts, are defined instead of using the carrier-aggregation framework only. To some extent carrier aggregation could have been used to handle devices with different bandwidth capabilities as well as bandwidth adaptation. However, from an RF perspective there is a significant difference. A component carrier is associated with various RF requirements such as out-of-band emission requirements as discussed in Chapter 18, but for a bandwidth part inside a carrier there is no such requirement—it is all handled by the requirements set on the carrier as such. Furthermore, from an MAC perspective there are also some differences in the handling of, for example, hybrid ARQ retransmissions which cannot move between component carriers.

7.5 FREQUENCY-DOMAIN LOCATION OF NR CARRIERS

In principle, an NR carrier could be positioned anywhere within the spectrum and, similarly to LTE, the basic NR physical-layer specification does not say anything about the exact frequency location of an NR carrier, including the frequency band. However, in practice, there is a need for restrictions on where an NR carrier can be positioned in the frequency domain to simplify RF implementation and to provide some structure to carrier assignments in a frequency band between different operators. In LTE, a 100 kHz carrier raster served this purpose and a similar approach has been taken in NR. However, the NR raster has a much finer granularity of 5 kHz up to 3 GHz carrier frequency, 15 kHz for 3–24.25 GHz, and 60 kHz above 24.25 GHz. This raster has the benefit of being a factor in the sub-carrier spacings relevant for each frequency range, as well as being compatible with the 100 kHz LTE raster in bands where LTE is deployed (below 3 GHz).

In LTE, this carrier raster also determines the frequency locations a device must search for as part of the initial access procedure. However, given the much

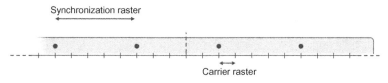

Carrier raster

FIGURE 7.9

NR carrier raster.

wider carriers possible in NR and the larger number of bands in which NR can be deployed, as well as the finer raster granularity, performing initial cell search on all possible raster positions would be too time consuming. Instead, to reduce the overall complexity and not spend an unreasonable time on cell search, NR also defines a sparser *synchronization raster*, which is what an NR device has to search upon initial access. A consequence of having a sparser synchronization raster than carrier raster is that, unlike LTE, the synchronization signals may not be centered in the carrier (see Fig. 7.9 and Chapter 16 for further details).

7.6 CARRIER AGGREGATION

The possibility of *carrier aggregation* is part of NR from the first release. Similar to LTE, multiple NR carriers can be aggregated and transmitted in parallel to/from the same device, thereby allowing for an overall wider bandwidth and correspondingly higher per-link data rates. The carriers do not have to be contiguous in the frequency domain but can be dispersed, both in the same frequency band as well as in different bands, resulting in three difference scenarios:

- Intraband aggregation with frequency-contiguous component carriers;
- Intraband aggregation with non-contiguous component carriers;
- Interband aggregation with non-contiguous component carriers.

Although the overall structure is the same for all three cases, the RF complexity can be vastly different.

Up to 16 carriers, possibly of different bandwidths and different duplex schemes, can be aggregated allowing for overall transmission bandwidths of up to $16 \cdot 400 \text{ MHz} = 6.4 \text{ GHz}$, which is far beyond typical spectrum allocations.

A device capable of carrier aggregation may receive or transmit simultaneously on multiple component carriers while a device not capable of carrier aggregation can access one of the component carriers. Thus, in most respects and unless otherwise mentioned, the physical-layer description in the following chapters applies to each component carrier separately in the case of carrier aggregation. It is worth noting that in the case of interband carrier aggregation of multiple half-duplex (TDD) carriers, the transmission direction on different carriers does not necessarily have to be the same. This implies that a

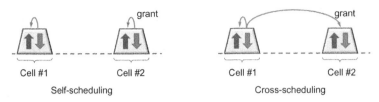

FIGURE 7.10

Self-scheduling and cross-scheduling.

carrier-aggregation-capable TDD device may need a duplex filter, unlike the typical scenario for a noncarrier-aggregation-capable device.

In the specifications, carrier aggregation is described using the term *cell*, that is, a carrier-aggregation-capable device is able to receive and transmit from/to multiple cells. One of these cells is referred to as the *primary* cell (PCell). This is the cell which the device initially finds and connects to, after which one or more *secondary* cells (SCells) can be configured once the device is in connected mode. The secondary cells can be rapidly activated or deceived to meet the variations in the traffic pattern. Different devices may have different cells as their primary cell—that is, the configuration of the primary cell is device-specific. Furthermore, the number of carriers (or cells) does not have to be the same in uplink and downlink. In fact, a typical case is to have more carriers aggregated in the downlink than in the uplink. There are several reasons for this. There is typically more traffic in the downlink that in the uplink. Furthermore, the RF complexity from multiple simultaneously active uplink carriers is typically larger than the corresponding complexity in the downlink.

Scheduling grants and scheduling assignments can be transmitted on either the same cell as the corresponding data, known as self-scheduling, or on a different cell than the corresponding data, known as cross-carrier scheduling, as illustrated in Fig. 7.10. In most cases, self-scheduling is sufficient.

7.6.1 **CONTROL SIGNALING**

Carrier aggregation uses L1/L2 control signaling for the same reason as when operating with a single carrier. The use of downlink controls signaling for scheduling information was touched upon in the previous section. There is also a need for uplink control signaling, for example, hybrid-ARQ acknowledgments to inform the gNB about the success or failure of downlink data reception. As baseline, all the feedback is transmitted on the primary cell, motivated by the need to support asymmetric carrier aggregation with the number of downlink carriers supported by a device unrelated to the number of uplink carriers. For a large number of downlink component carriers, a single uplink carrier may carry a large number of acknowledgments. To avoid overloading a single carrier, it is possible to configure two *PUCCH groups* where feedback relating to the first group is

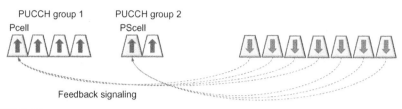

FIGURE 7.11

Multiple PUCCH groups.

transmitted in the uplink of the PCell and feedback relating to the other group of carriers is transmitted on the primary second cell (PSCell) (Fig. 7.11).

If carrier aggregation is used, the device may receive and transmit on multiple carriers, but reception on multiple carriers is typically only needed for the highest data rates. It is therefore beneficial to inactivate reception of carriers not used while keeping the configuration intact. Activation and inactivation of component carriers can be done through MAC signaling (more specifically, *MAC control elements*, discussed in Section 6.4.4.1) containing a bitmap where each bit indicates whether a configured SCell should be activated or deactivated.

7.7 SUPPLEMENTARY UPLINK

In addition to carrier aggregation, NR also supports so-called "*supplementary uplink*" (SUL). As illustrated in Fig. 7.12, SUL implies that a conventional downlink/uplink (DL/UL) carrier pair has an associated or supplementary uplink carrier with the SUL carrier typically operating in lower-frequency bands. As an example, a downlink/uplink carrier pair operating in the 3.5 GHz band could be complemented with a supplementary uplink carrier in the 800 MHz band. Although Fig. 7.12 seems to indicate that the conventional DL/UL carrier pair operates on paired spectra with frequency separation between the downlink and uplink carriers, it should be understood that the conventional carrier pair could equally well operate in unpaired spectra with downlink/uplink separation by means of TDD. This would, for example, be the case in an SUL scenario where the conventional carrier pair operates in the unpaired 3.5 GHz band.

While the main aim of carrier aggregation is to enable higher peak data rates by increasing the bandwidth available for transmission to/from a device, the typical aim of SUL is to extend uplink coverage, that is, to provide higher uplink data rates in power-limited situations, by utilizing the lower path loss at lower frequencies. Furthermore, in an SUL scenario the non-SUL uplink carrier is typically significantly more wideband compared to the SUL carrier. Thus, under good channel conditions such as the device located relatively close to the cell site, the non-SUL carrier typically allows for substantially higher data rates compared to the SUL

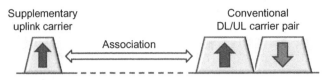

FIGURE 7.12

Supplementary uplink carrier complementing a conventional DL/UL carrier pair.

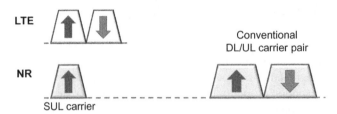

FIGURE 7.13

SUL carrier coexisting with LTE uplink carrier.

carrier. At the same time, under bad channel conditions, for example, at the cell edge, a lower-frequency SUL carrier typically allows for significantly higher data rates compared to the non-SUL carrier, due to the assumed lower path loss at lower frequencies. Hence, only in a relatively limited area do the two carriers provide similar data rates. As a consequence, *aggregating* the throughput of the two carriers has in most cases limited benefits. At the same time, scheduling only a single uplink carrier at a time simplifies transmission protocols and in particular the RF implementation as various intermodulation issues is avoided.

Note that for carrier aggregation the situation is different:

- The two (or more) carriers in a carrier-aggregation scenario are often of similar bandwidth and operating at similar carrier frequencies, making aggregation of the throughput of the two carriers more beneficial;
- Each uplink carrier in a carrier aggregation scenario is operating with its own downlink carrier, simplifying the support for simultaneous scheduling of multiple uplink transmissions in parallel.

Hence, only one of SUL and non-SUL is transmitting and simultaneous SUL and non-SUL transmission from a device is not possible.

One SUL scenario is when the SUL carrier is located in the uplink part of paired spectrum already used by LTE (see Fig. 7.13). In other words, the SUL carrier exists in an LTE/NR uplink coexistence scenario (see also Chapter 17). In many LTE deployments, the uplink traffic is significantly less than the corresponding downlink traffic. As a consequence, in many deployments, the uplink part of paired spectra is not fully utilized. Deploying an NR supplementary uplink carrier on top of the LTE uplink carrier in such a spectrum is a way to enhance the NR user experience with limited impact on the LTE network.

Finally, a supplementary uplink can also be used to reduce latency. In the case of TDD, the separation of uplink and downlink in the time domain may impose restrictions on when uplink data can be transmitted. By combining the TDD carrier with a supplementary carrier in paired spectra, latency-critical data can be transmitted on the supplementary uplink immediately without being restricted by the uplink−downlink partitioning on the normal carrier.

7.7.1 RELATION TO CARRIER AGGREGATION

Although SUL may appear similar to uplink carrier aggregation there are some fundamental differences.

In the case of carrier aggregation, each uplink carrier has its own associated downlink carrier. Formally, each such downlink carrier corresponds to a cell of its own and thus different uplink carriers in a carrier-aggregation scenario correspond to different cells (see left part of Fig. 7.14).

In contrast, in the case of SUL the supplementary uplink carrier does not have an associated downlink carrier of its own. Rather the supplementary carrier and the conventional uplink carrier share the same downlink carrier. As a consequence, the supplementary uplink carrier does not correspond to a cell of its own. Instead, in the SUL scenario there is a single cell with one downlink carrier and two uplink carriers (right part of Fig. 7.14).

It should be noted that in principle nothing prevents the combination of carrier aggregation, for example, a situation with carrier aggregation between two cells (two DL/UL carrier pairs) where one of the cells is an SUL cell with an additional supplementary uplink carrier. However, there are currently no band combinations defined for such carrier-aggregation/SUL combinations.

A relevant question is, if there is a *supplementary uplink*, is there such a thing as a *supplementary downlink*? The answer is yes—since the carrier aggregation framework allows for the number of downlink carriers to be larger than the number of uplink carriers, some of the downlink carriers can be seen as supplementary downlinks. One common scenario is to deploy an additional downlink carrier in unpaired spectra and aggregate it with a carrier in paired spectra to increase capacity and data rates. No additional mechanisms beyond carrier aggregation are needed and hence the term supplementary downlink is mainly used from a spectrum point of view as discussed in Chapter 3.

FIGURE 7.14

Carrier aggregation vs supplementary uplink.

7.7.2 CONTROL SIGNALING

In the case of supplementary-uplink operation, a device is explicitly configured (by means of RRC signaling) to transmit PUCCH on either the SUL carrier or on the conventional (non-SUL) carrier.

In terms of PUSCH transmission, the device can be configured to transmit PUSCH on the same carrier as PUCCH. Alternatively, a device configured for SUL operation can be configured for dynamic selection between the SUL carrier or the non-SUL carrier. In the latter case, the uplink scheduling grant will include an *SUL/non-SUL indicator* that indicates on what carrier the scheduled PUSCH transmission should be carried. Thus, in the case of supplementary uplink, a device will never transmit PUSCH *simultaneously* on both the SUL carrier and on the non-SUL carrier.

As described in Section 10.2, if a device is to transmit UCI on PUCCH during a time interval that overlaps with a scheduled PUSCH transmission on the same carrier, the device instead multiplexes the UCI onto PUSCH. The same rule is true for the SUL scenario, that is, there is not simultaneous PUSCH and PUCCH transmission even on different carriers. Rather, if a device is to transmit UCI on PUCCH one carrier (SUL or non-SUL) during a time interval that overlaps with a scheduled PUSCH transmission on either carrier (SUL or non SUL), the device instead multiplexes the UCI onto the PUSCH.

An alternative to supplementary uplink would be to rely on dual connectivity with LTE on the lower frequency and NR on the higher frequency. Uplink data transmission would in this case be handled by the LTE carrier with, from a data rate perspective, the benefits would be similar to supplementary uplink. However, in this case, the uplink control signaling related to NR downlink transmissions has to be handled by the high-frequency NR uplink carrier as each carrier pair has to be self-contained in terms of L1/L2 control signaling. Using a supplementary uplink avoids this drawback and allows L1/L2 control signaling to exploit the lower-frequency uplink. Another possibility would be to use carrier aggregation, but in this case a low-frequency downlink carrier has to be configured as well, something which may be problematic in the case of LTE coexistence.

7.8 DUPLEX SCHEMES

Spectrum flexibility is one of the key features of NR. In addition to the flexibility in transmission bandwidth, the basic NR structure also supports separation of uplink and downlink in time and/or frequency subject to either half duplex or full duplex operation, all using the same single frame structure. This provides a large degree of flexibility (Fig. 7.15):

• TDD—uplink and downlink transmissions use the same carrier frequency and are separated in time only;

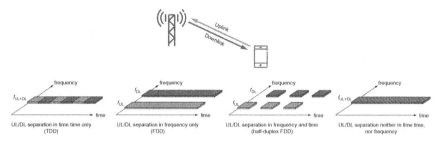

FIGURE 7.15

Duplex schemes.

- FDD—uplink and downlink transmissions use different frequencies but can occur simultaneously;
- Half-duplex FDD—uplink and downlink transmissions are separated in frequency *and* time, suitable for simpler devices operating in paired spectra.

In principle, the same basic NR structure would also allow full duplex operation with uplink and downlink separated neither in time, nor in frequency, although this would result in a significant transmitter-to-receiver interference problem whose solution is still in the research stage and left for the future.

LTE also supported both TDD and FDD, but unlike the *single* frame structure used in NR, LTE used two *different* frame structure types used.[4] Furthermore, unlike LTE where the uplink–downlink allocation does not change over time,[5] the TDD operation for NR is designed with *dynamic* TDD as a key technology component.

7.8.1 TIME-DIVISION DUPLEX (TDD)

In the case of TDD operation, there is a single carrier frequency and uplink and downlink transmissions are separated in the time domain on a cell basis. Uplink and downlink transmissions are nonoverlapping in time, both from a cell and a device perspective. TDD can therefore be classified as half-duplex operation.

In LTE, the split between uplink and downlink resources in the time domain was semistatically determined and essentially remained constant over time. NR, on the other hand, uses *dynamic TDD* as the basis where (parts of) a slot can be dynamically allocated to either uplink or downlink as part of the scheduler decision. This enables following rapid traffic variations which are particularly pronounced in dense deployments with a relatively small number of users per base station. Dynamic TDD is particularly useful in small-cell and/or isolated cell

---

[4]Originally, LTE supported frame structure type 1 for FDD and frame structure type 2 for TDD, but in later releases frame structure type 3 was added to handle operation in unlicensed spectra.

[5]In LTE Rel-12 the eIMTA feature provides some support for time-varying uplink–downlink allocation.

deployments where the transmission power of the device and the base station is of the same order and the intersite interference is reasonable. If needed, the scheduling decisions between the different sites can be coordinated. It is much simpler to restrict the dynamics in the uplink−downlink allocation *when needed* and thereby have a more static operation than trying to add dynamics to a fundamentally static scheme, which was done when introducing eIMTA for LTE in release 12.

One example when intersite coordination is useful is a traditional macrodeployment. In such scenarios, a (more or less) static uplink−downlink allocation is a good choice as it avoids troublesome interference situations. Static or semistatic TDD operation is also necessary for handling coexistence with LTE, for example, when an LTE carrier and an NR carrier are using the same sites and the same frequency band. Such restrictions in the uplink−downlink allocation can easily be achieved as part of the scheduling implementation by using a fixed pattern in each base station. There is also a possibility to semistatically configure the transmission direction of some or all of the slots as discussed in Section 7.8.3, a feature that can allow for reduced device energy consumption as it is not necessary to monitor for downlink control channels in slots that are a priori known to be reserved for uplink usage.

An essential aspect of any TDD system, or half-duplex system in general, is the possibility to provide a sufficiently large *guard period* (or guard time), where neither downlink nor uplink transmissions occur. This guard period is necessary for switching from downlink to uplink transmission and vice versa and is obtained by using slot formats where the downlink ends sufficiently early prior to the start of the uplink. The required length of the guard period depends on several factors. First, it should be sufficiently large to provide the necessary time for the circuitry in base stations and the devices to switch from downlink to uplink. Switching is typically relatively fast, of the order of 20 µs or less, and in most deployments does not significantly contribute to the required guard time.

Second, the guard time should also ensure that uplink and downlink transmissions do not interfere at the base station. This is handled by advancing the uplink timing at the devices such that, at the base station, the last uplink subframe before the uplink-to-downlink switch ends before the start of the first downlink subframe. The uplink timing of each device can be controlled by the base station by using the timing advance mechanism, as will be elaborated upon in Chapter 15. Obviously, the guard period must be large enough to allow the device to receive the downlink transmission and switch from reception to transmission before it starts the (timing-advanced) uplink transmission (see Fig. 7.16). As the timing advance is proportional to the distance to the base station, a larger guard period is required when operating in large cells compared to small cells.

Finally, the selection of the guard period also needs to take interference between base stations into account. In a multicell network, intercell interference from downlink transmissions in neighboring cells must decay to a sufficiently low level before the base station can start to receive uplink transmissions. Hence, a

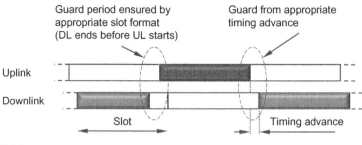

Guard period ensured by appropriate slot format (DL ends before UL starts)

Guard from appropriate timing advance

Uplink

Downlink

Slot

Timing advance

FIGURE 7.16

Creation of guard time for TDD operation.

larger guard period than is motivated by the cell size itself may be required as the last part of the downlink transmissions from distant base stations, otherwise it may interfere with uplink reception. The amount of guard period depends on the propagation environments, but in some macrocell deployments the interbase-station interference is a nonnegligible factor when determining the guard period. Depending on the guard period, some residual interference may remain at the beginning of the uplink period. Hence, it is beneficial to avoid placing interference-sensitive signals at the start of an uplink burst.

7.8.2 FREQUENCY-DIVISION DUPLEX (FDD)

In the case of FDD operation, uplink and downlink are carried on different carrier frequencies, denoted f_{UL} and f_{DL} in Fig. 7.15. During each frame, there is thus a full set of slots in both uplink and downlink, and uplink and downlink transmission can occur simultaneously within a cell. Isolation between downlink and uplink transmissions is achieved by transmission/reception filters, known as duplex filters, and a sufficiently large *duplex separation* in the frequency domain.

Even if uplink and downlink transmission can occur simultaneously within a cell in the case of FDD operation, a device may be capable of *full-duplex* operation or only *half-duplex* operation for a certain frequency band, depending on whether or not it is capable of simultaneous transmission/reception. In the case of full-duplex capability, transmission and reception may also occur simultaneously at a device, whereas a device capable of only half-duplex operation cannot transmit and receive simultaneously. Half-duplex operation allows for simplified device implementation due to relaxed or no duplex-filters. This can be used to reduce device cost, for example, for low-end devices in cost-sensitive applciations. Another example is operation in certain frequency bands with a very narrow duplex gap with correspondingly challenging design of the duplex filters. In this case, full duplex support can be *frequency-band-dependent* such that a device may support only half-duplex operation in certain frequency bands while being capable of full-duplex operation in the remaining supported bands. It should be noted that full/half-duplex capability is a property of the *device*; the base station can operate in full duplex irrespective of the device capabilities. For example, the base station can transmit to one device while simultaneously receiving from another device.

From a network perspective, half-duplex operation has an impact on the sustained data rates that can be provided to/from a single device as it cannot transmit in all uplink subframes. The cell capacity is hardly affected as typically it is possible to schedule different devices in uplink and downlink in a given subframe. No provisioning for guard periods is required from a network perspective as the network is still operating in full duplex and therefore is capable of simultaneous transmission and reception. The relevant transmission structures and timing relations are identical between full-duplex and half-duplex FDD and a single cell may therefore simultaneously support a mixture of full-duplex and half-duplex FDD devices. Since a half-duplex device is not capable of simultaneous transmission and reception, the scheduling decisions must take this into account and half-duplex operation can be seen as a scheduling restriction.

7.8.3 SLOT FORMAT AND SLOT-FORMAT INDICATION

Returning to the slot structure discussed in Section 7.2, it is important to point out that there is one set of slots in the uplink and another set of slots in the downlink, the reason being the time offset between the two as a function of timing advance. If both uplink and downlink transmission would be described using *the same* slot, which is often seen in various illustrations in the literature, it would not be possible to specify the necessary timing difference between the two.

Depending on the whether the device is capable of full duplex, as is the case for FDD, or half duplex only, as is the case for TDD, a slot may not be fully used for uplink or downlink transmission. As an example, the downlink transmission in Fig. 7.16 had to stop prior to the end of the slot in order to allow for sufficient time to switch to downlink reception. Since the necessary time between downlink and uplink depends on several factors, NR defines a wide range of *slot formats* defining which parts of a slot are used for uplink or downlink. Each slot format represents a combination of OFDM symbols denoted downlink, flexible, and uplink, respectively. The reason for having a third state, flexible, will be discussed further below, but one usage is to handle the necessary guard period in half-duplex schemes. A subset of the slot formats supported by NR are illustrated in Fig. 7.17. As seen in the figure, there are downlink-only and uplink-only slot formats which are useful for full-duplex operation (FDD), as well as partially filled uplink and downlink slots to handle the case of half-duplex operation (TDD).

The name slot format is somewhat misleading as there are separate slots for uplink and downlink transmissions, each filled with data in such a way that there is no simultaneous transmission and reception in the case of TDD. Hence, the slot format for a downlink slot should be understood as downlink transmissions can

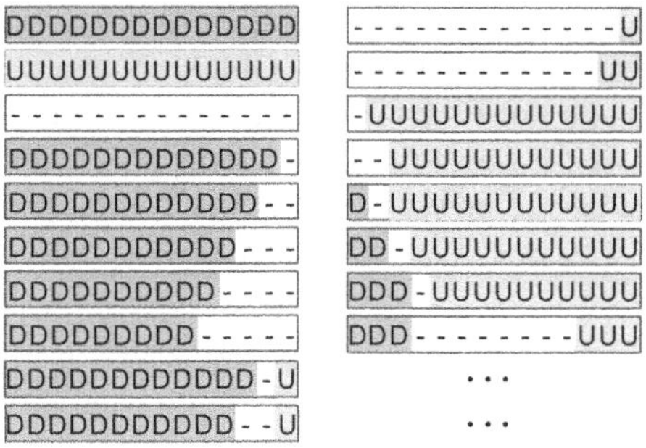

FIGURE 7.17

A subset of the possible slot formats in NR ("D" is downlink, "U" is uplink, and "−" is flexible).

only occur in "downlink" or "flexible" symbols, and in an uplink slot, uplink transmissions can only occur in "uplink" or "flexible" symbols. Any guard period necessary for TDD operation is taken from the flexible symbols.

One of the key features of NR is, as already mentioned, the support for *dynamic TDD* where the scheduler dynamically determines the transmission direction. Since a half-duplex device cannot transmit and receive simultaneously, there is a need to split the resources between the two directions. In NR, three different signaling mechanisms provide information to the device on whether the resources are used for uplink or downlink transmission:

- Dynamic signaling for the scheduled device;
- Semistatic signaling using RRC; and
- Dynamic slot-format indication shared by a group of devices.

Some or all of these mechanisms are used in combination to determine the instantaneous transmission direction as will be discussed below. Although the description below uses the term dynamic TDD, the framework can in principle be applied to half-duplex operation in general, including half-duplex FDD.

The first mechanism and the basic principle is for the device to monitor for control signaling in the downlink and transmit/receive according to the received scheduling grants/assignments. In essence, a half-duplex device would view each OFDM symbol as a downlink symbol unless it has been instructed to transmit in the uplink. It is up to the scheduler to ensure that a half-duplex device is not requested to simultaneously receive and transmit. For a full-duplex-capable device (FDD), there is obviously no such restriction and the scheduler can independently schedule uplink and downlink.

The general principle above is simple and provides a flexible framework. However, if the network knows a priori that it will follow a certain uplink−downlink allocation, for example, in order to provide coexistence with some other TDD technology or to fulfill some spectrum regulatory requirement, it can be advantageous to provide this information to the device. For example, if it is known to a device that a certain set of OFDM symbols is assigned to uplink transmissions, there is no need for the device to monitoring for downlink control signaling in the part of the downlink slots overlapping with these symbols. This can help reducing the device power consumption. NR therefore provides the possibility to optionally signal the uplink−downlink allocation through RRC signaling.

The RRC-signaled pattern classifies OFDM symbols as "downlink," "flexible," or "uplink." For a half-duplex device, a symbol classified as "downlink" can only be used for downlink transmission with no uplink transmission in the same period of time. Similarly, a symbol classified as "uplink" means that the device should not expect any overlapping downlink transmission. "Flexible" means that the device cannot make any assumptions on the transmission direction. Downlink control signaling should be monitored and if a scheduling message is found, the device should transmit/receive accordingly. Thus, the fully dynamic scheme outlined above is equivalent to semistatically declaring all symbols as "flexible."

The RRC-signaled pattern is expressed as a concatenation of up to two sequences of downlink−flexible−uplink, together spanning a configurable period from 0.5 ms up to 10 ms. Furthermore, *two* patterns can be configured, one cell-specific provided as part of system information and one signaled in a device-specific manner. The resulting pattern is obtained by combining these two where the dedicated pattern can further restrict the flexible symbols signaled in the cell-specific pattern to be either downlink or uplink. Only if both the cell-specific pattern and the device-specific pattern indicate flexible should the symbols be for flexible use (Fig. 7.18).

The third mechanism is to dynamically signal the current uplink−downlink allocation to a group of devices monitoring a special downlink control message known as the *slot-format indicator* (SFI). Similar to the previous mechanism, the slot format can indicate the number of OFDM symbols that are downlink, flexible, or uplink, and the message is valid for one or more slots.

The SFI message will be received by a configured group of one or more devices and can be viewed as a pointer into an RRC-configured table where each row in the table is constructed from a set of predefined downlink/flexible/uplink patterns one slot in duration. Upon receiving the SFI, the value is used as an index into the SFI table to obtain the uplink−downlink pattern for one or more slots as illustrated in Fig. 7.19. The set of predefined downlink/flexible/uplink patterns is listed in the NR specifications and covers a wide range of possibilities, some examples of which can be seen in Fig. 7.17 and in the left part of Fig. 7.19. The SFI can also indicate the uplink−downlink situations for other cells (cross-carrier indication).

Since a dynamically scheduled device will know whether the carrier is currently used for uplink transmission or downlink transmission from its scheduling assignment/grant, the group-common SFI signaling is primarily intended for *non-scheduled* devices. In particular, it offers the possibility for the network to

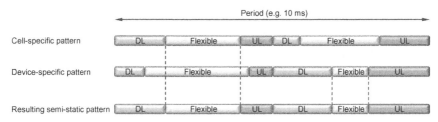

FIGURE 7.18

Example of cell-specific and device-specific uplink—downlink patterns.

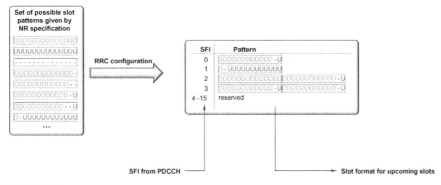

FIGURE 7.19

Example of configuring the SFI table.

overrule periodic transmissions of *uplink sounding signals* (SRS) or downlink measurements on *channel-state information reference signals* (CSI-RS). The SRS transmissions and CSI-RS measurements are used for assessing the channel quality as discussed in Chapter 8, and can be semi-statically configured. Overriding the periodic configuration can be useful in a network running with dynamic TDD (see Fig. 7.20 for an example illustration).

The SFI cannot override a semistatically configured uplink or downlink period, neither can it override a dynamically scheduled uplink or downlink transmission which takes place regardless of the SFI. However, the SFI can override a symbol period semistatically indicated as flexible by restricting it to be downlink or uplink. It can also be used to provide a reserved resource; if both the SFI and the semistatic signaling indicate a certain symbol to be flexible, then the symbol should be treated as reserved and not be used for transmission, nor should the device make any assumptions on the downlink transmission. This can be useful as a tool to reserve resource on an NR carrier, for example, used for other radio-access technologies or for features added to future releases of the NR standard.

The description above has focused on half-duplex devices in general and TDD in particular. However, the SFI can be useful also for full-duplex systems such as

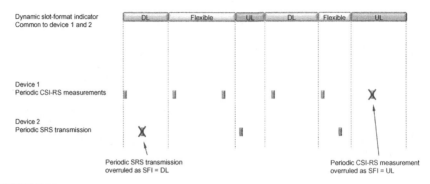

FIGURE 7.20

Controlling periodic CSI-RS measurements and SRS transmissions by using the SFI.

FDD, for example, to override periodic SRS transmissions. Since there are two independent "carriers" in this case, one for uplink and one for downlink, two SFIs are needed, one for each carrier. This is solved by using the multislot support in the SFI; one slot is interpreted as the current SFI for the downlink and the other as the current SFI for the uplink.

7.9 ANTENNA PORTS

Downlink multiantenna transmission is a key technology of NR. Signals transmitted from different antennas or signals subject to different, and for the receiver unknown, *multiantenna precoders* (see Chapter 9), will experience different "radio channels" even if the set of antennas are located at the same site.[6]

In general, it is important for a device to understand what it can assume in terms of the relationship between the radio channels experienced by different downlink transmissions. This is, for example, important in order for the device to be able to understand what reference signal(s) should be used for channel estimation for a certain downlink transmission. It is also important in order for the device to be able to determine relevant channel-state information, for example, for scheduling and link-adaptation purposes.

For this reason, the concept of *antenna port* is used in the NR, following the same principles as in LTE. An antenna port is defined such that the channel over which a symbol on the antenna port is conveyed can be inferred from the channel over which another symbol on the same antenna port is conveyed. Expressed differently, each individual downlink transmission is carried out from a specific antenna port, the identity of which is known to the device. Furthermore, the

---

[6]An unknown transmitter-side precoder needs to be seen as part of the overall radio channel.

Table 7.2 Antenna Ports in NR

Antenna Port	Uplink	Downlink
0-series	PUSCH and associated DM-RS	–
1000-series	SRS, precoded PUSCH	PDSCH
2000-series	PUCCH	PDCCH
3000-series	–	CSI-RS
4000-series	PRACH	SS block

device can assume that two transmitted signals have experienced the same radio channel *if and only if* they are transmitted from the same antenna port.[7]

In practice, each antenna port can, at least for the downlink, be seen as corresponding to a specific reference signal. A device receiver can then assume that this reference signal can be used to estimate the channel corresponding to the specific antenna port. The reference signals can also be used by the device to derive detailed channel-state information related to the antenna port.

The set of antenna ports defined in NR is outlined in Table 7.2. As seen in the table, there is a certain structure in the antenna port numbering such that antenna ports for different purposes have numbers in different ranges. For example, downlink antenna ports starting with 1000 are used for PDSCH. Different transmission layers for PDSCH can use antenna ports in this series, for example, 1000 and 1001 for a two-layer PDSCH transmission. The different antenna ports and their usage will be discussed in more detail in conjunction with the respective feature.

It should be understood that an antenna port is an abstract concept that does not necessarily correspond to a specific physical antenna:

- Two different signals may be transmitted in the same way from multiple physical antennas. A device receiver will then see the two signals as propagating over a single channel corresponding to the "sum" of the channels of the different antennas and the overall transmission could be seen as a transmission from a single antenna port being the same for the two signals.
- Two signals may be transmitted from the same set of antennas but with different, for the receiver unknown, antenna transmitter-side precoders. A receiver will have to see the unknown antenna precoders as part of the overall channel implying that the two signals will appear as having been transmitted from two different antenna ports. It should be noted that if the antenna precoders of the two transmissions would have been known to be the same, the transmissions could have been seen as originating from the same antenna port. The same would have been true if the precoders would have been known

---

[7]For certain antenna ports, more specifically those that correspond to so-called demodulation reference signals, the assumption of the same radio channel is only valid within a given scheduling occasion.

to the receiver as, in that case, the precoders would not need to be seen as part of the radio channel.

The last of these two aspects motivates the introduction of QCL framework as discussed in the next section.

7.10 **QUASI-COLOCATION**

Even if two signals have been transmitted from two different antennas, the channels experienced by the two signals may still have many *large-scale* properties in common. As an example, the channels experienced by two signals transmitted from two different antenna ports corresponding to different physical antennas at the same site will, even if being different in the details, typically have the same or at least similar large-scale properties, for example, in terms of Doppler spread/ shift, average delay spread, and average gain. It can also be expected that the channels will introduce similar average delay. Knowing that the radio channels corresponding to two different antenna ports have similar large-scale properties can be used by the device receiver, for example, in the setting of parameters for channel estimation.

In case of single-antenna transmission, this is straightforward. However, one integral part of NR is the extensive support for multiantenna transmission, beamforming, and simultaneous transmission from multiple geographically separates sites. In these cases, the channels of different antenna ports relevant for a device may differ even in terms of large-scale properties.

For this reason, the concept of *quasi-colocation* with respect to antenna ports is part of NR. A device receiver can assume that the radio channels corresponding to two different antenna ports have the same large-scale properties in terms of specific parameters such as average delay spread, Doppler spread/shift, average delay, average gain, and spatial Rx parameters if and only if the antenna ports are specified as being quasi-colocated. Whether or not two specific antenna ports can be assumed to be quasi-colocated with respect to a certain channel property is in some cases given by the NR specification. In other cases, the device may be explicitly informed by the network by means of signaling if two specific antenna ports can be assumed to be quasi-colocated or not.

The general principle of quasi-colocation is present already in the later releases of LTE when it comes to the temporal parameters. However, with the extensive support for beamforming in NR, the QCL framework has been extended to the spatial domain. Spatial quasi-colocation or, more formally, quasi-colocation with respect to RX parameters is a key part of beam management. Although somewhat vague in its formal definition, in practice spatial QCL between two different signals implies that they are transmitted from the same place and in the same beam. As a consequence, if a device knows that a certain receiver beam direction is good for one of the signals, it can assume that the same beam direction is suitable also for reception of the other signal.

In a typical situation, the NR specification states that certain transmissions, for example, PDSCH and PDCCH transmissions, are spatially quasi-colocated with specific reference signals, for example, CSI-RS or SS block. The device may have decided on a specific receiver beam direction based on measurements on the reference signal in question and the device can then assume that the same beam direction is a good choice also for the PDSCH/PDCCH reception.

Channel Sounding

8

Many transmission features in modern radio-access technologies are based on the availability of more or less detailed knowledge about different characteristics of the radio channel over which a signal is to be transmitted. This may range from rough knowledge of the radio-channel path loss for transmit-power adjustment to detailed knowledge about the channel amplitude and phase in the time, frequency, and/or spatial domain. Many transmission features will also benefit from knowledge about the interference level experienced at the receiver side.

Such knowledge about different channel characteristics can be acquired in different ways and by measurements on either the transmitter side or receiver side of a radio link. As an example, knowledge about downlink channel characteristics can be acquired by means of device measurements. The acquired information could then be reported to the network for the setting of different transmission parameters for subsequent downlink transmissions. Alternatively, if it can be assumed that the channel is reciprocal, that is, the channel characteristics of interest are the same in the downlink and uplink transmission directions, the network can, by itself, acquire knowledge about relevant downlink channel characteristics by estimating the same characteristics in the uplink direction.

The same alternatives exist when it comes to acquiring knowledge about uplink channel characteristics:

- The network may determine the uplink characteristics of interest and either provide the information to the device or directly control subsequent uplink transmissions based on the acquired channel knowledge;
- Assuming channel reciprocity, the device may, by itself, acquire knowledge about the relevant uplink channel characteristics by means of downlink measurements.

Regardless of the exact approach to acquire channel knowledge, there is typically a need for specific signals on which a receiver can measure/estimate channel characteristics of interest. This is often expressed as *channel sounding*.

This chapter will describe the NR support for such channel sounding. Especially, we will describe the specific reference signals, downlink *channel-state-information reference signals* (CSI-RS) and uplink *sounding reference signals* (SRS), on which channel sounding is typically based. We will also provide an overview of the NR framework for downlink physical-layer measurements and corresponding device reporting to the network.

8.1 DOWNLINK CHANNEL SOUNDING—CSI-RS

In the first release of LTE (release 8), channel knowledge for the downlink transmission direction was solely acquired by means of device measurements on the so-called *cell-specific reference signals* (CRS). The LTE CRS are transmitted over the entire carrier bandwidth within every LTE subframe of length 1 ms, and can be assumed to be transmitted over the entire cell area. Thus, a device accessing an LTE network can assume that CRS are always present and can be measured on.

In LTE release 10 the CRS were complemented by so-called CSI-RS. In contrast to CRS, the LTE CSI-RS are not necessarily transmitted continuously. Rather, an LTE device is explicitly configured to measure on a set of CSI-RS and does not make any assumptions regarding the presence of a CSI-RS unless it is explicitly configured for the device.

The origin for the introduction of CSI-RS was the extension of LTE to support spatial multiplexing with more than four layers, something which was not possible with the release-8 CRS. However, the use of CSI-RS was soon found to be an, in general, more flexible and efficient tool for channel sounding, compared to CRS. In later releases of LTE, the CSI-RS concept was further extended to also support, for example, interference estimation and multi-point transmission.

As already described, a key design principle for the development of NR has been to as much as possible avoid "always on" signals. For this reason, there are no CRS-like signals in NR. Rather, the only "always-on" NR signal is the so-called *SS block* (see Chapter 16) which is transmitted over a limited bandwidth and with a much larger periodicity compared to the LTE CRS. The SS block can be used for power measurements to estimate, for example, path loss and average channel quality. However, due to the limited bandwidth and low duty cycle, the SS block is not suitable for more detailed channel sounding aimed at tracking channel properties that vary rapidly in time and/or frequency.

Instead the concept of CSI-RS is reused in NR and further extended to, for example, provide support for beam management and mobility as a complement to SS block.

8.1.1 BASIC CSI-RS STRUCTURE

A configured CSI-RS may correspond to up to 32 different antenna ports, each corresponding to a channel to be sounded.

In NR, a CSI-RS is always configured on a per-device basis. It is important to understand though that configuration on a per-device basis does not necessarily mean that a transmitted CSI-RS can only be used by a single device. Nothing prevents identical CSI-RS using the same set of resource elements to be separately configured for multiple devices, in practice implying that a single CS-RS is shared between the devices

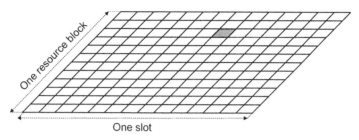

FIGURE 8.1

Single-port CSI-RS structure consisting of a single resource element within an RB/slot block.

As illustrated in Fig. 8.1, a single-port CSI-RS occupies a single resource element within a block corresponding to one resource block in the frequency domain and one slot in the time domain. In principle, the CSI-RS can be configured to occur anywhere within this block although in practice there are some restrictions to avoid collisions with other downlink physical channels and signals. Especially, a device can assume that transmission of a configured CSI-RS will not collide with:

- Any CORESET configured for the device;
- Demodulation reference signals associated with PDSCH transmissions scheduled for the device;
- Transmitted SS blocks.

A multi-port CSI-RS can be seen as multiple orthogonally transmitted per-antenna-port CSI-RS sharing the overall set of resource elements assigned for the configured multi-port CSI-RS. In the general case, this sharing is based on a combination of:

- *Code-domain sharing* (CDM), implying that different per-antenna-port CSI-RS are transmitted on the same set of resource elements with separation achieved by modulating the CSI-RS with different orthogonal patterns;
- *Frequency-domain sharing* (FDM), implying that different per-antenna-port CSI-RS are transmitted on different subcarriers within an OFDM symbol;
- *Time-domain sharing* (TDM), implying that different per-antenna-port CSI-RS are transmitted in different OFDM symbols within a slot.

Furthermore, as illustrated in Fig. 8.2, CDM between different per-antenna-port CSI-RS can be:

- In the frequency domain with CDM over two adjacent subcarriers ($2 \times$ CDM), allowing for code-domain sharing between two per-antenna-port CSI-RS;

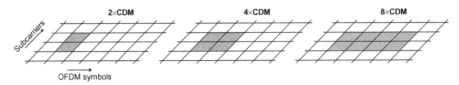

FIGURE 8.2

Different CDM structures for multiplexing per-antenna-port CSI-RS.

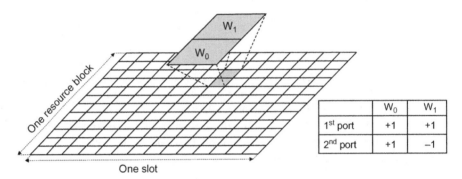

FIGURE 8.3

Structure of two-port CSI-RS based on 2 × CDM. The figure also illustrates the orthogonal patterns of each port.

- In the frequency and time domain with CDM over two adjacent subcarriers and two adjacent OFDM symbols (4 × CDM), allowing for code-domain sharing between up to four per-antenna-port CSI-RS;
- In the frequency and time domain with CDM over two adjacent subcarriers and four adjacent OFDM symbols (8 × CDM), allowing for code-domain sharing between up to eight per-antenna-port CSI-RS.

The different CDM alternatives of Fig. 8.2, in combination with FDM and/or TDM, can then be used to configure different multi-port CSI-RS structures where, in general, an N-port CSI-RS occupies a total of N resource elements within an RB/slot block.[1]

As a first example, Fig. 8.3 illustrates how a two-port CSI-RS consists of two adjacent resource elements in the frequency domain with sharing by means of CDM. In other words, the two-port CSI-RS has a structure identical to the basic 2 × CDM structure in Fig. 8.2.

In the case of CSI-RS corresponding to more than two antenna ports there is some flexibility in the sense that, for a given number of ports, there are multiple CSI-RS structures based on different combinations of CDM, TDM, and FDM.

---

[1]The "density-3" CSI-RS used for TRS (see Section 8.1.7) is an exception to this rule.

As an example, there are three different structures for an eight-port CSI-RS (see Fig. 8.4).

- Frequency-domain CDM over two resource elements ($2 \times$ CDM) in combination with four times frequency multiplexing (left part of Fig. 8.4). The overall CSI-RS resource thus consists of eight subcarriers within the same OFDM symbol.
- Frequency-domain CDM over two resource elements ($2 \times$ CDM) in combination with frequency and time multiplexing (middle part Fig. 8.4). The overall CSI-RS resource thus consists of four subcarriers within two OFDM symbols.
- Time/frequency-domain CDM over four resource elements ($4 \times$ CDM) in combination with two times frequency multiplexing. The overall CSI-RS resource thus once again consists of four subcarriers within two OFDM symbols.

Finally, Fig. 8.5 illustrates one out of three possible structures for a 32-port CSI-RS based on a combination of $8 \times$ CDM and four times frequency multiplexing. This example also illustrates that CSI-RS antenna ports separated in the frequency domain do not necessarily have to occupy consecutive subcarriers. Likewise, CSI-RS ports separated in the time domain do not necessarily have to occupy consecutive OFDM symbols.

In the case of a multi-port CSI-RS, the association between per-port CSI-RS and port number is done first in the CDM domain, then in the frequency domain, and finally in the time domain. This can, for example, be seen from the eight-port example of Fig. 8.4 where per-port CSI-RS separated by means of CDM correspond to consecutive port numbers. Furthermore, for the FDM + TDM case (center part of Fig. 8.4), port number zero to port number three are transmitted within the same OFDM symbol, while port number four to port number seven are jointly transmitted within another OFDM symbol. Port number zero to three and port number four to seven are thus separated by means of TDM.

8.1.2 FREQUENCY-DOMAIN STRUCTURE OF CSI-RS CONFIGURATIONS

A CSI-RS is configured for a given downlink bandwidth part and is then assumed to be confined within that bandwidth part and use the numerology of the bandwidth part.

The CSI-RS can be configured to cover the full bandwidth of the bandwidth part or just a fraction of the bandwidth. In the latter case, the CSI-RS bandwidth and frequency-domain starting position are provided as part of the CSI-RS configuration.

Within the configured CSI-RS bandwidth, a CSI-RS may be configured for transmission in every resource block, referred to as *CSI-RS density equal to one*.

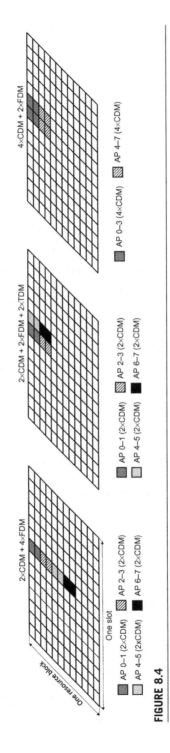

FIGURE 8.4

Three different structures for eight-port CSI-RS.

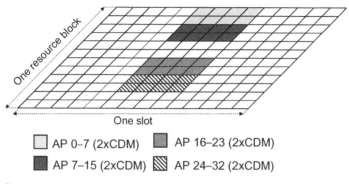

FIGURE 8.5

One structure (out of three supported structures) for a 32-port CSI-RS.

However, a CSI-RS may also be configured for transmission only in every second resource block, referred to as CSI-RS density equal to 1/2. In the latter case, the CSI-RS configuration includes information about the set of resource blocks (odd resource blocks or even resource blocks) within which the CSI-RS will be transmitted. CSI-RS density equal to 1/2 is not supported for CSI-RS with 4, 8, and 12 antenna ports.

There is also a possibility to configure a single-port CSI-RS *with a density of 3* in which case the CSI-RS occupies three subcarriers within each resource block. This CSI-RS structure is used as part of a so-called *Tracking Reference signal* (TRS) (see further details in Section 8.1.7).

8.1.3 TIME-DOMAIN PROPERTY OF CSI-RS CONFIGURATIONS

The per-resource-block CSI-RS structure outlined above describes the structure of a CSI-RS transmission, assuming the CSI-RS is actually transmitted in a given slot. In general, a CSI-RS can be configured for *periodic, semi-persistent*, or *aperiodic* transmission.

In the case of periodic CSI-RS transmission, a device can assume that a configured CSI-RS transmission occurs every N^{th} slot, where N ranges from as low as four, that is, CSI-RS transmissions every fourth slot, to as high as 640, that is, CSI-RS transmission only every 640th slot. In addition to the periodicity, the device is also configured with a specific slot offset for the CSI-RS transmission (see Fig. 8.6).

In the case of semi-persistent CSI-RS transmission, a certain CSI-RS periodicity and corresponding slot offset are configured in the same way as for periodic CSI-RS transmission. However, actual CSI-RS transmission can be activated/deactivated based on *MAC control elements* (MAC CE) (see Section 6.4.4). Once the CSI-RS transmission has been activated, the device can assume that the CSI-RS transmission will continue according to the configured periodicity until it is

FIGURE 8.6

Examples of CSI-RS periodicity and slot offset.

explicitly deactivated. Similarly, once the CSI-RS transmission has been deactivated, the device can assume that there will be no CSI-RS transmissions according to the configuration until it is explicitly re-activated.

In the case of aperiodic CSI-RS, no periodicity is configured. Rather, a device is explicitly informed ("triggered") about each CSI-RS transmission instant by means of signaling in the DCI.

It should be mentioned that the property of periodic, semi-persistent, or aperiodic is strictly speaking not a property of the CSI-RS itself but rather the property of a CSI-RS *resource set* (see Section 8.1.6). As a consequence, activation/deactivation and triggering of semi-persistent and aperiodic CSI-RS, respectively, is not done for a specific CSI-RS but for the set of CSI-RS within a resource set.

8.1.4 CSI-IM—RESOURCES FOR INTERFERENCE MEASUREMENTS

A configured CSI-RS can be used to derive information about the properties of the channel over which the CSI-RS is transmitted. A CSI-RS can also be used to estimate the interference level by subtracting the expected received signal from what is actually received on the CSI-RS resource.

However, the interference level can also be estimated from measurements on so-called *CSI-IM* (Interference Measurement) resources.

Fig. 8.7 illustrates the structure of a CSI-IM resource. As can be seen, there are two different CSI-IM structures, each consisting of four resource elements but with different time/frequency structures. Similar to CSI-RS, the exact location of the CSI-IM resource within the RB/slot block is flexible and part of the CSI-IM configuration.

The time-domain property of CSI-IM resources is the same as that of CSI-RS, that is, a CSI-IM resource could be periodic, semi-persistent (activation/deactivation by means of MAC CE), or aperiodic (triggered by DCI). Furthermore, for periodic and semi-persistent CSI-IM, the set of supported periodicities is the same as for CSI-RS.

In a typical case, a CSI-IM resource would correspond to resource elements where nothing is transmitted within the current cell while the activity within the CSI-IM resource in neighbor cells should correspond to normal activity of those cells. Thus, by measuring the receiver power within a CSI-IM resource, a device

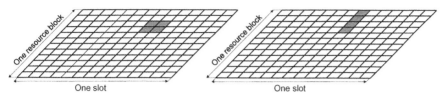

FIGURE 8.7

Alternative structures for a CSI-IM resource.

would get an estimate on the typical interference due to transmissions within other cells.

As there should be no transmissions on CSI-IM resources within the cell, devices should be configured with the corresponding resources as so-called ZP-CSI-RS resources (see below).

8.1.5 ZERO-POWER CSI-RS

The CSI-RS described above should more correctly be referred to as *non-zero-power* (NZP) CSI-RS to distinguish them from so-called *zero-power* (ZP) CSI-RS that can also be configured for a device.

If a device is scheduled for PDSCH reception on a resource that includes resource elements on which a configured CSI-RS is to be transmitted, the device can assume that the PDSCH rate matching and resource mapping avoid those resource elements. However, a device may also be scheduled for PDSCH reception on a resource that includes resource elements corresponding to a CSI-RS configured for a different device. The PDSCH must also in this case be rate matched around the resource elements used for CSI-RS. The configuration of a ZP-CSI-RS is a way to inform the device for which the PDSCH is scheduled about such rate matching.

A configured ZP-CSI-RS corresponds to a set of resource elements with the same structure as an NZP-CSI-RS. However, while a device can assume that an NZP-CI-RS is actually transmitted and is something on which a device can carry out measurements, a configured ZP-CSI-RS only indicates a set of resource blocks to which the device should assume that PDSCH is not mapped.

It should be emphasized that, despite the name, a device cannot assume that there are no transmissions (zero power) within the resource elements corresponding to a configured ZP-CSI-RS. As already mentioned, the resources corresponding to a ZP-CSI-RS may, for example, be used for transmission of NZP-CSI-RS configured for other devices. What the NR specification says is that a device cannot make *any* assumptions regarding transmissions on resources corresponding to a configured ZP-CSI-RS and that PDSCH transmission for the device is not mapped to resource elements corresponding to a configured ZP-CSI-RS.

8.1.6 CSI-RS RESOURCE SETS

In addition to being configured with CSI-RS, a device can be configured with one or several *CSI-RS resource sets*, officially referred to as *NZP-CSI-RS-ResourceSets*. Each such resource set includes one or several configured CSI-RS.[2] The resource set can then be used as part of *report configurations* describing measurements and corresponding reporting to be done by a device (see further details in Section 8.2). Alternatively, and despite the name, an *NZP-CSI-RS-ResourceSet* may include pointers to a set of SS blocks (see Chapter 16). This reflects the fact that some device measurements, especially measurements related to beam management and mobility, may be carried out on either CSI-RS or SS block.

Above it was described how a CSI RS could be configured for periodic, semi-persistent, or aperiodic transmission. As mentioned there, this categorization is strictly speaking not a property of the CSI-RS itself but a property of a resource set. Furthermore, all CSI-RS within a semi-persistent resource set are jointly activated/deactivated by means of a MAC CE command. Likewise, transmission of all CSI-RS within an aperiodic resource set is jointly triggered by means of DCI.

Similarly, a device may be configured with *CSI-IM resource sets*, each including a number of configured CSI-IM that can be jointly activated/deactivated (semi-persistent CSI-IM resource set) or triggered (aperiodic CSI-IM resource set).

8.1.7 TRACKING REFERENCE SIGNAL (TRS)

Due to oscillator imperfections, the device must track and compensate for variations in time and frequency to successfully receive downlink transmissions. To assist the device in this task, a *tracking reference signal* (TRS) can be configured. The TRS is not a CSI-RS. Rather a TRS is a *resource set* consisting of *multiple* periodic NZP-CSI-RS. More specifically a TRS consists of four one-port, density-3 CSI-RS located within two consecutive slots (see Fig. 8.8). The CRS-RS within the resource set, and thus also the TRS in itself, can be configured with a periodicity of 10, 20, 40, or 80 ms. Note that the exact set of resource elements (subcarriers and OFDM symbols) used for the TRS CSI-RS may vary. There is always a four-symbol time-domain separation between the two CSI-RS within a slot though. This time domain separation sets the limit for the frequency error that can be tracked. Likewise, the frequency-domain separation (four subcarriers) sets the limit for the timing error that can be tracked.

There is also an alternative TRS structure with the same per-slot structure as the TRS structure of Fig. 8.8 but only consisting of two CSI-RS *within a single slot*, compared to two consecutive slots for the TRS structure in Fig. 8.8

---

[2]Strictly speaking, the resource set includes *references* to configured CSI-RS.

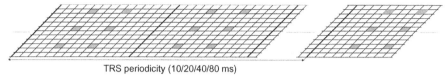

FIGURE 8.8

TRS consisting of four one-port, density-3 CSI-RS located within two consecutive slots.

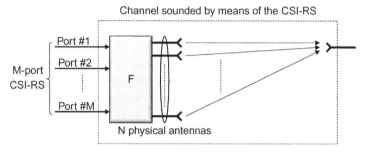

FIGURE 8.9

CSI-RS applied to spatial filter (F) before mapping to physical antennas.

For LTE, the CRS served the same purpose as the TRS. However, compared to the LTE CRS, the TRS implies much less overhead, only having one antenna port and only being present in two slots every TRS period.

8.1.8 MAPPING TO PHYSICAL ANTENNAS

In Chapter 7, the concept of antenna ports and the relation to reference signals were discussed. A multi-port CSI-RS corresponds to a set of antenna ports and the CSI-RS can be used for sounding of the channels corresponding to those antenna ports. However, a CSI-RS port is often not mapped directly to a physical antenna, implying that the channel being sounded based on a CSI-RS is often not the actual physical radio channel. Rather, more or less any kind of (linear) transformation or *spatial filtering*, labeled F in Fig. 8.9, may be applied to the CSI-RS before mapping to the physical antennas. Furthermore, the number of physical antennas (N in Fig. 8.9) to which the CSI-RS is mapped may very well be larger than the number of CSI-RS ports.[3] When a device does channel sounding based on the CSI-RS, neither the spatial filter F nor the N physical antennas will be explicitly visible. What the device will see is just the M "channels" corresponding to the M CSI-RS ports.

---

[3]Having N smaller than M does not make sense.

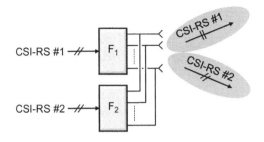

FIGURE 8.10

Different spatial filters applied to different CSI-RS.

The spatial filter F may very well be different for different CSI-RS. The network could, for example, map two different configured CSI-RS such that they are beam-formed in different directions (see Fig. 8.10). To the device this will appear as two CSI-RS transmitted over two different channels, despite the fact that they are transmitted from the same set of physical antennas and are propagating via the same set of physical channels.

Although the spatial filter F is not explicitly visible to the device, the device still has to make certain assumptions regarding F. Especially, F has a strong relation to the concept of antenna ports discussed in Chapter 7. In essence one can say that two signals are transmitted from the same antenna port if they are mapped to the same set of physical antennas by means of the same transformation F.

As an example, in the case of downlink multiantenna transmission (see Chapter 11), a device may measure on a CSI-RS and report a recommended *precoder matrix* to the network. The network may then use the recommended precoder matrix when mapping so-called transmission layers to antenna ports. When selecting a suitable precoder matrix the device will assume that the network, if using the recommended matrix, will map the output of the precoding to the antenna ports of the CSI-RS on which the corresponding device measurements were carried out. In other words, the device will assume that the precoded signal will be mapped to the physical antennas by means of the same spatial filter F as applied to the CSI-RS.

8.2 DOWNLINK MEASUREMENTS AND REPORTING

An NR device can be configured to carry out different measurements, in most cases with corresponding reporting to the network. In general, such a configuration of a measurement and corresponding reporting are done by means of a *report configuration*, in the 3GPP specifications [15] referred to as a *CSI-ReportConfig*.[4]

---

[4]Note that we are here talking about *physical-layer* measurements and reporting, to be distinguished from higher-layer reporting done by means of RRC signaling.

Each resource configuration describes/indicates:

- The specific quantity or set of quantities to be reported;
- The downlink resource(s) on which measurements should be carried out in order to derive the quantity or quantities to be reported;
- How the actual reporting is to be carried out, for example, when the reporting is to be done and what uplink physical channel to use for the reporting.

8.2.1 REPORT QUANTITY

A report configuration indicates a quantity or set of quantities that the device is supposed to report. The report could, for example, include different combinations of *channel-quality indicator* (CQI), *rank indicator* (RI), and *precoder-matrix indicator* (PMI), jointly referred to as *channel-state information* (CSI). Alternatively, the report configuration may indicate reporting of received signal strength, more formally referred to as *reference-signal received power* (RSRP). RSRP has historically been a key quantity to measure and report as part of higher-layer *radio-resource management* (RRM) and is so also for NR. However, NR also supports layer-1 reporting of RSRP, for example, as part of the support for beam management (see Chapter 12). What is then reported is more specifically referred to as *L1-RSRP*, reflecting the fact that the reporting does not include the more long-term ("layer-3") filtering applied for the higher-layer RSRP reporting.

8.2.2 MEASUREMENT RESOURCE

In addition to describing what quantity to report, a report configuration also describes the set of downlink signals or, more generally, the set of downlink resources on which measurements should be carried out in order to derive the quantity or quantities to be reported. This is done by associating the report configuration with one or several resource sets as described in Section 8.1.6.

A resource configuration is associated with at least one *NZP-CSI-RS-ResourceSet* to be used for measuring channel characteristics. As described in Section 8.1.6, a *NZP-CSI-RS-ResourceSet* may either contain a set of configured CSI-RS or a set of SS blocks. Reporting of, for example, L1-RSRP for beam management can thus be based on measurements on either a set of SS blocks or a set of CSI-RS.

Note that the resource configuration is associated with a resource *set*. Measurements and corresponding reporting are thus in the general case carried out on *a set* of CSI-RS or *a set* of SS blocks.

In some cases, the set will only include a single reference signal. An example of this is conventional feedback for link adaptation and multi-antenna precoding. In this case, the device would typically be configured with a resource set consisting of a single multi-port CSI-RS on which the device will carry out measurements to determine and report a combination of CQI, RI, and PMI.

On the other hand, in the case of beam management the resource set will typically consist of multiple CSI-RS, alternatively multiple SS blocks, where in practice each CSI-RS or SS block is associated with a specific beam. The device measures on the set of signals within the resource set and reports the result to the network as input to the beam-management functionality.

There are also situations when a device needs to carry out measurements without any corresponding reporting to the network. One such case is when a device should carry out measurements for receiver-side downlink beam forming. As will be described in Chapter 12, in such a case a device may measure on downlink reference signals using different receiver beams. However, the result of the measurement is not reported to the network but only used internally within the device to select a suitable receiver beam. At the same time the device needs to be configured with the reference signals to measure on. Such a configuration is also covered by report configurations for which, in this case, the quantity to be reported is defined as "None."

8.2.3 REPORT TYPES

In addition to the quantity to report and the set of resources to measure on, the report configuration also describes when and how the reporting should be carried out.

Similar to CSI-RS transmission, device reporting can be periodic, semi-persistent, or aperiodic.

As the name suggests, periodic reporting is done with a certain configured periodicity. Periodic reporting is always done on the PUCCH physical channel. Thus, in the case of periodic reporting, the resource configuration also includes information about a periodically available PUCCH resource to be used for the reporting.

In the case of semi-persistent reporting, a device is configured with periodically occurring reporting instances in the same way as for periodic reporting. However, actual reporting can be activated and deactivated by means of MAC signaling (MAC CE).

Similar to periodic reporting, semi-persistent reporting can be done on a periodically assigned PUCCH resource. Alternatively, semi-persistent reporting can be done on a semi-persistently allocated PUSCH. The latter is typically used for larger reporting payloads.

Aperiodic reporting is explicitly triggered by means of DCI signaling, more specifically within a CSI-request field within the uplink scheduling grant (DCI format 0-1). The DCI field may consist of up to 6 bits with each configured aperiodic report associated with a specific bit combination. Thus, up to 63 different aperiodic reports can be triggered.[5]

---

[5]The all-zero value indicates "no triggering."

Table 8.1 Allowed Combinations of Report Type and Resource Type

Report Type	Resource Type		
	Periodic	**Semi-persistent**	**Aperidic**
Periodic	Yes	–	–
Semi-persistent	Yes	Yes	–
Aperiodic	Yes	Yes	Yes

Aperiodic reporting is always done on the scheduled PUSCH and thus requires an uplink scheduling grant. This is the reason why the triggering of aperiodic reporting is only included in the uplink scheduling grant and not in other DCI formats.

It should be noted that, in the case of aperiodic reporting, the report configuration could actually include multiple resource sets for channel measurements, each with its own set of reference signals (CSI-RS or SS block). Each resource set is associated with a specific value of the CSI-request field in the DCI. By means of the CSI request the network can, in this way, trigger the same type of reporting but based on different measurement resources. Note that the same could, in principle, have been done by configuring the device with multiple report configurations, where the different resource configurations would specify the same reporting configuration and report type but different measurement resources.

Periodic, semi-persistent, and aperiodic reporting should not be mixed up with periodic, semi-persistent, and aperiodic CSI-RS as described in Section 8.1.3. As an example, aperiodic reporting and semi-persistent reporting could very well be based on measurements on periodic CSI-RS. On the other hand, periodic reporting can only be based on measurements on periodic CSI-RS but not on aperiodic and semi-persistent CSI-RS. Table 8.1 summarizes the allowed combinations of reporting type (periodic, semi-persistent, and aperiodic) and resource type (periodic, semi-persistent, and aperiodic).

8.3 UPLINK CHANNEL SOUNDING—SRS

To enable uplink channel sounding a device can be configured for transmission of *sounding reference signals* (SRS). In many respects SRS can be seen as the uplink equivalence to the downlink CSI-RS in the sense that both CSI-RS and SRS are intended for channel sounding, albeit in different transmission directions. Both CSI-RS and SRS can also serve as QCL references in the sense that other physical channels can be configured to be transmitted quasi-colocated with CSI-RS and SRS, respectively. Thus, given knowledge of a suitable receiver beam for the CSI-RS/SRS, the receiver knows that the same receiver beam should be suitable also for the physical channel in question.

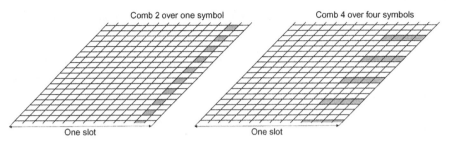

FIGURE 8.11

Examples of SRS time/frequency structures.

However, on a more detailed level, the structure of SRS is quite different from CSI-RS.

- SRS is limited to a maximum of four antenna ports, while CSI-RS supports up to 32 antenna ports.
- Being an uplink signal, SRS is designed to have low cubic-metric [60] enabling high device power-amplifier efficiency.

The basic time/frequency structure of an SRS is exemplified in Fig. 8.11. In the general case, an SRS spans one, two, or four consecutive OFDM symbols and is located somewhere within the last six symbols of a slot. In the frequency domain, an SRS has a so-called "comb" structure, implying that an SRS is transmitted on every Nth subcarrier where N can take the values two or four ("comb-2" and "comb-4," respectively).

SRS transmissions from different devices can be frequency multiplexed within the same frequency range by being assigned different combs corresponding to different frequency offsets. For comb-2, that is, when SRS is transmitted on every second subcarrier, two SRS can be frequency multiplexed. In the case of comb-4, up to four SRS can be frequency multiplexed. Fig. 8.12 illustrates an example of SRS multiplexing assuming a comb-2 SRS spanning two OFDM symbols.

8.3.1 SRS SEQUENCES AND ZADOFF–CHU SEQUENCES

The sequences applied to the set of SRS resource elements are partly based on so-called *Zadoff–Chu* sequences [25]. Due to their specific properties, Zadoff–Chu sequences are used at several places within the NR specifications, especially in the uplink transmission direction. Zadoff–Chu sequences are also extensively used in LTE [28].

A Zadoff–Chu sequence of length M is given by the following expression:

$$z_i^u = e^{-j\frac{\pi u i(i+1)}{M}} \; ; \; 0 \le i < M \tag{8.1}$$

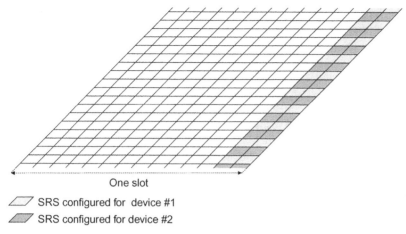

One slot

SRS configured for device #1
SRS configured for device #2

FIGURE 8.12

Comb-based frequency multiplexing of SRS from two different devices assuming comb-2.

As can be seen from Eq. (8.1), a Zadoff–Chu sequence has a characterizing parameter u, referred to as the *root index* of the Zadoff–Chu sequence. For a given sequence length M, the number of root indices generating unique Zadoff–Chu sequences equals the number of integers that are relative prime to M. For this reason, Zadoff–Chu sequences of prime length are of special interest as they maximize the number of available Zadoff–Chu sequences. More specifically, assuming the sequence length M being a prime number there are $M-1$ unique Zadoff–Chu sequences.

A key property of Zadoff–Chu sequences is that the discrete Fourier transform of a Zadoff–Chu sequence is also a Zadoff–Chu sequence.[6] From Eq. (8.1) it is obvious that a Zadoff–Chu sequence has constant time-domain amplitude, making it good from a power-amplifier-efficiency point of view. As the Fourier transform of a Zadoff–Chu sequence is also a Zadoff–Chu sequence, there would then also be constant power in the frequency domain, that is, in addition to constant time-domain amplitude, Zadoff–Chu sequences also have flat spectra. As a flat spectrum is equivalent to zero cyclic autocorrelation for any non-zero cyclic shift, this implies that two different time-domain cyclic shifts of the same Zadoff–Chu sequence are orthogonal to each other. Note that a cyclic shift in the time domain corresponds to applying a continuous phase rotation in the frequency domain.

Although Zadoff–Chu sequences of prime length are preferred in order to maximize the number of available sequences, SRS sequences are not of prime length. The SRS sequences are therefore *extended* Zadoff–Chu sequences based

---

[6]The inverse obviously holds as well, that is, the inverse DFT of a Zadoff–Chu sequence is also a Zadoff–Chu sequence.

on the longest prime-length Zadoff−Chu sequence with a length M smaller or equal to the desired SRS-sequence length. The sequence is then cyclically extended in the frequency domain up to the desired SRS sequence length. As the extension is done in the frequency domain, the extended sequence still has a constant spectrum, and thus "perfect" cyclic autocorrelation, but the time-domain amplitude will vary somewhat.

Extended Zadoff−Chu sequences will be used as SRS sequences for sequence lengths of 36 or larger, corresponding to an SRS extending over 6 and 12 resource blocks in the cases of comb-2 and comb-4, respectively. For shorter sequence lengths, special flat-spectrum sequences with good time-domain envelope properties have been found from computer search. The reason is that, for shorter sequences, there would not be a sufficient number of Zadoff−Chu sequences available.

The same principle will be used also for other cases where Zadoff−Chu sequences are used within the NR specifications, for example, for uplink DMRS (see Section 9.11.1).

8.3.2 MULTIPORT SRS

In the case of an SRS supporting more than one antenna port, the different ports share the same set of resource elements and the same basic SRS sequence. Different phase rotations are then applied to separate the different ports as illustrated in Fig. 8.13.

As described above, applying a phase rotation in the frequency domain is equivalent to applying a cyclic shift in the time domain. In the NR specification the above operation is actually referred to as "cyclic shift," although it is mathematically described as a frequency-domain phase shift.

8.3.3 TIME-DOMAIN STRUCTURE OF SRS

Similar to CSI-RS, an SRS can be configured for *periodic*, *semi-persistent*, or *aperiodic* transmission:

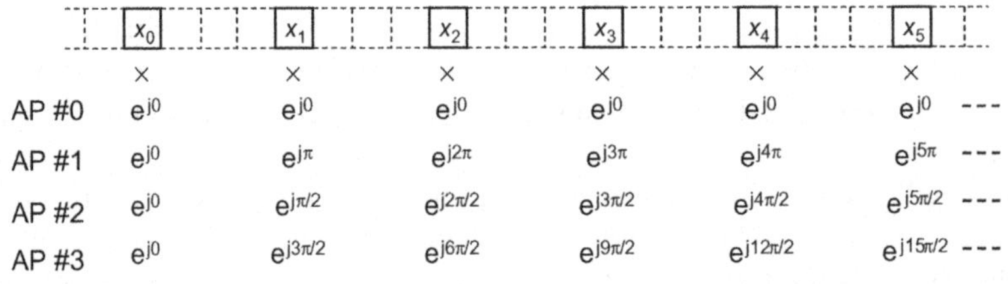

	x_0	x_1	x_2	x_3	x_4	x_5	
	×	×	×	×	×	×	
AP #0	e^{j0}	e^{j0}	e^{j0}	e^{j0}	e^{j0}	e^{j0}	---
AP #1	e^{j0}	$e^{j\pi}$	$e^{j2\pi}$	$e^{j3\pi}$	$e^{j4\pi}$	$e^{j5\pi}$	---
AP #2	e^{j0}	$e^{j\pi/2}$	$e^{j2\pi/2}$	$e^{j3\pi/2}$	$e^{j4\pi/2}$	$e^{j5\pi/2}$	---
AP #3	e^{j0}	$e^{j3\pi/2}$	$e^{j6\pi/2}$	$e^{j9\pi/2}$	$e^{j12\pi/2}$	$e^{j15\pi/2}$	---

FIGURE 8.13

Separation of different SRS antenna ports by applying different phase shifts to the basic frequency-domain SRS sequence $x_0, x_1, x_2, \ldots$. The figure assumes a comb-4 SRS.

- A periodic SRS is transmitted with a certain configured periodicity and a certain configured slot offset within that periodicity;
- A semi-persistent SRS has a configured periodicity and slot offset in the same way as a periodic SRS. However, actual SRS transmission according to the configured periodicity and slot offset is activated and deactivated by means of MAC CE signaling;
- An aperiodic SRS is only transmitted when explicitly triggered by means of DCI.

It should be pointed out that, similar to CSI-RSI, activation/deactivation and triggering for semi-persistent and aperiodic SRS, respectively, is actually not done for a specific SRS but rather done for a so-called *SRS resource set* which, in the general case, included multiple SRS (see below).

8.3.4 SRS RESOURCE SETS

Similar to CSI-RS, a device can be configured with one or several *SRS resource sets*, where each resource set includes one or several configured SRS. As described above, a SRS can be configured for periodic, semi-persistent, or aperiodic transmission. All SRS included within a configured SRS resource set have to be of the same type. In other words, periodic, semi-persistent, or aperiodic transmission can also be seen as a property of an SRS resource set.

A device can be configured with multiple SRS resource sets that can be used for different purposes, including both downlink and uplink multiantenna precoding and downlink and uplink beam management.

The transmission of aperiodic SRS, or more accurately, transmission of the set of configured SRS included in an aperiodic SRS resource set, is triggered by DCI. More specifically, DCI format 0-1 (uplink scheduling grant) and DCI format 1-1 (downlink scheduling assignment) include a 2-bit *SRS-request* that can trigger the transmission of one out of three different aperiodic SRS resource sets configured for the device (the fourth bit combination corresponds to "no triggering").

8.3.5 MAPPING TO PHYSICAL ANTENNAS

Similar to CSI-RS, SRS ports are often not mapped directly to the device physical antennas but via some spatial filter F that maps M SRS ports to N physical channels (see Fig. 8.14).

In order to provide connectivity regardless of the rotational direction of the device, NR devices supporting high-frequency operation will typically include multiple antenna panels pointing in different directions. The mapping of SRS to one such panel is an example of a transformation F from SRS antenna ports to the set of physical antennas. Transmission from different panels will then correspond to different spatial filters F as illustrated in Fig. 8.15.

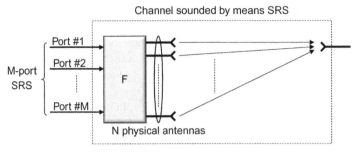

FIGURE 8.14

SRS applied to spatial filter (F) before mapping to physical antennas.

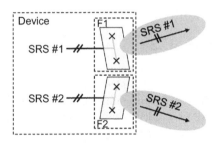

FIGURE 8.15

Different spatial filters applied to different SRS.

Similar to the downlink, the spatial filtering F has a real impact despite the fact that it is never explicitly visible to the network receiver but just seen as an integrated part of the overall channel. As an example, the network may sound the channel based on a device-transmitted SRS and then decide on a precoder matrix that the device should use for uplink transmission. The device is then assumed to use that precoder matrix in combination with the spatial filter F applied to the SRS. In other cases, a device may be explicitly scheduled for data transmission using the antenna ports defined by a certain SRS. In practice this implies that the device is assumed to transmit using the same spatial F that has been used for the SRS transmission. In practice, this may imply that the device should transmit using the same beam or panel that has been used for the SRS transmission.

Transport-Channel Processing

9

This chapter will provide a more detailed description of the downlink and uplink physical-layer functionality such as coding, modulation, multi-antenna precoding, resource-block mapping, and reference signal structure.

9.1 OVERVIEW

The physical layer provides services to the MAC layer in the form of transport channels as described in Section 6.4.5. In the downlink, there are three different types of transport channels defined for NR: the Downlink Shared Channel (DL-SCH), the Paging Channel (PCH), and the Broadcast Channel (BCH), although the latter two are not used in the non-standalone operation. In the uplink, there is only one uplink transport-channel type carrying transport blocks in NR,[1] the Uplink Shared Channel (UL-SCH). The overall transport channel processing for NR follows a similar structure as for LTE (see Fig. 9.1). The processing is mostly similar in uplink and downlink and the structure in Fig. 9.1 is applicable for the DL-SCH, BCH, and PCH in the downlink, and the UL-SCH in the uplink. The part of the BCH that is mapped to the PBCH follows a different structure, described in Section 16.1, as does the RACH.

Within each *transmission time interval* (TTI), up to two transport blocks of dynamic size are delivered to the physical layer and transmitted over the radio interface for each component carrier. Two transport blocks are only used in the case of spatial multiplexing with more than four layers, which is only supported in the downlink direction and mainly useful in scenarios with very high signal-to-noise ratios. Hence, at most a single transport block per component carrier and TTI is a typical case in practice.

A CRC for error-detecting purposes is added to each transport block, followed by error-correcting coding using LDPC codes. Rate matching, including physical-layer hybrid-ARQ functionality, adapts the number of coded bits to the scheduled resources. The code bits are scrambled and fed to a modulator, and finally the

---

[1]Strictly speaking, the Random-Access Channel is also defined as a transport-channel type (see Chapter 16). However, RACH only includes a layer-1 preamble and carries no data in the form of transport blocks.

5G NR: The Next Generation Wireless Access Technology. DOI: https://doi.org/10.1016/B978-0-12-814323-0.00009-0

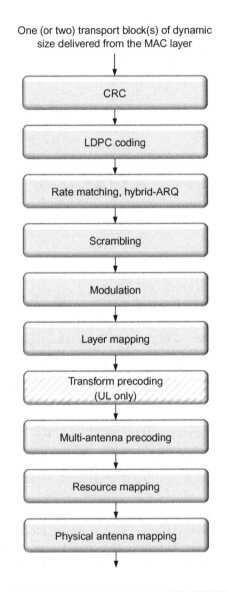

One (or two) transport block(s) of dynamic size delivered from the MAC layer

FIGURE 9.1

General transport-channel processing.

modulation symbols are mapped to the physical resources, including the spatial domain. For the uplink there is also a possibility of a DFT-precoding. The differences between uplink and downlink is, apart from DFT-precoding being possible in the uplink only, mainly around antenna mapping and associated reference signals.

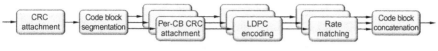

FIGURE 9.2

Channel coding.

In the following, each of the processing steps will be discussed in more detail. For carrier aggregation, the processing steps are duplicated for each of the carriers and the description herein is applicable to each of the carriers. Since most of the processing steps are identical for uplink and downlink, the processing will be described jointly and any differences between uplink and downlink explicitly mentioned when relevant.

9.2 CHANNEL CODING

An overview of the channel coding steps is provided in Fig. 9.2 and described in more detail in the following sections. First, a CRC is attached to the transport block to facilitate error detection, followed by code block segmentation. Each code block is LDPC-encoded and rate matched separately, including physical-layer hybrid-ARQ processing, and the resulting bits are concatenated to form the sequence of bits representing the coded transport block.

9.2.1 CRC ATTACHMENT PER TRANSPORT BLOCK

In the first step of the physical-layer processing, a CRC is calculated for and appended to each transport block. The CRC allows for receiver-side detection of errors in the decoded transport block and can, for example, be used by the hybrid-ARQ protocol as a trigger for requesting retransmissions.

The size of the CRC depends on the transport-block size. For transport blocks larger than 3824 bits, a 24-bit CRC is used, otherwise a 16-bit CRC is used to reduce overhead.

9.2.2 CODE-BLOCK SEGMENTATION

The LDPC coder in NR is defined up to a certain code-block size (8424 bits for base graph 1 and 3840 bits for base graph 2). To handle transport block sizes larger than this, code-block segmentation is used where the transport block, including the CRC, is split into multiple equal-sized[2] code blocks as illustrated in Fig 9.3.

---

[2]The set of possible transport-block sizes are such that it is always possible to split a too large transport block into smaller equal-sized code-blocks.

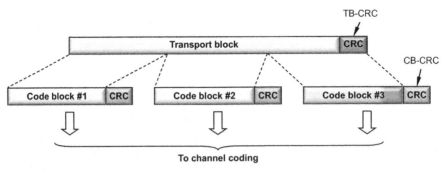

FIGURE 9.3

Code block segmentation.

As can be seen in Fig. 9.3, code-block segmentation also implies that an additional CRC (also of length 24 bits but different compared to the transport-block CRC described above) is calculated for and appended to each code block. In the case of a single code-block transmission no additional code-block CRC is applied.

One could argue that, in the case of code-block segmentation, the transport-block CRC is redundant and implies unnecessary overhead as the set of code-block CRCs should indirectly provide information about the correctness of the complete transport block. However, to handle *code-block group* (CBG) retransmissions as discussed in Chapter 13, a mechanism to detect errors per code block is necessary. CBG retransmission means that only the erroneous code-block groups are retransmitted instead of the complete transport block to improve the spectral efficiency. The per-CB CRC can also be used for the device to limit decoding in case of a retransmission only to those CBs whose CRCs did not check even if per-CBG retransmission is not configured. This helps reducing the device processing load. The transport-block CRC also adds an extra level of protection in terms of error detection. Note that code-block segmentation is only applied to large transport blocks for which the relative extra overhead due to the additional transport-block CRC is small.

9.2.3 CHANNEL CODING

Channel coding is based on LDPC codes, a code design which was originally proposed in the 1960s [34] but forgotten for many years. They were "rediscovered" in the 1990s [59] and found to be an attractive choice from an implementation perspective. From an error-correcting capability point of view, turbo codes, as used in LTE, can achieve similar performance, but LDPC codes can offer lower complexity, especially at higher code rates, and were therefore chosen for NR.

The basis for LDPC codes is a sparse (low-density) parity check matrix H where for each valid code word c the relation $Hc^T = 0$ holds. Designing a good

LDPC code to a large extent boils down to finding a good parity check matrix **H** which is sparse (the sparseness implies relatively simple decoding). It is common to represent the parity-check matrix by a graph connecting n variable nodes at the top with $(n-k)$ constraint nodes at the bottom of the graph, a notation that allows a wide range of properties of an (n, k) LDPC code to be analyzed. This explains why the term *base graph* is used in the NR specifications. A detailed description of the theory behind LDPC codes is beyond the scope of this book, but there is a rich literature in the field (for example, see [68]).

Quasi-cyclic LDPC codes with a dual-diagonal structure of the kernel part of the parity check matrix are used in NR, which gives a decoding complexity which is linear in the number of coded bits and enables a simple encoding operation. Two base graphs are defined, BG1 and BG2, representing the two base matrices. The reason for two base graphs instead of one is to handle the wide range of payload sizes and code rates in an efficient way. Supporting a very large payload size at a medium to high code rate, which is the case for very high data rates, using a code designed to support a very low code rate is not efficient. At the same time, the lowest code rates are necessary to provide good performance in challenging situations. In NR, BG1 is designed for code rates from 1/3 to 22/24 (approximately 0.33−0.92) and BG 2 from 1/5 to 5/6 (approximately 0.2−0.83). Through puncturing, the highest code rate can be increased somewhat, up to 0.95, beyond which the device is not required to decode. The choice between BG1 and BG2 is based on the transport block size and code rate targeted for the first transmission (see Fig. 9.4).

The base graphs, and the corresponding base matrices, define the general structure of the LDPC code. To support a range of payload sizes, 51 different *lifting sizes* and sets of *shift coefficients* are defined and applied to the base matrices.

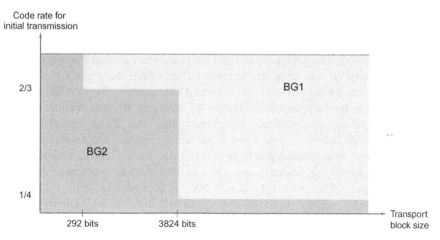

FIGURE 9.4

Selection of base graph for the LDPC code.

In short, for a given lifting size Z, each "1" in the base matrix is replaced by the $Z \times Z$ identify matrix circularly shifted by the corresponding shift coefficient and each "0" in the base matrix is replaced by the $Z \times Z$ all-zero matrix. Hence, a relatively large number of parity-check matrices can be generated to support multiple payload sizes while maintaining the general structure of the LDPC code. To support payload sizes that are not a native payload size of one of the 51 defined parity check matrices, known filler bits can be appended to the code block before encoding. Since the NR LDPC codes are systematic codes, the filler bits can be removed before transmission.

9.3 RATE MATCHING AND PHYSICAL-LAYER HYBRID-ARQ FUNCTIONALITY

The rate-matching and physical-layer hybrid-ARQ functionality serves two purposes, namely to extract a suitable number of coded bits to match the resources assigned for transmission and to generate different redundancy versions needed for the hybrid-ARQ protocol. The number of bits to transmit on the PDSCH or PUSCH depends on a wide range of factors, not only the number of resource blocks and the number of OFDM symbols scheduled, but also on the amount of overlapping resource elements used for other purposes and such as reference signals, control channels, or system information. There is also a possibility to, in the downlink, define *reserved resources* as a tool to provide future compatibility (see Section 9.10), which affects the number of resource elements usable for the PDSCH.

Rate matching is performed separately for each code block. First, a fixed number of the systematic bits are punctured. The fraction of systematic bits punctured can be relatively high, up to 1/3 of the systematic bits, depending on the code-block size. The remaining coded bits are written into a circular buffer, starting with the non-punctured systematic bits and continuing with parity bits as illustrated in Fig. 9.5. The selection of the bits to transmit is based on reading the required number of bits from the circular buffer where the exact set of bits to transmit depends on the *redundancy version* (RV) corresponding to different starting positions in the circular buffer. Hence, by selecting different redundancy versions, different sets of coded bits representing the same set of information bits can be generated, which is used when implementing hybrid-ARQ with incremental redundancy. The starting points in the circular buffer are defined such that both RV0 and RV3 are self-decodable, that is, includes the systematic bits under typical scenarios. This is also the reason RV3 is located after "nine o'clock" in Fig. 9.5 as this allows more of the systematic bits to be included in the transmission.

In the receiver, *soft combining* is an important part of the hybrid-ARQ functionality as described in Section 13.1. The soft values representing the received coded bits are buffered and, if a retransmission occurs, decoding is performed

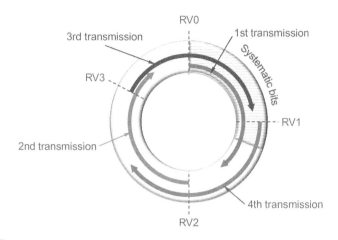

FIGURE 9.5

Example of circular buffer for incremental redundancy.

using the buffered bits combined with the retransmitted coded bits. In addition to a gain in accumulated received E_b/N_0, with different coded bits in different transmission attempts, additional parity bits are obtained and the resulting code rate after soft combining is lower with a corresponding coding gain obtained.

Soft combining requires a buffer in the receiver. Typically, a fairly high probability of successful transmission on the first attempt is targeted and hence the soft buffer remains unused most of the time. Since the soft buffer size is fairly large for the largest transport block sizes, requiring the receiver to buffer all soft bits even for the largest transport block sizes is suboptimal from a cost−performance tradeoff perspective. Hence, limited-buffer rate-matching is supported as illustrated in Fig. 9.6. In principle, only bits the device can buffer are kept in the circular buffer, that is, the size of the circular buffer is determined based on the receiver's soft buffering capability.

For the downlink, the device is not required to buffer more soft bits than corresponding to the largest transport block size coded at rate 2/3. Note that this only limits the soft buffer capacity for the highest transport block sizes, that is, the highest data rates. For smaller transport block sizes, the device is capable of buffering all soft bits down to the mother code rate.

For the uplink, full-buffer rate matching, where all soft bits are buffered irrespective of the transport block size, is supported given sufficient gNB memory. Limited-buffer rate matching using the same principles as for the downlink can be configured using RRC signaling.

The final step of the rate-matching functionality is to interleave the bits using a block interleaver and to collect the bits from each code block. The bits from the circular buffer are written row-by-row into a block interleaver and read out column-by-column. The number of rows in the interleaver is given by the

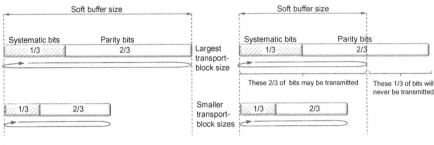

FIGURE 9.6

Limited-buffer rate matching.

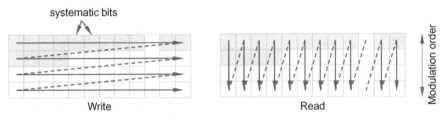

FIGURE 9.7

Bit interleaver (16QAM assumed in this example).

modulation order and hence the bits in one column correspond to one modulation symbol[3] (see Fig. 9.7). This results in the systematic bits spread across the modulation symbols, which improves performance. Bit collection concatenates the bits for each code block.

9.4 SCRAMBLING

Scrambling is applied to the block of coded bits delivered by the hybrid-ARQ functionality by multiplying the sequence of coded bits with a bit-level *scrambling sequence*. Without scrambling, the channel decoder at the receiver could, at least in principle, be equally matched to an interfering signal as to the target signal, thus being unable to properly suppress the interference. By applying different scrambling sequences for neighboring cells in the downlink or for different devices in the uplink, the interfering signal(s) after descrambling is (are) randomized, ensuring full utilization of the processing gain provided by the channel code.

The scrambling sequence in both downlink (PDSCH) and uplink (PUSCH) depends on the identity of the device, that is, the C-RNTI, and a *data scrambling*

---

[3]This structure improves the performance for higher-order modulation.

identity configured in each device. If no data scrambling identity is configured, the physical layer cell identity is used as a default value to ensure that neighboring devices, both in the same cell and between cells, use different scrambling sequences. Furthermore, in the case of two transport blocks being transmitted in the downlink to support more than four layers, different scrambling sequences are used for the two transport blocks.

9.5 MODULATION

The modulation step transforms the block of scrambled bits to a corresponding block of complex modulation symbols. The modulation schemes supported include QPSK, 16QAM, 64QAM, and 256QAM in both uplink and downlink. In addition, for the uplink $\pi/2$-BPSK is supported in the case the DFT-precoding is used, motivated by a reduced cubic metric [60] and hence improved power-amplifier efficiency, in particular for coverage limited scenarios. Note that $\pi/2$-BPSK is neither supported nor useful in the absence of DFT-precoding as the cubic metric in this case is dominated by the OFDM waveform.

9.6 LAYER MAPPING

The purpose of the layer-mapping step is to distribute the modulation symbols across the different transmission layers. This is done in a similar way as for LTE; every *n*th symbol is mapped to the *n*th layer. One coded transport block can be mapped on up to four layers. In the case of five to eight layers, supported in the downlink only, a second transport block is mapped to layers five to eight following the same principle as for the first transport block.

Multi-layer transmission is only supported in combination with OFDM, the baseline waveform in NR. With DFT-precoding in the uplink, only a single transmission layer is supported. This is motivated both by the receiver complexity, which in the case of multi-layer transmission would be significantly higher with a DFT-precoder than without, and the use case originally motivating the additional support of DFT-precoding, namely handling of coverage-limited scenarios. In such a scenario, the received signal-to-noise ratio is too low for efficient usage of spatial multiplexing and there is no need to support spatial multiplexing to a single device.

9.7 UPLINK DFT PRECODING

DFT precoding can be configured in the uplink only. In the downlink, as well as the case of OFDM in the uplink, the step is transparent.

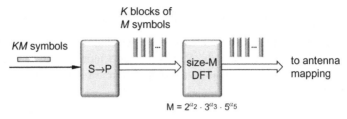

FIGURE 9.8

DFT-precoding.

In the case that DFT-precoding is applied in the uplink, blocks of M symbols are fed through a size-M DFT as illustrated in Fig. 9.8, where M corresponds to the number of subcarriers assigned for the transmission. The reason for the DFT pre-coding is to reduce the cubic metric for the transmitted signal, thereby enabling higher power-amplifier efficiency. From an implementation complexity point of view the DFT size should preferably be constrained to a power of 2. However, such a constraint would limit the scheduler flexibility in terms of the amount of resources that can be assigned for an uplink transmission. Rather, from a flexibility point of view all possible DFT sizes should preferably be allowed. For NR, the same middle-way as for LTE has been adopted where the DFT size, and thus also the size of the resource allocation, is limited to products of the integers 2, 3, and 5. Thus, for example, DFT sizes of 60, 72, and 96 are allowed but a DFT size of 84 is not allowed.[4] In this way, the DFT can be implemented as a combination of rela-tively low-complex radix-2, radix-3, and radix-5 FFT processing.

9.8 MULTI-ANTENNA PRECODING

The purpose of multi-antenna precoding is to map the different transmission layers to a set of antenna ports using a precoder matrix. In NR, the precoding and multi-antenna operation differs between downlink and uplink and the codebook-based precoding step is, except for CSI reporting, only visible in the uplink direc-tion. For a detailed discussion on how the precoding step is used to realize beam-forming and different multi-antenna schemes see Chapters 11 and 12.

9.8.1 DOWNLINK PRECODING

In the downlink, the demodulation reference signal (DMRS) used for channel esti-mation is subject to the same precoding as the PDSCH (see Fig. 9.9). Thus, the precoding is not explicitly visible to the receiver but is seen as part of the overall

---

[4]As uplink resource assignments are always done in terms of resource blocks of size 12 subcarriers, the DFT size is always a multiple of 12.

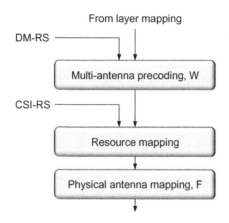

FIGURE 9.9

Downlink precoding.

channel. This is similar to the receiver-transparent spatial filtering discussed in the context of CSI-RS and SRS in Chapter 8. In essence, in terms of actual downlink transmission, any multi-antenna precoding can be seen as part of such, to the device, transparent spatial filtering.

However, for the purpose of CSI reporting, the device may assume that a specific precoding matrix W is applied at the network side. The device is then assuming that the precoder maps the signal to the antenna ports of the CSI-RS used for the measurements on which the reporting was done. The network is still free to use whatever precoder it finds advantageous for data transmission.

To handle receiver-side beamforming, or in general multiple reception antennas with different spatial characteristics, QCL relations between a DM-RS port group, which is the antenna ports used for PDSCH transmission,[5] and the antenna ports used for CSI-RS or SS block transmission can be configured. The *Transmission Configuration Index* (TCI) provided as part of the scheduling assignment indicates the QCL relations to use, or in other words, which reception beam to use. This is described in more detail in Chapter 12.

Demodulation reference signals are, as discussed in Section 9.11, transmitted in the scheduled resource blocks and it is from those reference signals that the device can estimate the channel, including any precoding W and spatial filtering F applied for PDSCH. In principle, knowledge about the correlation between reference signal transmissions, both in terms of correlation introduced by the radio channel itself and correlation in the use of precoder, is useful to know and can be exploited by the device to improve the channel estimation accuracy.

---

[5]The specification is prepared to handle two DM-RS port groups as part of multi-TRP schemes not part of release 15 but planned for later releases. In that case, some of the PDSCH layers belong to one DM-RS port group and the other layers to the other DM-RS port group.

In the time domain, the device is not allowed to make any assumptions on the reference signals being correlated between PDSCH scheduling occasions. This is necessary to allow full flexibility in terms of beamforming and spatial processing as part of the scheduling process.

In the frequency domain, the device can be given some guidance on the correlation. This is expressed in the form of *physical resource-block groups* (PRGs). Over the frequency span of one PRG, the device may assume the downlink precoder remains the same and may exploit this in the channel-estimation process, while the device may not make any assumptions in this respect between PRGs. From this it can be concluded that there is a trade-off between the precoding flexibility and the channel-estimation performance—a large PRG size can improve the channel-estimation accuracy at the cost of precoding flexibility and vice versa. Hence, the gNB may indicate the PRG size to the device where the possible PRG sizes are two resource blocks, four resource blocks, or the scheduled bandwidth as shown in the bottom of Fig. 9.10. A single value may be configured, in which case this value is used for the PDSCH transmissions. It is also possible to dynamically, through the DCI, indicate the PRG size used. In addition, the device can be configured to assume that the PRG size equals the scheduled bandwidth in the case that the scheduled bandwidth is larger than half the bandwidth part.

9.8.2 UPLINK PRECODING

Similar to the downlink, uplink demodulation reference signals used for channel estimation are subject to the same precoding as the uplink PUSCH. Thus, also for the uplink the precoding is not directly visible from a receiver perspective but is seen as part of the overall channel (see Fig. 9.11).

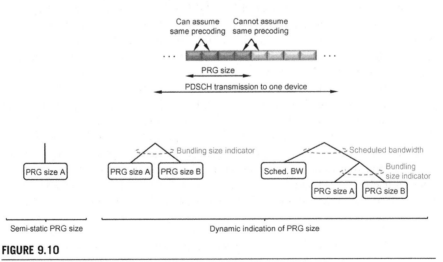

FIGURE 9.10

Physical resource-block groups (top) and indication thereof (bottom).

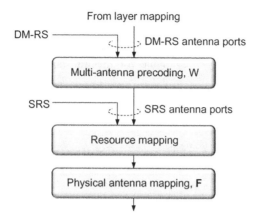

FIGURE 9.11

Uplink precoding.

However, from a scheduling point of view, the multi-antenna precoding of Fig. 9.1 is visible in the uplink as the network may provide the device with a specific precoder matrix W the receiver should use for the PUSCH transmission. This is done through the *precoding information* and *antenna port* fields in the DCI. The precoder is then assumed to map the different layers to the antenna ports of a configured SRS indicated by the network. In practice this will be the same SRS as the network used for the measurement on which the precoder selection was made. This is known as *codebook-based* precoding since the precoder W to use is selected from a codebook of possible matrices and explicitly signaled. Note that the spatial filter F selected by the device also can be seen as a precoding operation, although not explicitly controlled by the network. The network can however restrict the freedom in the choice of F through the *SRS resource indicator* (SRI) provided as part of the DCI.

There is also a possibility for the network to operate with *non-codebook-based* precoding. In this case W is equal to the identity matrix and precoding is handled solely by the spatial filter F based on recommendations from the device.

Both codebook-based and non-codebook-based precoding are described in detail in Chapter 11.

9.9 RESOURCE MAPPING

The resource-block mapping takes the modulation symbols to be transmitted on each antenna port and maps them to the set of available resource elements in the set of resource blocks assigned by the MAC scheduler for the transmission. As described in Section 7.3, a resource block is 12 subcarriers wide and typically multiple OFDM symbols, and resource blocks, are used for the transmission. The set of time—frequency resources used for transmission is determined by the scheduler. However, some or all of the resource elements within the scheduled resource blocks may not be available for the transport-channel transmission as they are used for:

- Demodulation reference signals (potentially including reference signals for *other* coscheduled devices in the case of multi-user MIMO) as described in Section 9.11;
- Other types of reference signals such as CSI-RS and SRS (see Chapter 8);
- Downlink L1/L2 control signaling (see Chapter 10);
- Synchronization signals and system information as described in Chapter 16;
- Downlink reserved resources as a means to provide forward compatibility as described in Section 9.10.

The time—frequency resources to be used for transmission are signaled by the scheduler as a set of *virtual resource blocks* and a set of OFDM symbols. To these scheduled resources, the modulation symbols are mapped to resource elements in a frequency-first, time-second manner. The frequency-first, time-second mapping is chosen to achieve low latency and allows both the transmitter and receiver to process the data "on the fly". For high data rates, there are multiple code blocks in each OFDM symbol and the device can decode those received in one symbol while receiving the next OFDM symbol. Similarly, assembling an OFDM symbol can take place while transmitting the previous symbols, thereby enabling a pipelined implementation. This would not be possible in the case of a time-first mapping as the complete slot needs to be prepared before the transmission can start.

The virtual resource blocks containing the modulation symbols are mapped to *physical resource blocks* in the bandwidth part used for transmission. Depending on the bandwidth part used for transmission, the *carrier resource blocks* can be determined and the exact frequency location on the carrier determined (see Fig. 9.12 for an illustration). The reason for this, at first sight somewhat complicated mapping process with both virtual and physical resource blocks is to be able to handle a wide range of scenarios.

There are two methods for mapping virtual resource blocks to physical resource blocks, non-interleaved mapping (Fig. 9.12: top) and interleaved mapping (Fig. 9.12: bottom). The mapping scheme to use can be controlled on a dynamic basis using a bit in the DCI scheduling the transmission.

Non-interleaved mapping means that a virtual resource block in a bandwidth part maps directly to the physical resource block in the same bandwidth part. This is useful in cases when the network tries to allocate transmissions to physical resource with instantaneously favorable channel conditions. For example, the scheduler might have determined that physical resource blocks six to nine in Fig. 9.12 have favorable radio channel properties and are therefore preferred for transmission and a non-interleaved mapping is used.

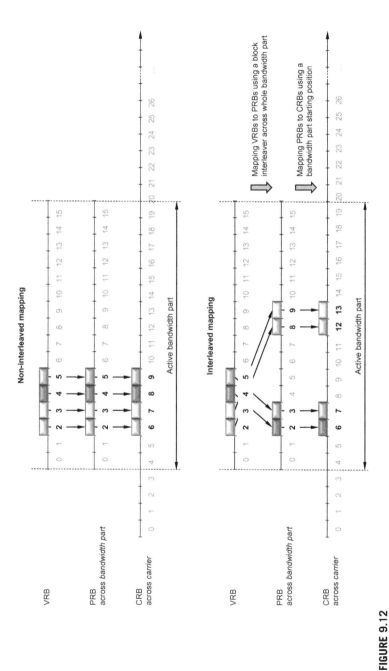

FIGURE 9.12

Mapping from virtual to physical to carrier resource blocks.

Interleaved mapping maps virtual resource blocks to physical resource blocks using an interleaver spanning the whole bandwidth part and operating on pairs or quadruplets of resource blocks. A block interleaver with two rows is used, with pairs/quadruplets of resource blocks written column-by-column and read out row-by-row. Whether to use pairs or quadruplets of resource blocks in the interleaving operation is configurable by higher-layer signaling.

The reason for interleaved resource-block mapping is to achieve frequency diversity, the benefits of which can be motivated separately for small and large resource allocations.

For small allocations, for example voice services, channel-dependent scheduling may not be motivated from an overhead perspective due to the amount of feedback signaling required, or may not be possible due to channel variations not being possible to track for a rapidly moving device. Frequency diversity by distributing the transmission in the frequency domain is in such cases an alternative way to exploit channel variations. Although frequency diversity could be obtained by using resource *allocation type 0* (see Section 10.1.10), this resource allocation scheme implies a relatively large control signaling overhead compared to the data payload transmitted as well as limited possibilities to signal very small allocations. Instead, by using the more compact *resource allocation type 1*, which is only capable of signaling contiguous resource allocations, combined with an interleaved virtual to physical resource block mapping, frequency diversity can be achieved with a small relative overhead. This is very similar to the distributed resource mapping in LTE. Since resource allocation type 0 can provide a high degree of flexibility in the resource allocation, interleaved mapping is supported for resource allocation type 1 only.

For larger allocations, possibly spanning the whole bandwidth part, frequency diversity can still be advantageous. In the case of a large transport block, that is, at very high data rates, the coded data are split into multiple code blocks as discussed in Section 9.2.2. Mapping the coded data directly to physical resource blocks in a frequency-first manner (remember, frequency-first mapping is beneficial from an overall latency perspective) would result in each code block occupying only a fairly small number of contiguous physical resource blocks. Hence, if the channel quality varies across the frequency range used for transmission, some code blocks may suffer worse quality than other code blocks, possibly resulting in the overall transport block failing to decode despite almost all code blocks being correctly decoded. The quality variations across the frequency range may occur even if the radio channel is flat due to imperfections in RF components. If an interleaved resource-block mapping is used, one code block occupying a contiguous set of virtual resource blocks would be distributed in the frequency domain across multiple, widely separated physical resource blocks, similarly to what is the case for the small allocations discussed in the previous paragraph. The result of the interleaved VRB-to-PRB mapping is a quality-averaging effect across the code blocks, resulting in a higher likelihood of correctly decoding very large transport blocks. This aspect of resource block mapping was not present in LTE, partially because the data rates were not as high as in NR, partly because the code blocks in LTE are interleaved.

The discussion above holds in general and for the downlink. In the uplink, release 15 only specifies RF requirements for contiguous allocations and therefore interleaved mapping is only supported for downlink transmissions. To obtain frequency diversity also in the uplink, frequency hopping can be used where the data in the first set of OFDM symbols in the slot are transmitted on the resource block as indicated by the scheduling grant. In the remaining OFDM symbols, data are transmitted on a different set of resource blocks given by a configurable offset from the first set. Uplink frequency hopping can be dynamically controlled using a bit in the DCI scheduling the transmission.

9.10 DOWNLINK RESERVED RESOURCES

One of the key requirements on NR was to ensure forward compatibility, that is, to allow future extensions and technologies to be introduced in a simple way without causing backward-compatibility problems with, at that point in time, already-deployed NR networks. Several NR technology components contribute to meeting this requirement, but the possibility to define *reserved resources* in the downlink is one of the more important tools. Reserved resources are semistatically configured time–frequency resources around which the PDSCH can be rate-matched.

Reserved resources can be configured in three different ways:

- By referring to an LTE carrier configuration, thereby allowing for transmissions on an NR carrier deployed on top of an LTE carrier (LTE/NR spectrum coexistence) to avoid the cell-specific reference signals of the LTE carrier (see further details in Chapter 17);
- By referring to a CORESET;
- By configuring resource sets using a set of bitmaps.

There are no reserved resources in the uplink; avoiding transmission on certain resources can be achieved through scheduling.[6]

Configuring reserved resources by referring to a configured CORESET is used to dynamically control whether control signaling resources can be reused for data or not (see Section 10.1.2). In this case the reserved resource is identical to the CORESET configured and the gNB may dynamically indicate whether these resources are usable for PDSCH or not. Thus, reserved resources do not have to be periodically occurring but can be used when needed.

---

[6]One reason is that only frequency-contiguous allocations are supported in the uplink in release 15, resulting in "bitmap-1" being unable to be used as this may result in non-contiguous frequency-domain allocations.

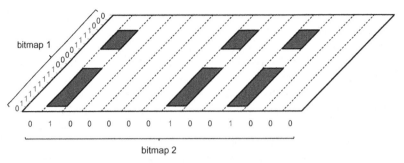

FIGURE 9.13

Configuring reserved resources.

The third way to configure reserved resources is based on bitmaps. The basic building block for a resource-set configuration covers one or two slots in the time domain and can be described by two bitmaps as illustrated in Fig. 9.13:

- A first time-domain bitmap, which in the NR specifications is referred to as "bitmap-2," indicates a set of OFDM symbols within the slot (or within a pair two slots).
- Within the set of OFDM symbols indicated by bitmap-2, an arbitrary set of resource blocks, that is, blocks of 12 resource elements in the frequency domain, may be reserved. The set of resource blocks is indicated by a second bitmap, in the NR specifications referred to as "bitmap-1."

If the resource set is defined on a carrier level, bitmap-1 has a length corresponding to the number of resource blocks within the carrier. If the resource set is bandwidth-part specific, the length of bitmap-1 is given by the bandwidth of the bandwidth part.

The same bitmap-1 is valid for all OFDM symbols indicated by bitmap-2. In other words, the same set of resource elements are reserved in all OFDM symbols indicated by bitmap-2. Furthermore, the frequency-domain granularity of the resource-set configuration provided by bitmap-1 is one resource block. In other words, all resource elements within a (frequency-domain) resource block are either reserved or not reserved.

Whether or not the resources configured as reserved resources are actually reserved or can be used for PDSCH can either be semistatically or dynamically controlled.

In the case of semistatic control, a third bitmap (bitmap-3) determines whether or not the resource-set defined by the bitmap-1/bitmap-2 pair or the CORSET is valid for a certain slot or not. The bitmap-3 has a granularity equal to the length of bitmap-2 (either one or two slots) and a length of 40 slots. In other words, the overall time-domain periodicity of a semistatic resource set defined by the triplet {bitmap-1, bitmap-2, bitmap-3} is 40 slots in length.

In the case of dynamic activation/deactivation of a rate-matching resource set, an indicator in the scheduling assignment indicates if the semistatically configured pattern is valid or not for a certain dynamically scheduled transmission. Note that, although Fig. 9.14 assumes scheduling on a slot basis, dynamic indication is equally applicable to transmission durations shorter than a slot. The indicator in the DCI should not be seen as corresponding to a certain slot. Rather, it should be seen as corresponding to a certain scheduling assignment. What the indicator does is simply indicate if, for a given scheduling assignment defined by a given DCI, a configured resource set should be assumed active or not during the time over which the assignment is valid.

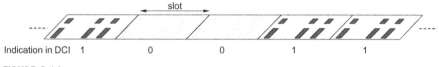

FIGURE 9.14

Dynamic activation/deactivation of a resource set by means of a DCI indicator.

In the general case, a device can be configured with up to eight different resource sets. Each resource set is configured either by referring to a CORSEST or by using the bitmap approach described above. By configuring more than one resource-set configuration, more elaborate patterns of reserved resources can be realized, as illustrated in Fig. 9.15.

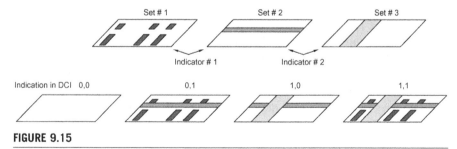

FIGURE 9.15

Dynamic activation/deactivation in the case of multiple configured resource sets.

Although a device can be configured with up to eight different resource-set configurations, each of which can be configured for dynamic activation, the configurations cannot be independently activated in the scheduling assignment. Rather, to maintain a reasonable overhead, the scheduling assignment includes at most two indicators. Each resource set configured for dynamic activation/deactivation is assigned to either one or both of these indications and jointly activates/deactivates or disables all resource sets assigned to that indicator. Fig. 9.15

illustrates an example with three configured resource sets, where resource set #1 and resource set #2 are assigned to indicator #1 and indicator #2, respectively, while resource set #2 is assigned to both indicators. Note that the patterns in Fig. 9.15 are not necessarily realistic, but rather chosen for illustrative purposes.

9.11 REFERENCE SIGNALS

Reference signals are predefined signals occupying specific resource elements within the downlink time—frequency grid. The NR specification includes several types of reference signals transmitted in different ways and intended to be used for different purposes by a receiving device.

Unlike LTE, which relies heavily on always-on, cell-specific reference signals in the downlink for coherent demodulation, channel quality estimation for CSI reporting, and general time—frequency tracking, NR uses different downlink reference signals for different purposes. This allows for optimizing each of the reference signals for their specific purpose. It is also in line with the overall principle of ultra-lean transmission as the different reference signals can be transmitted only when needed. Later release of LTE took some steps in this direction, but NR can exploit this to a much larger degree as there are no legacy NR devices to cater for.

The NR reference signals include:

- *Demodulation reference signals* (DM-RS) for PDSCH are intended for channel estimation at the device as part of coherent demodulation. They are present only in the resource blocks used for PDSCH transmission. Similarly, the DM-RS for PUSCH allows the gNB to coherently demodulate the PUSCH. The DM-RS for PDSCH and PUSCH is the focus of this section; DM-RS for PDCCH and PBCH are described in Chapters 10 and 16, respectively.
- *Phase-tracking reference signals* (PT-RS) can be seen as an extension to DM-RS for PDSCH/PUSCH and are intended for phase-noise compensation. The PT-RS is denser in time but sparser in frequency than the DM-RS, and, if configured, occurs only in combination with DM-RS. A discussion of the phase-tracking reference signal is found later in this chapter.
- *CSI reference signals* (CSI-RS) are downlink reference signals intended to be used by devices to acquire downlink channel-state information (CSI). Specific instances of CSI reference signals can be configured for time/frequency tracking and mobility measurements. CSI reference signals are described in Section 8.1.
- *Tracking reference signals* (TRS) are sparse reference signals intended to assist the device in time and frequency tracking. A specific CSI-RS configuration serves the purpose of a TRS (see Section 8.1.7).
- *Sounding reference signals* (SRS) are uplink reference signals transmitted by the devices and used for uplink channel-state estimation at the base stations. Sounding reference signals are described in Section 8.3.

In the following, the demodulation reference signals intended for coherent demodulation of PDSCH and PUSCH are described in more detail, starting with the reference signal structure used for OFDM. The same DM-RS structure is used for both downlink and uplink in the case of OFDM. For DFT-spread OFDM in the uplink, a reference signal based on Zadoff—Chu sequences as in LTE is used to improve the power-amplifier efficiency but supporting contiguous allocations and single-layer transmission only as discussed in a later section. Finally, a discussion on the phase-tracking reference signal is provided.

9.11.1 DEMODULATION REFERENCE SIGNALS FOR OFDM-BASED DOWNLINK AND UPLINK

The DM-RS in NR provides quite some flexibility to cater for different deployment scenarios and use cases: a front-loaded design to enable low latency, support for up to 12 orthogonal antenna ports for MIMO, transmissions durations from 2 to 14 symbols, and up to four reference-signal instances per slot to support very high-speed scenarios.

To achieve low latency, it is beneficial to locate the demodulation reference signals early in the transmission, sometimes known as front-loaded reference signals. This allows the receiver to obtain a channel estimate early and, once the channel estimate is obtained, process the received symbols on the fly without having to buffer a complete slot prior to data processing. This is essentially the same motivation as for the frequency-first mapping of data to the resource elements.

Two main time-domain structures are supported, differencing in the location of the first DM-RS symbol:

- *Mapping type A*, where the first DM-RS is located in symbol 2 or 3 *of the slot* and the DM-RS is mapped relative to the start of the slot boundary, regardless of where in the slot the actual data transmission starts. This mapping type is primarily intended for the case where the data occupy (most of) a slot. The reason for symbol 2 or 3 in the downlink is to locate the first DM-RS occasion after a CORESET located at the beginning of a slot.
- *Mapping type B*, where the first DM-RS is located in the first symbol *of the data allocation*, that is, the DM-RS location is not given relative to the slot boundary but rather relative to where the data are located. This mapping is originally motivated by transmissions over a small fraction of the slot to support very low latency and other transmissions that benefit from not waiting until a slot boundary starts but can be used regardless of the transmission duration.

The mapping type for PDSCH transmission can be dynamically signaled as part of the DCI (see Section 9.11 for details), while for the PUSCH the mapping type is semistatically configured.

Although front-loaded reference signals are beneficial from a latency perspective, they may not be sufficiently dense in the time domain in the case of rapid channel variations. To support high-speed scenarios, it is possible to configure up to three additional DM-RS occasions in a slot. The channel estimator in the receiver can use these additional occasions for more accurate channel estimation, for example, to use interpolation between the occasions within a slot. It is not possible to interpolate between slots, or in general different transmission occasions, as different slots may be transmitted to different devices and/or in different beam directions. This is a difference compared to LTE, where interslot interpolation of the channel estimates is possible but also restricts the multi-antenna and beamforming flexibility in LTE compared to NR.

The different time-domain allocations for DM-RS are illustrated in Fig. 9.16, including both single-symbol and double-symbol DM-RS. The purpose of the double-symbol DM-RS is primarily to provide a larger number of antenna ports than what is possible with a single-symbol structure as discussed below. Note that the time-domain location of the DM-RS depends on the scheduled data duration. Furthermore, not all patterns illustrated in Fig. 9.16 are applicable to the PDSCH (for example, mapping type B for PDSCH only supports duration 2, 4, and 7.

Multiple orthogonal reference signals can be created in each DM-RS occasion. The different reference signals are separated in the frequency and code domains, and, in the case of a double-symbol DM-RS, additionally in the time domain. Two different types of demodulation reference signals can be configured, type 1 and type 2, differing in the mapping in the frequency domain and the maximum number of orthogonal reference signals. Type 1 can provide up to four orthogonal signals using a single-symbol DM-RS and up to eight orthogonal reference signals using a double-symbol DM-RS. The corresponding numbers for type 2 are six and twelve. The reference signal types (1 or 2) should not be confused with the mapping types (A or B); different mapping types can be combined with different reference signal types.

Reference signals should preferably have small power variations in the frequency domain to allow for a similar channel-estimation quality for all frequencies spanned by the reference signal. Note that this is equivalent to a well-focused time-domain autocorrelation of the transmitted reference signal. For OFDM-based modulation, a pseudo-random sequence is used, more specifically a length $2^{31}-1$ Gold sequence, which fulfills the requirements on a well-focused autocorrelation. The sequence is generated across all the common resource blocks (CRBs) in the frequency domain but transmitted only in the resource blocks used for data transmission as there is no reason for estimating the channel outside the frequency region used for transmission. Generating the reference signal sequence across all the resource blocks ensures that the same underlying sequence is used for multiple devices scheduled on overlapping time—frequency resources in the case of multi-user MIMO (see Fig. 9.17) (orthogonal sequences are used on top of the pseudo-random sequence to obtain multiple orthogonal reference signals from the same pseudo-random sequence as discussed later). If the underlying pseudo-random sequence would differ between different co-scheduled devices, the resulting reference signals would not be orthogonal. The pseudo-random sequence is

FIGURE 9.16

Time-domain location of DM-RS.

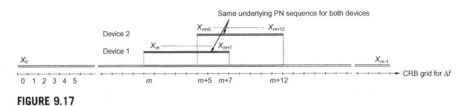

FIGURE 9.17

Generating DM-RS sequences based on common resource block 0.

generated using a configurable identity, similar to the virtual cell ID in LTE. If no identity has been configured, it defaults to the physical-layer cell identity.

Returning to the type 1 reference signals, the underlying pseudo-random sequence is mapped to every second subcarrier in the frequency domain in the OFDM symbol used for reference signal transmission, see Fig. 9.18 for an illustration assuming only front-loaded reference signals are being used. Antenna ports[7] 1000 and 1001 use even-numbered subcarriers in the frequency domain and are separated from each other by multiplying the underlying pseudo-random sequence with different length-2 orthogonal sequences in the frequency domain, resulting in transmission of two orthogonal reference signals for the two antenna ports. As long as the radio channel is flat across four consecutive subcarriers, the two reference signals will be orthogonal also at the receiver. Antenna ports 1000 and 1001 are said to belong to *CDM group 0* as they use the same subcarriers but are separated in the code-domain using different orthogonal sequences. Reference signals for antenna ports 1002 and 1003 belong to CDM group 1 and are generated in the same way using odd-numbered subcarriers, that is, separated in the code domain within the CDM group and in the frequency domain between CDM groups. If more than four orthogonal antenna ports are needed, two consecutive OFDM symbols are used instead. The structure above is used in each of the OFDM symbols and a length-2 orthogonal sequence is used to extend the code-domain separation to also include the time domain, resulting in up to eight orthogonal sequences in total.

Demodulation reference signals type 2 (see Fig. 9.19) have a similar structure to type 1, but there are some differences, most notably the number of antenna ports supported. Each CDM group for type 2 consists of two neighboring subcarriers over which a length-2 orthogonal sequence is used to separate the two antenna ports sharing the same set of subcarriers. Two such pairs of subcarriers are used in each resource block for one CDM group. Since there are 12 subcarriers in a resource block, up to three CDM groups with two orthogonal reference signals each can be created using one resource block in one OFDM symbol. By using a second OFDM symbol and a time-domain length-2 sequence in the same was as for type 1, a maximum of 12 orthogonal reference signals can be created

---

[7]The downlink antenna port numbering is assumed in this example. The uplink structure is similar but with different antenna port numbers.

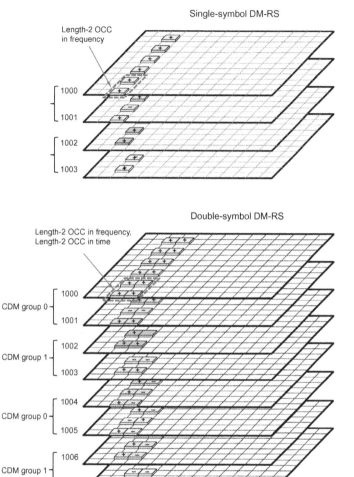

FIGURE 9.18

Demodulation reference signals type 1.

with type 2. Although the basic structures of type 1 and type 2 have many similarities, there are also differences. Type 1 is denser in the frequency domain, while type 2 trades the frequency-domain density for a larger multiplexing capacity, that is, a larger number of orthogonal reference signals. This is motivated by the support for multi-user MIMO with simultaneous transmission to a relatively large number of devices.

The reference signal structure to use is determined based on a combination of dynamic scheduling and higher-layer configuration. If a double-symbol reference signal is configured, the scheduling decision, conveyed to the device using the

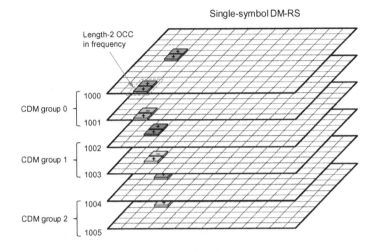

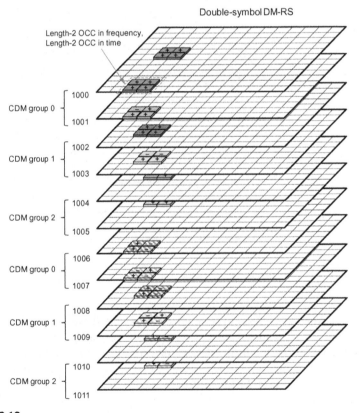

FIGURE 9.19

Demodulation reference signals type 2.

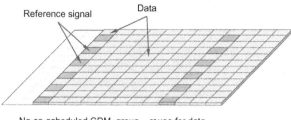

No co-scheduled CDM group – reuse for data

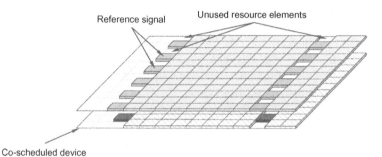

Co-scheduled CDM groups – leave resource elements empty

FIGURE 9.20

Rate matching data around co-scheduled CDM groups.

downlink control information, indicates to the device whether to use single-symbol or double-symbol reference signals. The scheduling decision also contains information for the device which reference signals (more specifically, which CDM groups) that are intended for other devices (see Fig. 9.20). The scheduled device maps the data around both its own reference signals as well as the reference signals intended for another device. This allows for a dynamic change of the number of coscheduled devices in the case of multi-user MIMO. In the case of spatial multiplexing (also known as single-user MIMO) of multiple layers for the same device, the same approach is used—each layer leaves resource elements corresponding to another CDM group intended for the same device unused. This is to avoid interlayer interference for the reference signals.

The reference signal description above is applicable to both uplink and downlink. Note though, that for precoder-based uplink transmissions, the uplink reference signal is applied *before* the precoder (see Fig. 9.11). Hence, the reference signal transmitted is not the structure above, but the precoded version of it.[8]

---

[8]In general, the reference signal transmitted is in addition subject to any implementation-specific multi-antenna processing, captured by the spatial filter F in Section 9.8, and the word "transmitted" should be understood from a specification perspective.

9.11.2 DEMODULATION REFERENCE SIGNALS FOR DFT-PRECODED OFDM UPLINK

DFT-precoded OFDM supports single-layer transmission only and is primarily designed with coverage-challenged situations in mind. Due to the importance of low cubic metric and corresponding high power-amplifier efficiency for uplink DFT-precoded OFDM, the reference signal structure is somewhat different compared to the OFDM case. In essence, transmitting reference signals frequency multiplexed with other uplink transmissions from the same device is not suitable for the uplink as that would negatively impact the device power-amplifier efficiency due to increased cubic metric. Instead, certain OFDM symbols within a slot are used exclusively for DM-RS transmission—that is, the reference signals are *time multiplexed* with the data transmitted on the PUSCH from the same device. The structure of the reference signal itself then ensures a low cubic metric within these symbols as described below.

In the time domain, the reference signals follow the same mapping as configuration type 1. As DFT-precoded OFDM is capable of single-layer transmission only and DFT-precoded OFDM is primarily intended for coverage-challenged situations, there is no need to support configuration type 2 and its capability of handling a high degree of multi-user MIMO. Furthermore, since multi-user MIMO is not a targeted scenario for DFT-precoded OFDM, there is no need to define the reference signal sequence across all common resource blocks as for the corresponding OFDM case, but it is sufficient to define the sequence for the transmitted physical resource blocks only.

Uplink reference signals should preferably have small power variations in the frequency domain to allow for similar channel-estimation quality for all frequencies spanned by the reference signal. As already discussed, for OFDM transmission it is fulfilled by using a pseudo-random sequence with good autocorrelation properties. However, for the case of DFT-precoded OFDM, limited power variations as a function of time are also important to achieve a low cubic metric of the transmitted signal. Furthermore, a sufficient number of reference-signal sequences of a given length, corresponding to a certain reference-signal bandwidth, should be available in order to avoid restrictions when scheduling multiple devices in different cells. A type of sequence fulfilling these two requirements is the Zadoff–Chu sequence, discussed in Chapter 8. From a Zadoff–Chu sequence with a given group index and sequence index, additional reference-signal sequences can be generated by applying different linear phase rotations in the frequency domain, as illustrated in Fig. 9.21. This is the same principle as used in LTE.

9.11.3 PHASE-TRACKING REFERENCE SIGNALS (PT-RS)

Phase-tracking reference signals (PT-RS) can be seen as an extension to DM-RS, intended for tracking phase variations across the transmission duration, for

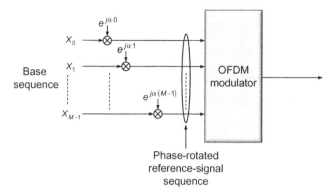

FIGURE 9.21

Generation of uplink reference-signal sequence from phase-rotated base sequence.

example, one slot. These phase variations can come from phase noise in the oscillators, primarily at higher carrier frequencies where the phase noise tends to be higher. It is an example of a reference signal type existing in NR but with no corresponding signal in LTE. This is partially motivated by the lower carrier frequencies used in LTE, and hence less problematic phase noise situation, and partly it is motivated by the presence of cell-specific reference signals in LTE which can be used for tracking purposes. Since the main purpose is to track phase noise, the PT-RS needs to be dense in time but can be sparse in frequency. The PT-RS only occurs in combination with DM-RS and only if the network has configured the PT-RS to be present. Depending on whether OFDM or DFTS-OFDM is used, the structure differs.

For OFDM, the first reference symbol (prior to applying any orthogonal sequence) in the PDSCH/PUSCH allocation is repeated every Lth OFDM symbol, starting with the first OFDM symbol in the allocation. The repetition counter is reset at each DM-RS occasion as there is no need for PT-RS immediately after a DM-RS. The density in the time-domain is linked to the scheduled MCS in a configurable way.

In the frequency domain, phase-tracking reference signals are transmitted in every second or fourth resource block, resulting in a sparse frequency domain structure. The density in the frequency domain is linked to the scheduled transmission bandwidth such that the higher the bandwidth, the lower the PT-RS density in the frequency domain. For the smallest bandwidths, no PT-RS is transmitted.

To reduce the risk of collisions between phase-tracking reference signals associated with different devices scheduled on overlapping frequency-domain resources, the subcarrier number and the resource blocks used for PT-RS transmission are determined by the C-RNTI of the device. The antenna port used for PT-RS transmission is given by the lowest numbered antenna port in the DM-RS antenna port group. Some examples of PT-RS mappings are given in Fig. 9.22.

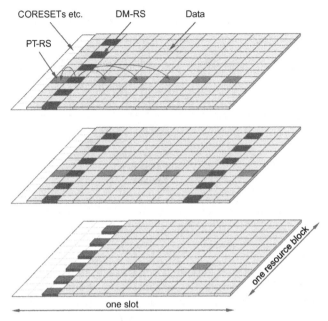

FIGURE 9.22

Examples of PT-RS mapping in one resource block and one slot.

For DFT-precoded OFDM in the uplink, the samples representing the phase-tracking reference signal are inserted prior to DFT precoding. The time domain mapping follows the same principles as the pure OFDM case.

Physical-Layer Control Signaling

To support the transmission of downlink and uplink transport channels, there is a need for certain *associated control signaling*. This control signaling is often referred to as L1/L2 *control signaling*, indicating that the corresponding information partly originates from the physical layer (layer 1) and partly from MAC (layer 2).

In this chapter, the downlink control signaling, including scheduling grants and assignments, will be described, followed by the uplink control signaling carrying the necessary feedback from the device.

10.1 DOWNLINK

Downlink L1/L2 control signaling consists of downlink scheduling assignments, including information required for the device to be able to properly receive, demodulate, and decode the DL-SCH on a component carrier, and uplink scheduling grants informing the device about the resources and transport format to use for uplink (UL-SCH) transmission. In addition, the downlink control signaling can also be used for special purposes such as conveying information about the symbols used for uplink and downlink in a set of slots, preemption indication, and power control.

In NR, there is only a single control channel, the *physical downlink control channel* (PDCCH). On a high level, the principles of the PDCCH processing in NR are similar to LTE, namely that the device tries to blindly decode candidate PDCCHs transmitted from the network using one or more search spaces. However, there are some differences compared to LTE based on the different design targets for NR as well as experience from LTE deployments:

- The PDCCH in NR does not necessarily span the full carrier bandwidth, unlike the LTE PDCCH. This is a natural consequence of the fact that not all NR devices may be able to receive the full carrier bandwidth as discussed in Chapter 5, and led to the design of a more generic control channel structure in NR.
- The PDCCH in NR is designed to support device-specific beamforming, in line with the general beam-centric design of NR and a necessity when operating at very high carrier frequencies with a corresponding challenging link budget.

5G NR: The Next Generation Wireless Access Technology. DOI: https://doi.org/10.1016/B978-0-12-814323-0.00010-7

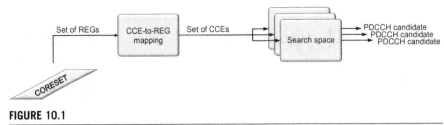

FIGURE 10.1

Overview of PDCCH processing in NR.

These two aspects were to some extent addressed in the LTE EPDCCH design in release 11, although in practice EPDCCH has not been used extensively except as a basis for the control signaling for eMTC.

Two other control channels present in LTE, the PHICH and the PCFICH, are not needed in NR. The former is used in LTE to handle uplink retransmissions and is tightly coupled to the use of a synchronous hybrid-ARQ protocol, but since the NR hybrid-ARQ protocol is asynchronous in both uplink and downlink the PHICH is not needed in NR. The latter channel, the PCFICH, is not necessary in NR as the size of the *control resource sets* (CORESETs) does not vary dynamically and reuse of control resources for data is handled in a different way than in LTE, as discussed further below.

In the following sections, the NR downlink control channel, the PDCCH, will be described, including the notion of a CORESETs, the time−frequency resources upon which the PDCCH is transmitted. First, the PDCCH processing including coding and modulation will be discussed, followed by a discussion on the CORESETs structure. There can be multiple CORESETs on a carrier and part of the control resource set is the mapping from resource elements to *control channel elements* (CCEs). One or more CCEs from one control resource set are aggregated to form the resources used by one PDCCH. Blind detection, the process where the device attempts to detect if there are any PDCCHs transmitted to the device, is based on search spaces. There can be multiple search spaces using the resources in a single CORESET, as illustrated in Fig. 10.1. Finally, the contents of the *downlink control information* (DCI) will be described.

10.1.1 PHYSICAL DOWNLINK CONTROL CHANNEL

The PDCCH processing steps are illustrated in Fig. 10.2. At a high level, the PDCCH processing in NR is more similar to the LTE EPDCCH than the LTE PDCCH in the sense that each PDCCH is processed independently.

The payload transmitted on a PDCCH is known as *Downlink Control Information* (DCI) to which a 24-bit CRC is attached to detect transmission errors and to aid the decoder in the receiver. Compared to LTE, the CRC size has been increased to reduce the risk of incorrectly received control information and to assist early termination of the decoding operation in the receiver.

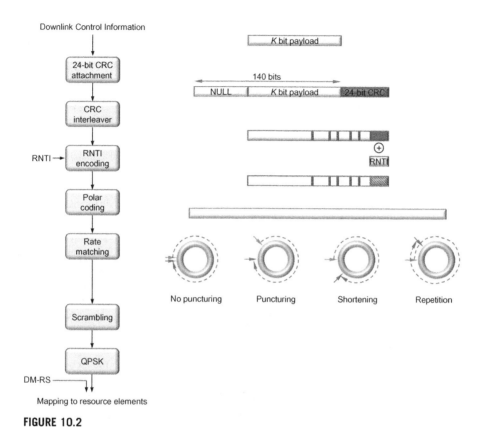

FIGURE 10.2

PDCCH processing.

Similarly to LTE, the device identity modifies the CRC transmitted through a scrambling operation. Upon receipt of the DCI, the device will compute a scrambled CRC on the payload part using the same procedure and compare it against the received CRC. If the CRC checks, the message is declared to be correctly received and intended for the device. Thus, the identity of the device that is supposed to receive the DCI message is implicitly encoded in the CRC and not explicitly transmitted. This reduces the number of bits necessary to transmit on the PDCCH as, from a device point of view, there is no difference between a corrupt message whose CRC will not check, and a message intended for another device. Note that the RNTI does not necessarily have to be the identity of the device, the C-RNTI, but can also be different types of group or common RNTIs, for example, to indicate paging or a random-access response.

Channel coding of the PDCCH is based on Polar codes, a relatively new form of channel coding. The basic idea behind Polar codes is to transform several instances of the radio channel into a set of channels that are either noiseless or completely noisy and then transmit the information bits on the noiseless channels. Decoding can be done in several ways, but a typical approach is to use successive cancellation and list decoding. List decoding uses the CRC as part of the decoding process, which means that the error-detecting capabilities are reduced. For example, list decoding of size eight results in a loss of three bits from an error-detecting perspective, resulting in the 24-bits CRC providing error-detecting capabilities corresponding to a 21-bit CRC. This is part of the reason for the larger CRC size compared to LTE.

Unlike the tailbiting convolutional codes used in LTE, which can handle any number of information bits, Polar codes need to be designed with a maximum number of bits in mind. In NR, the Polar code has been designed to support 512 coded bits (prior to rate matching) in the downlink. Up to 140 information bits can be handled, which provides a sufficient margin for future extensions as the DCI payload size in release 15 is significantly less. To assist early termination in the decoding process, the CRC is not attached at the end of the information bits, but inserted in a distributed manner, after which the Polar code is applied. Early termination can also be achieved by exploiting the path metric in the decoder.

Rate matching is used to match the number of coded bits to the resources available for PDCCH transmission. This is a somewhat intricate process and is based on one of shortening, puncturing, or repetition of the coded bits after sub-block interleaving of 32 blocks. The set of rules selecting between shortening, puncturing, and repetition, as well as when to use which of the schemes, is designed to maximize performance.

Finally, the coded and rate-matched bits are scrambled, modulated using QPSK, and mapped to the resource elements used for the PDCCH, the details of which will be discussed below. Each PDCCH has its own reference signal, which means that the PDCCH can make full use of the antenna setup, for example, be beamformed in a particular direction. The complete PDCCH processing chain is illustrated in Fig. 10.2.

The mapping of the coded and modulated DCI to resource elements is subject to a certain structure, based on *control-channel elements* (CCEs) and *resource-element groups* (REGs). Although the names are borrowed from LTE, the size of the two differ from their LTE counterparts, as does the CCE-to-REG mapping.

A PDCCH is transmitted using 1, 2, 4, 8, or 16 contiguous control-channel elements with the number known as the *aggregation level*. The control-channel element is the unit upon which the search spaces for blind decoding are defined as will be discussed in Section 10.1.3. A control-channel element consists of six resource-element groups, each of which is equal to one resource block in one OFDM symbol. After accounting for the DM-RS overhead, there are 54 resource elements (108 bits) available for PDCCH transmission in one control-channel element.

The CCE-to-REG mapping can be either interleaved or non-interleaved. The motivation for having two different mapping schemes is, similarly to the case of the LTE EPDCCH, to be able to provide frequency diversity by using an interleaved mapping or to facilitate interference coordination and frequency-selective transmission of control channels by using non-interleaved mapping. The details of the CCE-to-REG mapping will be discussed in the next section as part of the overall CORESET structure.

10.1.2 CONTROL RESOURCE SET

Central to downlink control signaling in NR is the concept of CORESETs. A control resource set is a time–frequency resource in which the device tries to decode candidate control channels using one or more search spaces. The size and location of a CORESET in the time–frequency domain is semistatically configured by the network and can thus be set to be smaller than the carrier bandwidth. This is especially important in NR as a carrier can be very wide, up to 400 MHz, and it is not reasonable to assume all devices can receive such a wide bandwidth.

In LTE, the concept of a CORESET is not explicitly present. Instead, downlink control signaling in LTE uses the full carrier bandwidth in the first 1–3 OFDM symbols (four for the most narrowband case). This is known as the control region in LTE and in principle this control region would correspond to the "LTE CORESET" if that term would have been used. Having the control channels spanning the full carrier bandwidth was well motivated by the desire for frequency diversity and the fact that all LTE devices support the full 20 MHz carrier bandwidth (at least at the time of specifying release 8). However, in later LTE releases this lead to complications when introducing support for devices not supporting the full carrier bandwidth, for example, the eMTC devices introduced in release 12. Another drawback of the LTE approach is the inability to handle frequency-domain interference coordination between cells for the downlink control channels. To some extent, these drawbacks with the LTE control channel design were addressed with the introduction of the EPDCCH in release 11, but the EPDCCH feature has so far not been widely deployed in practice as an LTE network still needs to provide PDCCH support for initial access and to handle non-EPDCCH-capable LTE devices. Therefore, a more flexible structure is used in NR from the start.

A CORESET can occur at any position within a slot and anywhere in the frequency range of the carrier (see Fig. 10.3). However, a device is not expected to handle CORESETs outside its active bandwidth part. The reason for configuring CORESETs on the cell level and not per bandwidth part is to facilitate reuse of CORSETs between bandwidth parts, for example, when operating with bandwidth adaptation as discussed in Section 14.1.1.

The first CORSET, CORESET 0, is provided by the master information block (MIB) as part of the configuration of the initial bandwidth part to be able to receive the remaining system information and additional configuration information from the network. After connection setup, a device can be configured with multiple, potentially overlapping, CORESETs in addition to using RRC signaling.

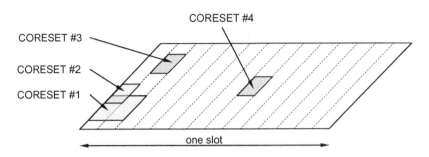

FIGURE 10.3

Examples of CORESET configurations.

In the time domain, a CORESET can be up to three OFDM symbols in duration and located anywhere within a slot, although a common scenario, suitable for traffic scenarios when a scheduling decision is taken once per slot, is to locate the CORESET at the beginning of the slot. This is similar to the LTE situation with control channels at the beginning of each LTE subframe. However, configuring a CORESET at other time instances can be useful, for example to achieve very low latency for transmissions occupying only a few OFDM symbols without waiting for the start of the next slot. It is important to understand that a CORESET is defined from a device perspective and only indicates where a device may receive PDCCH transmissions. It does not say anything on whether the gNB actually transmits a PDCCH or not.

Depending on where the front-loaded DM-RS for PDSCH are located, in the third or fourth OFDM symbol of a slot (see Section 9.11.1), the maximum duration for a CORESET is two or three OFDM symbols. This is motivated by the typical case of locating the CORESET before the start of downlink reference signals and the associated data. In the frequency domain, a CORESET is defined in multiples of six resource blocks up to the carrier bandwidth.

Unlike LTE, where the control region can vary dynamically in length as indicated by a special control channel (the PCFICH), a CORESET in NR is of fixed size. This is beneficial from an implementation perspective, both for the device and the network. From a device perspective, a pipelined implementation is simpler if the device can directly start to process the PDCCH without having to first decode another channel like the PCFICH in LTE. Having a streamlined and implementation-friendly structure of the PDCCH is important in order to realize the very low latency possible in NR. However, from a spectral efficiency point of view, it is beneficial if resources can be shared flexibly between control and data in a dynamic manner. Therefore, NR provides the possibility to start the PDSCH data before the end of a CORESET. It is also possible to, for a given device, reuse unused CORESET resources as illustrated in Fig. 10.4. To handle this, the general mechanism of reserved resources is used (see Section 9.10). Reserved

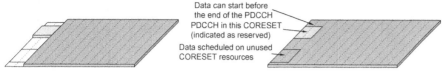

Data can start before the end of the PDCCH

PDCCH in this CORESET (indicated as reserved)

Data scheduled on unused CORESET resources

FIGURE 10.4

No reuse (left) and reuse (right) of CORESET resources for data transmission (the device is configured with two CORESETs in this example).

resources that overlap with the CORESET are configured and information in the DCI indicates to the device whether the reserved resources are usable by the PDSCH or not. If they are indicated as reserved, the PDSCH is rate-matched around the reserved resources overlapping with the CORESET, and if the resources are indicated as available, the PDSCH uses the reserved resources for data except for the resources used by the PDCCH upon which the device received the DCI scheduling the PDSCH.

For each CORESET there is an associated CCE-to-REG mapping, a mapping that is described using the term REG bundle. A REG bundle is a set of REGs across which the device can assume the precoding is constant. This property can be exploited to improve the channel-estimation performance in a similar way as resource-block bundling for the PDSCH.

As already mentioned, the CCE-to-REG mapping can be either interleaved or non-interleaved, depending on whether frequency-diverse or frequency-selective transmission is desired. There is only one CCE-to-REG mapping for a given CORESET, but since the mapping is a property of the CORESET, multiple CORESETs can be configured with different mappings, which can be useful. For example, one or more CORESETs configured with non-interleaved mapping to benefit from frequency-dependent scheduling, and one or more configured with interleaved mapping to act as a fallback in case the channel-state feedback becomes unreliable due to the device moving rapidly.

The non-interleaved mapping is straightforward. The REG bundle size is six for this case, that is, the device may assume the precoding is constant across a whole CCE. Consecutive bundles of six REGs are used to form a CCE.

The interleaved case is a bit more intricate. In this case, the REG bundle size is configurable between two alternatives. One alternative is six, applicable to all CORESET durations, and the other alternative is, depending on the CORSET duration, two or three. For a duration of one or two OFDM symbols, the bundle size can be two or six, and for a duration of three OFDM symbols, the bundle size can be three or six. In the interleaved case, the REG bundles constituting a CCE are obtained using a block interleaver to spread out the different REG bundles in frequency, thereby obtaining frequency diversity. The number of rows in the block interleaver is configurable to handle different deployment scenarios (Fig. 10.5).

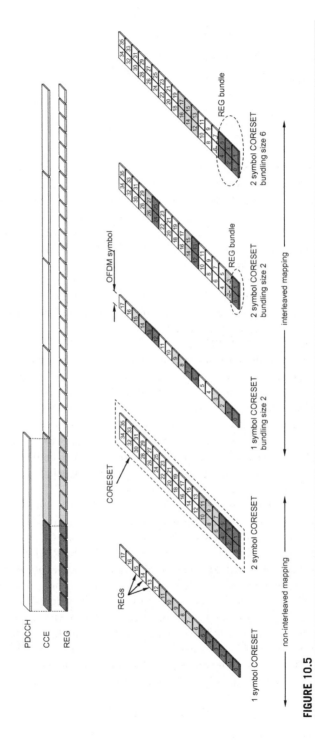

FIGURE 10.5

Examples of CCE-to-REG mapping.

As part of the PDCCH reception process, the device needs to form a channel estimate using the reference signals associated with the PDCCH candidate being decoded. A single antenna port is used for the PDCCH, that is, any transmit diversity or multi-user MIMO scheme is handled in a device-transparent manner.

The PDCCH has its own demodulation reference signals, based on the same type of pseudo-random sequence as the PDSCH—the pseudo-random sequence is generated across all the common resource blocks in the frequency domain but transmitted only in the resource blocks used for the PDCCH (with one exception as discussed below). However, during initial access, the location for the common resource blocks is not yet known as it is signaled as part of the system information. Hence, for CORESET 0 configured by the PBCH, the sequence is generated starting from the first resource block in the CORESET instead.

Demodulation reference-signals specific for a given PDCCH candidate are mapped onto every fourth subcarrier in a resource-element group, that is, the reference signal overhead is 1/4. This is a denser reference signal pattern than in LTE, which uses a reference signal overhead of 1/6, but in LTE the device can interpolate channel estimates in time and frequency as a consequence of LTE using a cell-specific reference signal common to all devices and present regardless of whether a control-channel transmission takes place or not. The use of a dedicated reference signal per PDCCH candidate is beneficial, despite the slightly higher overhead, as it allows for different types of device-transparent beamforming. By using a beamformed control channel, the coverage and performance can be enhanced compared to the non-beamformed control channels in LTE.[1] This is an essential part of the beam-centric design of NR.

When attempting to decode a certain PDCCH candidate occupying a certain set of CCEs, the device can compute the REG bundles that constitute the PDCCH candidate. Channel estimation must be performed per REG bundle as the network may change precoding across REG bundles. In general, this results in sufficiently accurate channel estimates for good PDCCH performance. However, there is also a possibility to configure the device to assume the same precoding across contiguous resource blocks in a CORESET, thereby allowing the device to do frequency-domain interpolation of the channel estimates. This also implies that the device may use reference signals outside the PDCCH it is trying to detect, sometimes referred to as wideband reference signals (see Fig. 10.6 for an illustration). In some sense this gives the possibility to partially mimic the LTE cell-specific reference signals in the frequency domain, of course with a corresponding limitation in terms of beamforming possibilities.

Related to channel estimation are, as has been discussed for other channels, the quasi-colocation relations applicable to the reference signals. If the device knows that two reference signals are quasi-collocated, this knowledge can be exploited to improve the channel estimation and, more importantly for the PDCCH, to manage different reception beams at the device (see Chapter 12 for a

---

[1]The LTE EPDCCH introduced device-specific reference signals in order to allow beamforming.

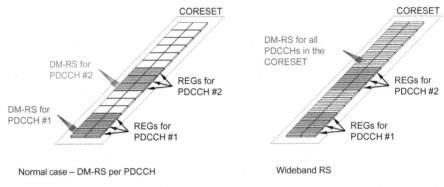

Normal case – DM-RS per PDCCH   Wideband RS

FIGURE 10.6

Normal RS structure (left) and wideband RS structure (right).

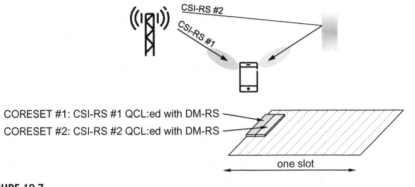

FIGURE 10.7

Example of QCL relation for PDCCH beam management.

detailed discussion on beam management and spatial quasi-colocation). To handle this, each CORESET can be configured with a *transmission configuration indication* (TCI) state, that is, providing information of the antenna ports with which the PDCCH antenna ports are quasi-colocated. If the device is a certain CORESET spatially colocated with a certain CSI-RS, the device can determine which reception is appropriate when attempting to receive PDCCHs in this CORESET, as illustrated in Fig. 10.7. In this example, two CORESETs have been configured in the device, one CORESET with spatial QCL between DM-RS and CSI-RS #1, and one CORESET with spatial QCL between DM-RS and CSI-RS #2. Based on CSI-RS measurements, the device has determined the best reception beam for each of the two CSI-RS:es. When monitoring CORESET #1 for possible PDCCH transmissions, the device knows the spatial QCL relation and uses the appropriate reception beam (similarly for CORESET #2). In this way, the device can handle multiple reception beams as part of the blind decoding framework.

If no quasi-colocation is configured for a CORESET the device assumes the PDCCH candidates to be quasi-colocated with the SS block with respect to delay spread, Doppler spread, Doppler shift, average delay, and spatial Rx parameters. This is a reasonable assumption as the device has been able to receive and decode the PBCH in order to access the system.

10.1.3 BLIND DECODING AND SEARCH SPACES

As described above, different DCI formats can be used for transmission on a PDCCH and the format used is a priori unknown to the device. Therefore, the device needs to blindly detect the DCI format. In LTE, the format was tightly coupled to the DCI size and monitoring for a certain DCI format in most cases implied monitoring for a new DCI size.

In NR, the coupling between DCI formats and DCI sizes is less pronounced. Different formats could still have different DCI sizes, but several formats share the same DCI size. This allows adding more formats in the future without increasing the number of blind decodings. An NR device needs to monitor for up to four different DCI sizes; one size used for the fallback DCI formats, one for downlink scheduling assignments, and (unless the uplink downlink non-fallback formats are size-aligned) one for uplink scheduling grants. In addition, a device may need to monitor for slot-format indication and preemption indication using a fourth size, depending on the configuration.

The CCE structure described in the previous section helps in reducing the number of blind decoding attempts but is not sufficient. Hence, it is required to have mechanisms to limit the number of PDCCH candidates that the device is supposed to monitor. Clearly, from a scheduling point of view, restrictions in the allowed aggregations are undesirable as they may reduce the scheduling flexibility and require additional processing at the transmitter side. At the same time, requiring the device to monitor all possible CCE aggregations in all configured CORESETs is not attractive from a device-complexity point of view. To impose as few restrictions as possible on the scheduler while at the same time limiting the maximum number of blind decoding attempts in the device, NR defines so-called *search spaces*. A search space is a set of candidate control channels formed by CCEs at a given aggregation level, which the device is supposed to attempt to decode. As there are multiple aggregation levels a device can have multiple search spaces. There can be multiple search spaces using the same CORESET and, as already described, there can be multiple CORESETs configured for a device. A device is not supposed for PDCCH outside its active bandwidth part, which follows from the overall purpose of a bandwidth part. Furthermore, the monitoring instance of a search space is configurable as illustrated in Fig. 10.8.

At a configured monitoring occasion for a search space, the devices will attempt to decode the candidate PDCCHs for that search space. Five different aggregation levels corresponding to 1, 2, 4, 8, and 16 CCEs, respectively, can be configured. The highest aggregation level, 16, is not supported in LTE and was

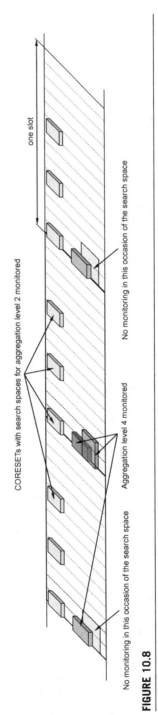

FIGURE 10.8

Example of PDCCH monitoring configuration.

added to NR in case of extreme coverage requirements. The number of PDCCH candidates can be configured per search space (and thus also per aggregation level). Hence NR has a more flexible way of spending the blind decoding attempts across aggregation levels than LTE, where the number of blind decodes at each aggregation level was fixed. This is motivated by the wider range of deployments expected for NR. For example, in a small-cell scenario the highest aggregation levels may not be used, and it is better to spend the limited number of blind decoding attempts the device is dimensioned for on the lower aggregation levels than on blind decoding on an aggregation level that is never used.

Upon attempting to decode a candidate PDCCH, the content of the control channel is declared as valid for this device if the CRC checks and the device processes the information (scheduling assignment, scheduling grants, etc.). If the CRC does not check, the information is either subject to uncorrectable transmission errors or intended for another device and in either case the device ignores that PDCCH transmission.

The network can only address a device if the control information is transmitted on a PDCCH formed by the CCEs in one of the device's search spaces. For example, device A in Fig. 10.9 cannot be addressed on a PDCCH starting at CCE number 20, whereas device B can. Furthermore, if device A is using CCEs 16−23, device B cannot be addressed on aggregation level 4 as all CCEs in its level-4 search space are blocked by use for other devices. From this it can be intuitively understood that for efficient utilization of the CCEs in the system, the search spaces should differ between devices. Each device in the system can therefore have one or more *device-specific* search spaces configured. As a device-specific search space is typically smaller than the number of PDCCHs the network could transmit at the corresponding aggregation level, there must be a mechanism determining the set of CCEs in a device-specific search space.

One possibility would be to let the network configure the device-specific search space in each device, similar to the way the CORESETs are configured. However, this would require explicit signaling to each of the devices and possibly reconfiguration at handover. Instead, the device-specific search spaces for PDCCH are defined without explicit signaling through a function of the device identity unique in the cell, that is, the C-RNTI. Furthermore, the set of CCEs the device should monitor for a certain aggregation level also varies as a function of time to avoid two devices constantly blocking each other. If they collide at one time instant, they are not likely to collide at the next time instant. In each of these search spaces, the device is attempting to decode the PDCCHs using the device-specific C-RNTI identity.[2] If valid control information is found, for example, a scheduling grant, the device acts accordingly.

However, there is also information intended for a group of devices. Furthermore, as part of the random-access procedure, it is necessary to transmit

---

[2]There is sometimes an additional device-specific identity, the CS-RNTI, used for semipersistent scheduling, as discussed in Chapter 14.

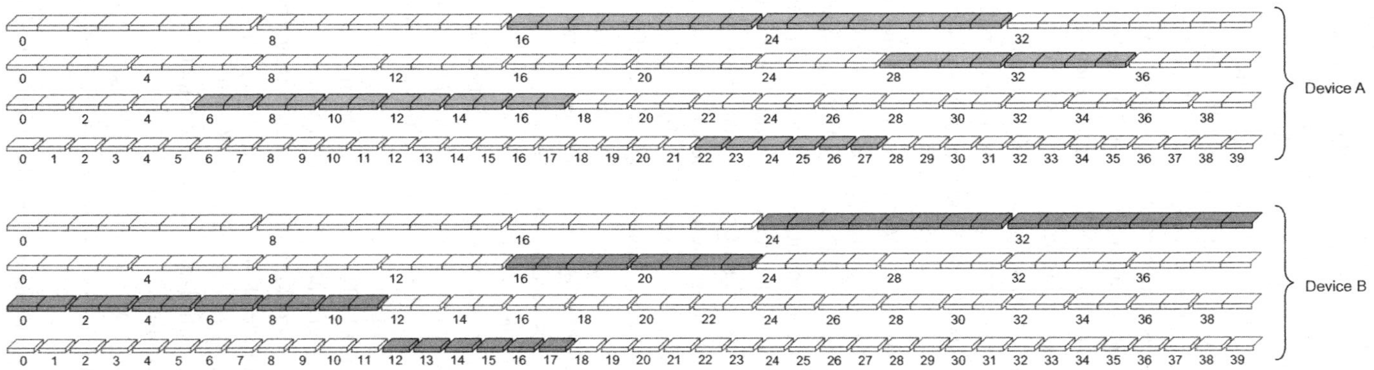

FIGURE 10.9

Example of search spaces for two different devices.

information to a device before it has been assigned a unique identity. These messages are scheduled with different predefined RNTIs, for example, the SI-RNTI for scheduling system information, the P-RNTI transmission of a paging message, the RA-RNTI for transmission of the random-access, and TPC-RNTI for uplink power control response. Other examples are the INT-RNTI used for preemption indication and the SFI-RNTI used for conveying slot-related information. These types of information cannot rely on a device-specific search space as different devices would monitor different CCEs despite the message being intended for all of them. Hence, NR also defines *common search spaces*.[3] A common search space is similar in structure to a device-specific search space with the difference that the set of CCEs is predefined and hence known to all devices, regardless of their own identity.

The number of blind decoding attempts depends on the subcarrier spacing (and hence the slot duration). For 15/30/60/120 kHz subcarrier spacing, up to 44/36/22/20 blind decoding attempts per slot can be supported across all DCI payload sizes—a number selected to offer a good tradeoff between device complexity and scheduling flexibility. However, the number of blind decoded is not the only measure of complexity but also channel estimation needs to be accounted for. The number of channel estimates for subcarrier spacings of 15/30/60/120 kHz has been limited to 56/56/48/32 CCEs across all CORESETs in a slot. Depending on the configuration, the number of PDCCH candidate may be limited either by the number of blind decodes, or by the number of channel estimates. Finally, to limit the device complexity, a "3 + 1" DCI size budget is defined, meaning that a device at most monitors three different DCI sizes using the C-RNTI (and hence being time-critical) and one DCI size using other RNTIs (and hence less time critical).

In the case of carrier aggregation, the general blind decoding operation described above is applied per component carrier. The total number of channel estimates and blind decoding attempts is increased compared to the single carrier case, but not in direct proportion to the number of aggregated carriers.

10.1.4 DOWNLINK SCHEDULING ASSIGNMENTS—DCI FORMATS 1–0 AND 1–1

Having described the transmission of DCI on PDCCH, the detailed contents of the control information can be discussed, starting with the downlink scheduling assignments. Downlink scheduling assignments use DCI format 1−1, the non-fallback format, or DCI format 1−0, also known as the fallback format.

The non-fallback format 1−1 supports all NR features. Depending on the features that are configured in the system, some information fields may or may not be present. For example, if carrier aggregation is not configured, there is no need

---

[3]The NR specifications defined different types of common search spaces depending on the RNTI monitored, but this is not important for understanding the general principle of search spaces.

to include carrier-aggregation-related information in the DCI. Hence the DCI size for format 1−1 depends on the overall configuration, but as long as the device knows which features are configured, it also knows the DCI size and blind detection can be performed.

The fallback format 1−0 is smaller in size, supports a limited set of NR functionality, and the set of information fields is in general not configurable, resulting in a (more or less) fixed DCI size. One use case of the fallback format is to handle periods of uncertainty in the configuration of a device as the exact time instant when a device applies the configuration information is not known to the network, for example due to transmission errors. Another reason for using the fallback DCI is to reduce control signaling overhead. In many cases the fallback format provides sufficient flexibility for scheduling smaller data packets.

Parts of the contents are the same for the different DCI formats, as seen in Table 10.1, but there are also differences due to the different capabilities. The information in the DCI formats used for downlink scheduling can be organized into different groups, with the fields present varying between the DCI formats. The content of DCI formats for downlink scheduling assignments is described below:

- Identifier of DCI format (1 bit). This is a header to indicate whether the DCI is a downlink assignment or an uplink grant, which is important in case the payload sizes of multiple DCI formats are aligned and the size cannot be used to differentiate the DCI formats (one example hereof is the fallback formats 0−0 and 1−0 which are of equal size).
- Resource information, consisting of:
  - Carrier indicator (0 or 3 bit). This field is present if cross-carrier scheduling is configured and is used to indicate the component carrier the DCI relates to. The carrier indicator is not present in the fallback DCI for example used for common signaling to multiple devices, as not all devices may be configured with (or capable of) carrier aggregation.
  - Bandwidth-part indicator (0−2 bit), used to activate one of up to four bandwidth parts configured by higher-layer signaling. Not present in the fallback DCI.
  - Frequency-domain resource allocation. This field indicates the resource blocks on one component carrier upon which the device should receive the PDSCH. The size of the field depends on the size of the bandwidth and on the resource allocation type, type 0 only, type 1 only, or dynamic switching between the two as discussed in Section 10.1.10. Format 1−0 supports resource allocation type 1 only as the full flexibility in resource allocation is not needed in this case.
  - Time-domain resource allocation (1−4 bit). This field indicates the resource allocation in the time domain as described in Section 10.1.11
  - VRB-to-PRB mapping (0 or 1 bit) to indicate whether interleaved or non-interleaved VRB-to-PRB mapping should be used as described in Section 9.9. Only present for resource allocation type 1.

Table 10.1 DCI Formats 1−0 and 1−1 for Downlink Scheduling

Field		Format 1−0	Format 1−1
Format identifier		•	•
Resource information	CFI		•
	BWP indicator		•
	Frequency domain allocation	•	•
	Time-domain allocation	•	•
	VRB-to-PRB mapping	•	•
	PRB bundling size indicator		•
	Reserved resources		•
	Zero-power CSI-RS trigger		•
Transport-block related	MCS	•	•
	NDI	•	•
	RV	•	•
	MCS, 2nd TB		•
	NDI, 2nd TB		•
	RV, 2nd TB		•
Hybrid-ARQ related	Process number	•	•
	DAI	•	•
	PDSCH-to-HARQ feedback timing	•	•
	CBGTI		•
	CBGFI		•
Multi-antenna related	Antenna ports		•
	TCI		•
	SRS request		•
	DM-RS sequence initialization		•
PUCCH-related information	PUCCH power control	•	•
	PUCCH resource indicator		•

- PRB size indicator (0 or 1 bit), used to indicate the PDSCH bundling size as described in Section 9.9.
- Reserved resources (0−2 bit), used to indicate to the device if the reserved resources can be used for PDSCH or not as described in Section 9.10.
- Zero-power CSI-RS trigger (0−2 bit), see Section 8.1 for a discussion on CSI reference signals.
- Transport-block-related information:
  - Modulation-and-coding scheme (5 bit), used to provide the device with information about the modulation scheme, the code rate, and the transport-block size, as described further below.
  - New-data indicator (1 bit), used to clear the soft buffer for initial transmissions as discussed in Section 13.1.

- Redundancy version (2 bit) (see Section 13.1).
- If a second transport block is present (only if more than four layers of spatial multiplexing are supported in DCI format 1−1), the three fields above are repeated for the second transport block.
- Hybrid-ARQ-related information:
  - Hybrid-ARQ process number (4 bit), informing the device about the hybrid-ARQ process to use for soft combining.
  - Downlink assignment index (DAI, 0, 2, or 4 bit), only present in the case of a dynamic hybrid-ARQ codebook as described in Section 13.1.5. DCI format 1−1 supports 0, 2, or 4 bits, while DCI format 1−0 uses 2 bits.
  - HARQ feedback timing (3 bit), providing information on *when* the hybrid-ARQ acknowledgment should be transmitted relative to the reception of the PDSCH.
  - CBG transmission indicator (CBGTI, 0, 2, 4, 6, or 8 bit), indicating the code block groups retransmitted as described in Section 13.1.2. Only present in DCI format 1−1 and only if CBG retransmissions are configured.
  - CBG flush information (CBGFI, 0−1 bit), indicating soft buffer flushing as described in Section 13.1.2. Only present in DCI format 1−1 and only if CBG retransmissions are configured.
- Multi-antenna-related information (present in DCI format 1−1 only):
  - Antenna ports (4−6 bit), indicating the antenna ports upon which the data are transmitted as well as antenna ports scheduled for other users as discussed in Chapters 9 and 11.
  - Transmission configuration indication (TCI, 0 or 3 bit), used to indicate the QCL relations for downlink transmissions as described in Chapter 12.
  - SRS request (2 bit), used to request transmission of a sounding reference signal as described in Section 8.3.
  - DM-RS sequence initialization (0 or 1 bit), used to select between two preconfigured initialization values for the DM-RS sequence.
- PUCCH-related information:
  - PUCCH power control (2 bit), used to adjust the PUCCH transmission power.
  - PUCCH resource indicator (3 bit), used to select the PUCCH resource from a set of configured resources (see Section 10.2.7).

10.1.5 UPLINK SCHEDULING GRANTS—DCI FORMATS 0−0 AND 0−1

Uplink scheduling grants use one of DCI formats 0−1, the non-fallback format, or DCI format 0−0, also known as the fallback format. The reason for having both a fallback and a non-fallback format is the same as for the downlink, namely to handle uncertainties during RRC reconfiguration and to provide a low-overhead format for transmissions not exploiting all uplink features. As for the uplink, the information fields present in the non-fallback format depend on the features that are configured.

The DCI sizes for the uplink DCI format $0-1$ and downlink DCI format $1-1$ are aligned with padding added to the smaller of the two in order to reduce the number of blind decodes.

Parts of the contents are the same for the different DCI formats, as seen in Table 10.2, but there are also differences due to the different capabilities. The information in the DCI formats used for uplink scheduling can be organized into different groups, with the fields present varying between the DCI formats. The content of DCI formats $0-1$ and $0-0$ is described below:

- Identifier of DCI format (1 bit), a header to indicate whether the DCI is a downlink assignment or an uplink grant.
- Resource information, consisting of:
  - Carrier indicator (0 or 3 bit). This field is present if cross-carrier scheduling is configured and is used to indicate the component carrier the DCI relates to. The carrier indicator is not present in DCI format $0-0$.

Table 10.2 DCI Formats $0-0$ and $0-1$ for Uplink Scheduling

Field		Format 0−0	Format 0−1
Identifier		•	•
Resource information	CFI		•
	UL/SUL	•	•
	BWP indicator		•
	Frequency domain allocation	•	•
	Time-domain allocation	•	•
	Frequency hopping	•	•
Transport-block-related	MCS	•	•
	NDI	•	•
	RV	•	•
Hybrid-ARQ-related	Process number	•	•
	DAI		•
	CBGTI		•
Multi-antenna-related	DM-RS sequence initialization		•
	Antenna ports		•
	SRI		•
	Precoding information		•
	PTRS−DMRS association		•
	SRS request		•
	CSI request		•
Power control	PUSCH power control	•	•
	Beta offset		•

- UL/SUL indicator (0 or 1 bit), used to indicate whether the grant relates to the supplementary uplink or the ordinary uplink (see Section 7.7). Only present if a supplementary uplink is configured as part of the system information.
- Bandwidth-part indicator (0—2 bit), used to activate one of up to four bandwidth parts configures by higher-layer signaling. Not present in DCI format 0—0.
- Frequency-domain resource allocation. This field indicates the resource blocks on one component carrier upon which the device should transmit the PUSCH. The size of the field depends on the size of the bandwidth and on the resource allocation type, type 0 only, type 1 only, or dynamic switching between the two as discussed in Section 10.1.10. Format 0—0 supports resource allocation type 1 only.
- Time-domain resource allocation (0—4 bit). This field indicates the resource allocation in the time domain as described in Section 10.1.11.
- Frequency-hopping flag (0 or 1 bit), used to handle frequency hopping for resource allocation type 1.

- Transport-block-related information:
  - Modulation-and-coding scheme (5 bit), used to provide the device with information about the modulation scheme, the code rate, and the transport-block size, as described further below.
  - New-data indicator (1 bit), used to indicate whether the grant relates to retransmission of a transport block or transmission of a new transport block.
  - Redundancy version (2 bit).

- Hybrid-ARQ-related information:
  - Hybrid ARQ process number (4 bit), informing the device about the hybrid-ARQ process to (re)transmit.
  - Downlink assignment index (DAI), used for handling of hybrid-ARQ codebooks in case of UCI transmitted on PUSCH. Not present in DCI format 0—0.
  - CBG transmission indicator (CBGTI, 0, 2, 4, or 6 bit), indicating the code block groups to retransmit as described in Section 13.1. Only present in DCI format 0—1 and only if CBG retransmissions are configured.

- Multi-antenna-related information (present in DCI format 1—1 only):
  - DMRS sequence initialization (1 bit), used to select between two preconfigured initialization values for the DM-RS sequence.
  - Antenna ports (2—5 bit), indicating the antenna ports upon which the data are transmitted as well as antenna ports scheduled for other users as discussed in Chapters 9 and 11.
  - SRS resource indicator (SRI), used to determine the antenna ports and uplink transmission beam to use for PUSCH transmission as described in Section 11.3. The number of bits depends on the number of SRS groups configured and whether codebook-based or non-codebook-based precoding is used.

- Precoding information (0—6 bit), used to select the precoding matrix **W** and the number of layers for codebook-based precoding as described in Section 11.3. The number of bits depends on the number of antenna ports and the maximum rank supported by the device.
- PTRS-DMRS association (0 or 2 bit), used to indicate the association between the DM-RS and PT-RS ports.
- SRS request (2 bit), used to request transmission of a sounding reference signal as described in Section 8.3.
- CSI request (0—6 bit), used to request transmission of a CSI report as described in Section 8.1.
- Power-control-related information:
  - PUSCH power control (2 bit), used to adjust the PUSCH transmission power.
  - Beta offset (0 or 2 bit), used to control the amount of resources used by UCI on PUSCH in case dynamic beta offset signaling is configured for DCI format 0—1 as discussed in Section 10.2.8.

10.1.6 SLOT FORMAT INDICATION—DCI FORMAT 2—0

DCI format 2—0, if used, is used to signal the slot format information (SFI) to the device as discussed in Section 7.8.3. The SFI is transmitted using the regular PDCCH structure and using the SFI-RNTI, common to multiple devices. To assist the device in the blind decoding process, the device is configured with information about the up to two PDCCH candidates upon which the SFI can be transmitted.

10.1.7 PREEMPTION INDICATION—DCI FORMAT 2—1

DCI format 2—1 is used to signal the preemption indicator to the device. It is transmitted using the regular PDCCH structure, using the INT-RNTI which can be common to multiple devices. The details and usage of the preemption indicator are discussed in Section 14.1.2.

10.1.8 UPLINK POWER CONTROL COMMANDS—DCI FORMAT 2—2

As a complement to the power-control commands provided as part of the downlink scheduling assignments and the uplink scheduling grants, there is the potential to transmit a power-control command using DCI format 2—2. The main motivation for DCI format 2—2 is to support power control for semipersistent scheduling. In this case there is no dynamic scheduling assignment or scheduling grant which can include the power control information for the PUCCH and PUSCH, respectively. Consequently, another mechanism is needed and DCI format 2—2 fulfills this need. The power-control message is directed to a group of devices using an RNTI specific for that group and each device is configured with the power control bits in the join message it should follow. DCI format $2-2$ is aligned with the size of DCI formats $0-0/1-0$ to reduce the blind decoding complexity.

10.1.9 SRS CONTROL COMMANDS—DCI FORMAT 2-3

DCI format $2-3$ is used for power control of uplink sounding reference signals for devices which have not coupled the SRS power control to the PUSCH power control, either because independent control is desirable or because the device is configured without PUCCH and PUSCH. The structure is similar to DCI format $2-2$, but with the possibility to, for each device, configure two bits for SRS request in addition to the two power control bits. DCI format $2-2$ is aligned with the size of DCI formats $0-0/1-0$ to reduce the blind decoding complexity.

10.1.10 SIGNALING OF FREQUENCY-DOMAIN RESOURCES

To determine the frequency-domain resources to transmit or receive upon, two fields are of interest: the resource-block allocation field and the bandwidth part indicator.

The resources allocation fields determine the resources blocks in the active bandwidth part upon which data are transmitted. There are two different possibilities for signaling the resources-block allocation, type 0 and type 1, both inherited from LTE where they are known as downlink resource allocation type 0 and type 2. In LTE, the resource-block allocation signaled the allocation across the carrier. However, in NR the indication is for the active bandwidth part.

Type 0 is a bitmap-based allocation scheme. The most flexible way of indicating the set of resource blocks the device is supposed to receive the downlink transmission upon is to include a bitmap with size equal to the number of resource blocks in the bandwidth part. This would allow for an arbitrary combination of resource blocks to be scheduled for transmission to the device but would, unfortunately, also result in a very large bitmap for the larger bandwidths. For example, in the case of a bandwidth part of 100 resource blocks, the downlink PDCCH would require 100 bits for the bitmap alone, to which the other pieces of information need to be added. Not only would this result in a large control-signaling overhead, but it could also result in downlink coverage problems as more than 100 bits in one OFDM symbol correspond to a data rate exceeding 1.4 Mbit/s for 15 kHz subcarrier spacing and even higher for the higher subcarrier spacings. Consequently, there is a need to reduce the bitmap size while keeping sufficient allocation flexibility. This can be achieved by pointing not to individual resource blocks in the frequency domain, but to groups of contiguous resource blocks, as shown in at the top of Fig. 10.10. The size of such a resource-block group is determined by the size of the bandwidth part. Two different configurations are possible for each size of the bandwidth parts, possibly resulting in different resource-block-group sizes for a given size of the bandwidth part.

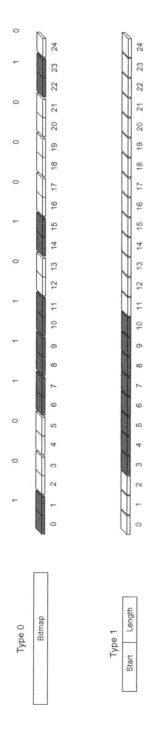

FIGURE 10.10

Illustration of resource-block allocation types (a bandwidth part of 25 resource blocks is used in this example).

Resource allocation type 1 does not rely on a bitmap. Instead, it encodes the resource allocation as a start position and length of the resource-block allocation. Thus, it does not support arbitrary allocations of resource blocks but only frequency-contiguous allocations, thereby reducing the number of bits required for signaling the resource-block allocation.

The resource allocation scheme to use is configured according to three alternatives: type 0, type 1, or dynamic selection between the two using a bit in the DCI. For the fallback DCIs, only resource block allocation type 1 is supported as a small overhead is more important than the flexibility to configure non-contiguous resources.

Both resource-allocation types refer to *virtual* resource blocks (see Section 7.3 for a discussion of resource-block types). For resource-allocation types 0, a non-interleaved mapping from virtual to physical resource blocks is used, meaning that the virtual resource blocks are directly mapped to the corresponding physical resource blocks. For resource-allocation type 1, on the other hand, both interleaved and non-interleaved mapping is supported. The VRB-to-PRB mapping bit (if present, downlink only) indicates whether the allocation signaling uses interleaved or non-interleaved mapping.

Returning to the bandwidth part indicator, this field is used to switch the active bandwidth part. It can either point to the current active bandwidth part, or to another bandwidth part to activate. If the field points to the current active bandwidth part, the interpretation of the DCI content is straightforward—the resource allocation applies to the current active bandwidth part as described above.

However, if the bandwidth part indicator points to a different bandwidth part than the active bandwidth part, the handling becomes more intricate. Many transmission parameters in general are configured per bandwidth part. The DCI payload size therefore may differ between different bandwidth parts. The frequency-domain resource allocation field is an obvious example; the larger the bandwidth part, the larger the number of bits for frequency-domain resource allocation. At the same time, the DCI sizes assumed when performing blind detection were determined by the currently active bandwidth part, not the bandwidth part to which the bandwidth part index points. Requiring the device to performing blind detection of multiple DCI sizes matching all possible bandwidth part configurations would be too complex. Hence, the DCI information obtained under the assumption of the DCI format being given by the currently active bandwidth part must be transformed to the new bandwidth part, which may have not only a different size in general, but also be configured with a different set of transmission parameters, for example TCI states which are configured per bandwidth part. The transformation is done using padding/truncation for each DCI field to match the requirements of the targeted bandwidth part. Once this is done, the bandwidth part pointed to by the bandwidth part indicator becomes the new active bandwidth part and the scheduling grant is applied to this bandwidth part. Similar transformation is sometimes required for DCI formats 0−0 and 1−0 in situations where the "3 + 1" DCI size budget otherwise would be violated.

10.1.11 SIGNALING OF TIME-DOMAIN RESOURCES

The time-domain allocation for the data to be received or transmitted is dynamically signaled in the DCI, which is useful as the part of a slot available for downlink reception or uplink transmission may vary from slot to slot as a result of the use of dynamic TDD or the amount of resources used for uplink controls signaling. Furthermore, the slot in which the transmission occurs also needs to be signaled as part of the time-domain allocation. Although the downlink data in many cases are transmitted in the same slot as the corresponding assignment, this is frequently not the case for uplink transmissions.

One approach would be to separately signal the slot number, the starting OFDM symbol, and the number of OFDM symbols used for transmission or reception. However, as this would result in an unnecessarily large number of bits, NR has adopted an approach based on configurable tables. The time-domain allocation field in the DCI is used as an index into an RRC-configured table from which the time-domain allocation is obtained, as illustrated in Fig. 10.11.

There is one table for uplink scheduling grants and one table for downlink scheduling assignments. Up to 16 rows can be configured where each row contains:

- A slot offset, that is, the slot relative to the one where the DCI was obtained. For the downlink, slot offsets from 0 to 3 are possible, while for the uplink slot offsets from 0 to 7 can be used. The larger uplink range can be motivated by the need for scheduling uplink transmissions further into the future for coexistence with (primarily) LTE TDD.
- The first OFDM symbol in the slot where the data are transmitted.
- The duration of the transmission in number of OFDM symbols in the slot. Not all combinations of start and length fit within one slot, for example, starting at OFDM symbol 12 and transmit during five OFDM symbols obviously results in crossing the slot boundary and represents an invalid combination. Therefore, the start and length are jointly encoded to cover only the valid combinations (although in Fig. 10.11 they are shown as two separate columns for illustrative reasons).
- For the downlink, the PDSCH mapping type, that is, the DM-RS location as described in Section 9.11, is also part of the table. This provides more flexibility compared to separately indicating the mapping type.

It is also possible to configure slot aggregation, that is, a transmission where the same transport block is repeated across up to eight slots. However, this is not

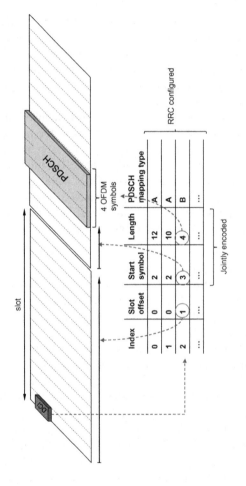

FIGURE 10.11

Signaling of time-domain allocation (downlink).

part of the dynamic signaling using a table but is a separate RRC configuration. Slot aggregation is primarily a tool to handle coverage-challenged deployments and thus there is less need for a fully dynamic scheme.

10.1.12 SIGNALING OF TRANSPORT-BLOCK SIZES

Proper reception of a downlink transmission requires, in addition to the set of resource blocks, knowledge about the modulation scheme and the transport-block size, information (indirectly) provided by the 5-bit MCS field. In principle, a similar approach as in LTE, namely to tabulate the transport block size as a function of the MCS field and the resource-block allocation would be possible. However, the significantly larger bandwidths supported in NR, together with a wide range of transmission durations and variations in the overhead depending on other features configured such as CSI-RS, would result in a large number of tables required to handle the large dynamic range in terms of transport block sizes. Such a scheme may also require modifications whenever some of these parameters change. Therefore, NR opted for a formula-based approach combined with a table for the smallest transport-block sizes instead of a purely table-based scheme to achieve the necessary flexibility.

The first step is to determine the modulation scheme and code rate from the MCS field. This is done using one of two tables, one table if 256QAM is not configured and another table if 256QAM is configured. Of the 32 combinations of the 5-bit MCS fields, 29 are used to signal the modulation-and-coding scheme, whereas three are reserved, the purpose of which is described later. Each of the 29 modulation-and-coding scheme entries represents a particular combination of modulation scheme and channel-coding rate or, equivalently, a certain spectral efficiency measured in the number of information bits per modulation symbol, ranging from approximately 0.2−5.5 bit/s/Hz. For devices configured with support for 256QAM, four of the 32 combinations are reserved and the remaining 28 combinations indicate a spectral efficiency in the range 0.2−7.4 bit/s/Hz.

Up to this point, the NR scheme is similar to the one used for LTE. However, to obtain a more flexible scheme, the following steps differ compared to LTE.

Given the modulation order, the number of resource blocks scheduled, and the scheduled transmission duration, the number of available resource elements can be computed. From this number the resource elements used for DM-RS are subtracted. A constant, configured by higher layers and modeling the overhead by other signals such as CSI-RS or SRS is also subtracted. The resulting estimate of resource elements available for data is then, together with the number of transmission layers, the modulation order, and the code rate obtained from the MCS, used to calculate an intermediate number of information bits. This intermediate number is then quantized to obtain the final transport block size while at the same time ensuring byte-aligned code blocks, and that no filler bits are needed in the LDPC coding. The quantization also results in the same transport block size being obtained, even if there are small variations in the amount of resources allocated, a

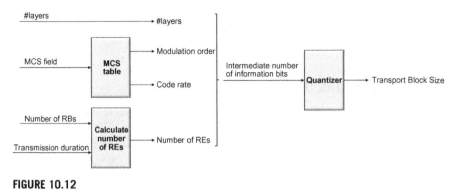

FIGURE 10.12

Calculating the transport block size.

property that is useful when scheduling retransmissions on a different set of resources than the initial transmission.

Returning to the three or four reserved combinations in the modulation-and-coding field mentioned at the beginning of this section, those entries can be used for retransmissions only. In the case of a retransmission, the transport-block size is, by definition, unchanged and fundamentally there is no need to signal this piece of information. Instead, the three or four reserved values represent the modulation scheme—QPSK, 16QAM, 64QAM, or (if configured) 256QAM—which allows the scheduler to use an (almost) arbitrary combination of resource blocks for the retransmission. Obviously, using any of the three or four reserved combinations assumes that the device properly received the control signaling for the initial transmission; if this is not the case, the retransmission should explicitly indicate the transport-block size.

The derivation of the transport-block size from the modulation-and-coding scheme and the number of scheduled resource blocks is illustrated in Fig. 10.12.

10.2 UPLINK

Similar to LTE, there is also a need for uplink L1/L2 control signaling to support data transmission on downlink and uplink transport channels. Uplink L1/L2 control signaling consists of:

- Hybrid-ARQ acknowledgments for received DL-SCH transport blocks;
- Channel-state information (CSI) related to the downlink channel conditions, used to assist downlink scheduling, including multi-antenna and beamforming schemes; and
- Scheduling requests, indicating that a device needs uplink resources for UL-SCH transmission.

There is no UL-SCH transport-format information included in the uplink transmission. As mentioned in Section 6.4.4, the gNB is in complete control of the uplink UL-SCH transmissions and the device always follows the scheduling grants received from the network, including the UL-SCH transport format specified in those grants. Thus, the network knows the transport format used for the UL-SCH transmission in advance and there is no need for any explicit transport-format signaling on the uplink.

The *physical uplink control channel* (PUCCH) is the basis for transmission of uplink control. In principle, the UCI could be transmitted on the PUCCH regardless of whether the device is transmitting data on the PUSCH. However, especially if the uplink resources for the PUSCH and the PUCCH are on the same carrier (or, to be more precise, use the same power amplifier) but widely separated in the frequency domain, the device may need a relatively large power back-off to fulfill the spectral emission requirements with a corresponding impact on the uplink coverage. Hence, similarly to LTE, NR supports UCI on PUSCH as the basic way of handling simultaneous transmission of data and control. Thus, if the device is transmitting on the PUSCH the UCI is multiplexed with data on the granted resources instead of being transmitted on the PUCCH. Simultaneous PUSCH and PUCCH is not part of release 15 but may be introduced in a later release.

Beamforming can be applied to the PUCCH. This is realized by configuring one or more spatial relations between the PUCCH and downlink signals such as CSI-RS or SS block. In essence, such a spatial relation means that the device can transmit the uplink PUCCH using the same beam as it used for receiving the corresponding downlink signal. For example, if the spatial relation between PUCCH and SS block is configured, the device will transmit PUCCH using the same beam as it used for receiving the SS block. Multiple spatial relations can be configured and MAC control elements used to indicate which one to use.

In the case of carrier aggregation, the uplink control information is transmitted on the primary cell as a baseline. This is motivated by the need to support asymmetric carrier aggregation with the number of downlink carriers supported by a device that is unrelated to the number of uplink carriers. For a large number of downlink component carriers, a single uplink carrier may carry a large number of acknowledgments. To avoid overloading a single carrier, it is possible to configure two *PUCCH groups* where feedback relating to the first group is transmitted in the uplink of the PCell and feedback relating to the other group of carriers is transmitted on the primary second cell (PSCell), as illustrated in Fig. 10.13.

In the following section, the basic PUCCH structure and the principles for PUCCH control signaling are described, followed by control signaling on PUSCH.

10.2.1 BASIC PUCCH STRUCTURE

Uplink control information can be transmitted on PUCCH using several different formats.

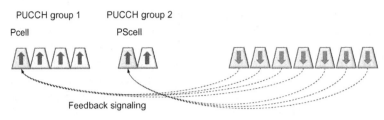

FIGURE 10.13

Multiple PUCCH groups.

Two of the formats, 0 and 2, are sometimes referred to as *short PUCCH formats*, as they occupy at most two OFDM symbols. In many cases the last one or two OFDM symbols in a slot are used for PUCCH transmission, for example, to transmit a hybrid-ARQ acknowledgment of the downlink data transmission. The short PUCCH formats include:

- PUCCH format 0, capable of transmitting at most two bits and spanning one or two OFDM symbols. This format can, for example, be used to transmit a hybrid-ARQ acknowledgment of a downlink data transmission, or to issue a scheduling request.
- PUCCH format 2, capable of transmitting more than two bits and spanning one or two OFDM symbols. This format can, for example, be used for CSI reports or for multi-bit hybrid-ARQ acknowledgments in the case of carrier aggregation or per-CBG retransmission.

Three of the formats, 1, 3, and 4, are sometimes referred to as *long PUCCH formats* as they occupy from 4 to 14 OFDM symbols. The reason for having a longer time duration than the previous two formats is coverage. If a duration of one or two OFDM symbols does not provide sufficient received energy for reliable reception, a longer time duration is necessary and one of the long PUCCH formats can be used. The long PUCCH formats include:

- PUCCH format 1, capable of transmitting at most two bits.
- PUCCH formats 3 and 4, both capable of transmitting more than two bits but differing in the multiplexing capacity, that is, how many devices that can use the same time−frequency resource simultaneously.

Since the PUSH uplink can be configured to use either OFDM or DFT-spread OFDM, one natural thought would be to adopt a similar approach for the PUCCH. However, to reduce the number of options to specify, this is not the case. Instead, the PUCCH formats are in general designed for low cubic metric, PUCCH format 2 being the exception and using pure OFDM only. Another choice made to simplify the overall design was to only support specification-transparent transmit diversity schemes. In other words, there is only a single antenna port specified for the PUCCH and if the device is equipped with multiple transmit antennas it is up to the device implementation how to exploit these antennas, for example by using some form of delay diversity. In the following, the detailed structure of each of these PUCCH formats will be described.

10.2.2 PUCCH FORMAT 0

PUCCH format 0, illustrated in Fig. 10.14, is one of the short PUCCH formats and is capable of transmitting up to two bits. It is used for hybrid-ARQ acknowledgments and scheduling requests.

Sequence selection is the basis for PUCCH format 0. For the small number of information bits supported by PUCCH format 0, the gain from coherent reception is not that large. Furthermore, multiplexing information and reference signals in one OFDM symbol while maintaining a low cubic metric is not possible. Therefore, a different structure where the information bit(s) selects the sequence to transmit is used. The transmitted sequence is generated by different phase rotations of the same underlying length-12 base sequence, where the base sequences are the same base sequences defined for generating the reference signal in the case of DFT-preceded OFDM as described in Section 9.11.2. Thus, the phase rotation applied to the base sequence carries the information. In other words, the information selects one of several phase-rotated sequences.

Twelve different phase rotations are defined for the same base sequence, providing up to 12 different orthogonal sequences from each base sequence. A linear phase rotation in the frequency domain is equivalent to applying a cyclic shift in the time domain, hence, the term "cyclic shift" is sometimes used with an implicit reference to the time domain.

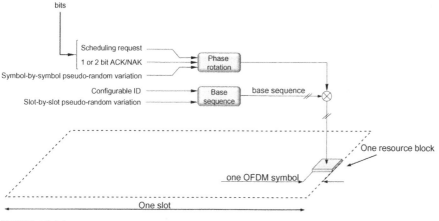

FIGURE 10.14

Example of PUCCH format 0.

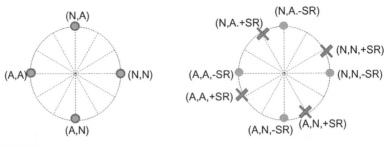

FIGURE 10.15

Examples of phase rotations as a function of hybrid-ARQ acknowledgments and scheduling request.

To maximize the performance, the phase rotations representing the different information bits are separated with $2\pi \cdot 6/12$ and $2\pi \cdot 3/12$ for one and two bits acknowledgments, respectively. In the case of a simultaneous scheduling request, the phase rotation is increased by $3\pi/12$ for one acknowledgment bit and by $2\pi/12$ for two bits, as illustrated in Fig. 10.15.

The phase rotation applied to a certain OFDM symbol carrying PUCCH format 0 depends not only on the information to be transmitted as already mentioned, but also on a reference rotation provided as part of the PUCCH resource allocation mechanism as discussed in Section 10.2.7. The intention with the reference rotation is to multiplex multiple devices on the same time—frequency resource. For example, two devices transmitting a single hybrid-ARQ acknowledgment can be given different reference phase rotations such that one device uses 0 and $2\pi \cdot 6/12$, while the other device uses $2\pi \cdot 3/12$ and $2\pi \cdot 9/12$. Finally, there is also a mechanism for cyclic shift hopping where a phase offset varying between different slots is added. The offset is given by a pseudo-random sequence. The underlying reason is to randomize interference between different devices.

The base sequence to use can be configured per cell using an identity provided as part of the system information. Furthermore, sequence hopping, where the base sequence used varies on a slot-by-slot basis, can be used to randomize the interference between different cells. As seen from this description many quantities are randomized in order to mitigate interference.

PUCCH format 0 is typically transmitted at the end of a slot as illustrated in Fig. 10.14. However, it is possible to transmit PUCCH format 0 also in other positions within a slot. One example is frequently occurring scheduling requests (as frequent as every second OFDM symbol can be configured). Another example when this can be useful is to acknowledge a downlink transmission on a downlink carrier at a high carrier frequency and, consequently, a correspondingly higher subcarrier spacing and shorter downlink slot duration. This is a relevant scenario in the case of carrier aggregation and supplementary uplink, as discussed in

Chapter 7. If low latency is important, the hybrid-ARQ acknowledgment needs to be fed back quickly after the end of the downlink slot, which is not necessarily at the end of the uplink slot if the subcarrier spacing differs between uplink and downlink.

In the case of two OFDM symbols used for PUCCH format 0, the same information is transmitted in both OFDM symbols. However, the reference phase rotation as well as the frequency-domain resources may vary between the symbols, essentially resulting in a frequency-hopping mechanism.

10.2.3 PUCCH FORMAT 1

PUCCH format 1 is to some extent the long PUCCH counterpart of format 0. It is capable of transmitting up to two bits, using from 4 to 14 OFDM symbols, each one resource block wide in frequency. The OFDM symbols used are split between symbols for control information and symbols for reference signals to enable coherent reception. The number of symbols used for control information and reference signal, respectively, is a trade-off between channel-estimation accuracy and energy in the information part. Approximately half the symbols for reference symbols were found to be a good compromise for the payloads supported by PUCCH format 2.

The one or two information bits to be transmitted are BPSK or QPSK modulated, respectively, and multiplied by the same type of length-12 low-PAPR sequence as used for PUCCH format 0. Similar to format 0, sequence and cyclic shift hopping can be used to randomize interference. The resulting modulated length-12 sequence is block-wise spread with an orthogonal DFT code of the same length as the number of symbols used for the control information. The use of the orthogonal code in the time domain increases the multiplexing capacity as multiple devices having the same base sequence and phase rotation still can be separated using different orthogonal codes.

The reference signals are inserted using the same structure, that is, an unmodulated length-12 sequence is block-spread with an orthogonal sequence and mapped to the OFDM symbols used for PUCCH reference-signal transmission. Thus, the length of the orthogonal code, together with the number of cyclic shifts, determines the number of devices that can transmit PUCCH format 1 on the same resource. An example is shown in Fig. 10.16 where nine OFDM symbols are used for PUCCH transmission, four carrying the information and five used for reference signals. Hence, up to four devices, determined by the shorter of the codes for the information part, can share the same cyclic shift of the base sequence, and a set of resources for PUCCH transmission in this particular example. Assuming a cell-specific base sequence and six out of the 12 cyclic shifts being useful from a delay-spread perspective, this results in a multiplexing capacity of at most 24 devices on the same time—frequency resources.

The longer transmission duration of the long PUCCH formats compared to a short single-symbol format opens the possibility for frequency hopping as a mean

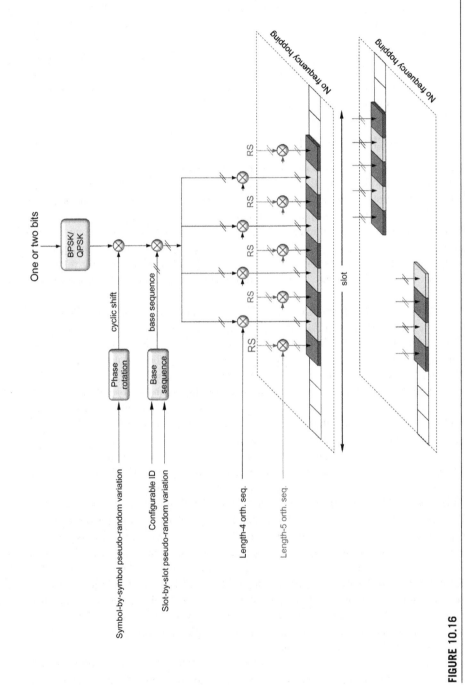

FIGURE 10.16

Example of PUCCH format 1 without frequency hopping (top) and with frequency hopping (bottom).

to achieve frequency diversity in a similar way as in LTE. However, unlike LTE, where hopping is always done at the slot boundary between the two slots used for PUCCH, additional flexibility is needed in NR as the PUCCH duration can vary depending on the scheduling decisions and overall system configuration. Furthermore, as the devices are supposed to transmit within their active bandwidth part only, hopping is typically not between the edges of the overall carrier bandwidth as in LTE. Therefore, whether to hop or not is configurable and determined as part of the PUCCH resource configuration. The position of the hop is obtained from the length of the PUCCH. If frequency hopping is enabled, one orthogonal block-spreading sequence is used per hop. An example is provided in Fig. 10.16 where, instead of a single set of length-4/length-5 orthogonal sequences, two sets of sequences length-2/length-2 and length-2/length-3, are used for the first and second hops, respectively.

10.2.4 **PUCCH FORMAT 2**

PUCCH format 2 is a short PUCCH format based on OFDM and used for transmission of more than two bits, for example, simultaneous CSI reports and hybrid-ARQ acknowledgments, or a larger number of hybrid-ARQ acknowledgments. A scheduling request can also be included in the bits jointly encoded. If the bits to be encoded are too large, the CSI report is dropped to preserve the hybrid-ARQ acknowledgments which are more important.

The overall transmission structure is straightforward. For larger payload sizes, a CRC is added. The control information (after CRC attachment) to be transmitted is coded, using Reed−Muller codes for payloads up to and including 11 bits and Polar[4] coding for larger payloads, followed by scrambling and QPSK modulation. The scrambling sequence is based on the device identity (the C-RNTI) together with the physical-layer cell identity (or a configurable virtual cell identity), ensuring interference randomization across cells and devices using the same set of time−frequency resources. The QPSK symbols are then mapped to subcarriers across multiple resource blocks using one or two OFDM symbols. A pseudo-random QPSK sequence, mapped to every third subcarrier in each OFDM symbol, is used as a demodulation reference signal to facilitate coherent reception at the base station.

The number of resource blocks used by PUCCH format 2 is determined by the payload size and a configurable maximum code rate. The number of resource blocks is thus smaller if the payload size is smaller, keeping the effective code rate roughly constant. The number of resource blocks used is upper bounded by a configurable limit.

PUCCH format 2 is typically transmitted at the end of a slot as illustrated in Fig. 10.17. However, similarly to format 0 and for the same reasons, it is possible to transmit PUCCH format 2 also in other positions within a slot.

---

[4]Polar coding is used for the DCI as well, but the details of the Polar coding for UCI are different.

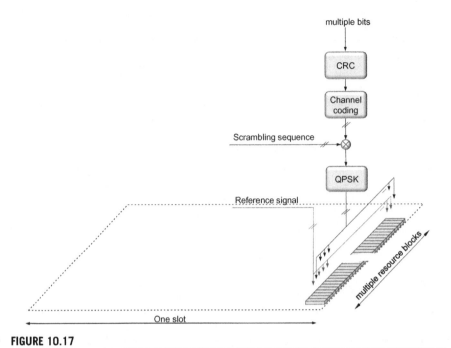

FIGURE 10.17

Example of PUCCH format 2 (the CRC is present only for larger payloads).

10.2.5 PUCCH FORMAT 3

PUCCH format 3 can be seen as the long PUCCH counterpart to PUCCH format 2. More than two bits can be transmitted using PUCCH format 3 using from 4 to 14 symbols, each of which can be multiple resource blocks wide. Thus, it is the PUCCH format with the largest payload capacity. Similar to PUCCH format 1, the OFDM symbols used are split between symbols for control information and symbols for reference signals to allow for a low cubic metric of the resulting waveform.

The control information to be transmitted is coded using Reed—Muller codes for 11 bits or less and Polar codes for large payloads, followed by scrambling and modulation. The scrambling sequence is based on the device identity (the C-RNTI) together with the physical-layer cell identity (or a configurable virtual cell identity), ensuring interference randomization across cells and devices using the same set of time—frequency resources. Following the principles of PUCCH format 2, a CRC is attached to the control information for the larger payloads. The modulation scheme used is QPSK but it is possible to optionally configure $\pi/2$-BPSK to lower the cubic metric at a loss in link performance.

The resulting modulation symbols are divided between the OFDM symbols. DFT precoding is applied to reduce the cubic metric and improve the power amplifier efficiency. The reference signal sequence is generated in the same way as for DFT-precoded PUSCH transmissions (see Section 9.11.2) for the same reason, namely to maintain a low cubic metric.

Frequency hopping can be configured for PUCCH format 3 as illustrated in Fig. 10.18, for example, to exploit frequency diversity, but it is also possible to operate without frequency hopping. The placements of the reference signal symbols depend on whether the frequency hopping is used or not and the length of the PUCCH transmission, as there must be at least one reference signal per hop. There is also a possibility to configure additional reference signal locations for the longer PUCCH durations to get two reference signal instances per hop.

The mapping of the UCI is such that the more critical bits, that is, hybrid-ARQ acknowledgments, scheduling request, and CSI part 1, are jointly coded and mapped close to the DM-RS locations, while the less critical bits are mapped in the remaining positions.

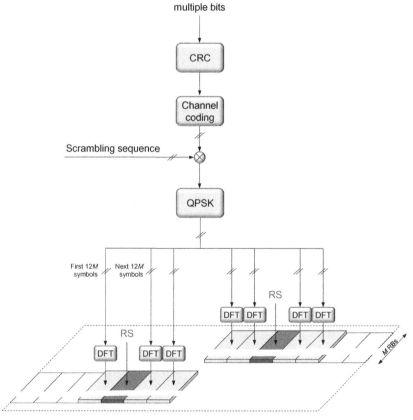

FIGURE 10.18

Example of PUCCH format 3 (the CRC is present for large payload sizes only).

10.2.6 PUCCH FORMAT 4

PUCCH format 4 (see Fig. 10.19) is in essence the same as PUCCH format 3 but with the possibility to code-multiplex multiple devices in the same resource and using at most one resource block in the frequency domain. Each control-information-carrying OFDM symbol carries $12/N_{SF}$ unique modulation symbols. Prior to DFT-precoding, each modulation symbol is block-spread with an orthogonal sequence of length N_{SF}. Spreading factors two and four are supported, implying a multiplexing capacity of two or four devices on the same set of resource blocks.

10.2.7 RESOURCES AND PARAMETERS FOR PUCCH TRANSMISSION

In the discussion of the different PUCCH formats above, a number of parameters were assumed to be known. For example, the resource blocks to map the transmitted signal to, the initial phase rotation for PUCCH format 0, whether to use

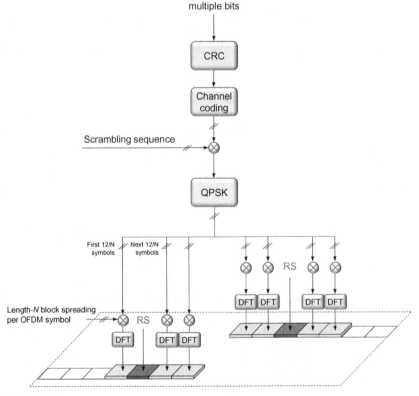

FIGURE 10.19

Example of PUCCH format 4.

frequency hopping or not, and the length in OFDM symbols for the PUCCH transmission. Furthermore, the device also needs to know which of the PUCCH formats to use, and which time−frequency resources to use.

In LTE, especially in the first releases, there is a fairly fixed linkage between the uplink control information, the PUCCH format, and the transmission parameters. For example, LTE PUCCH format 1a/1b is used for hybrid-ARQ acknowledgments and the time−frequency-code resources to use are given by a fixed time offset from the reception of the downlink scheduling assignment and the resources used for the downlink assignment. This is a low-overhead solution, but has the drawback of being inflexible and was extended to provide more flexibility in later releases of LTE supporting carrier aggregation and other more advanced features.

NR has adopted a more flexible scheme from the beginning, which is necessary given the very flexible framework with a wide range of service requirements in terms of latency and spectral efficiency, support of no predefined uplink−downlink allocation in TDD, different devices supporting aggregation of different number of carriers, and different antenna schemes requiring different amounts of feedback just to name some motivations. Central in this scheme is the notion of *PUCCH resource sets*. A PUCCH resource set contains at least four PUCCH resource configurations, where each resource configuration contains the PUCCH format to use and all the parameters necessary for that format. Up to four PUCCH resource sets can be configured, each of them corresponding to a certain range of UCI feedback to transmit. PUCCH resource set 0 can handle UCI payloads up to two bits and hence only contain PUCCH formats 0 and 1, while the remaining PUCCH resource sets may contain any PUCCH format except format 0 and 1.

When the device is about to transmit UCI, the UCI payload determines the PUCCH resource set and the ARI in the DCI determines the PUCCH resource configuration within the PUCCH resource set (see Fig. 10.20). Thus, the scheduler has control of where the uplink control information is transmitted. For periodic CSI reports and scheduling request opportunities, which both are semistatically configured, the PUCCH resources are provided as part of the CSI or SR configuration.

10.2.8 UPLINK CONTROL SIGNALING ON PUSCH

If the device is transmitting data on PUSCH—that is, has a valid scheduling grant—simultaneous control signaling could in principle remain on the PUCCH. However, as already discussed, this is not the case as in many cases it is preferable to multiplex data and control on PUSCH and avoid a simultaneous PUCCH. One reason is the increased cubic metric compared to UCI on PUSCH when using DFT-precoded OFDM. Another reason is the more challenging RF implementation if out-of-band emission requirements should be met at higher transmission powers and with PUSCH and PUCCH widely separate in the frequency domain. Hence, similar to LTE, UCI on PUSCH is the main mechanism for simultaneous

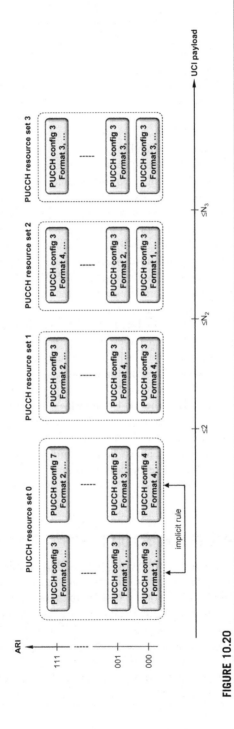

FIGURE 10.20

Example of PUCCH resource sets.

transmission of UCI and uplink data. The same principles are used for both OFDM and DFT-precoded OFDM in the uplink.

Only hybrid-ARQ acknowledgments and CSI reports are rerouted to the PUSCH. There is no need to request a scheduling grant when the device is already scheduled; instead, in-band buffer-status reports can be sent as described in Section 14.2.3.

In principle, the base station knows when to expect a hybrid-ARQ acknowledgment from the device and can therefore perform the appropriate demultiplexing of the acknowledgment and the data part. However, there is a certain probability that the device has missed the scheduling assignment on the downlink control channel. In this case the base station would expect a hybrid-ARQ acknowledgment while the device will not transmit one. If the rate-matching pattern would depend on whether an acknowledgment is transmitted or not, all the coded bits transmitted in the data part could be affected by a missed assignment and are likely to cause the UL-SCH decoding to fail.

One possibility to avoid this error is to puncture hybrid-ARQ acknowledgments onto the coded UL-SCH stream in which case the non-punctured bits are unaffected by the presence/absence of hybrid-ARQ acknowledgments. This is also the solution adopted in LTE. However, given the potentially large number of acknowledgment bits due to, for example, carrier aggregation of the use of codeblock group retransmissions, puncturing is less suitable as a general solution. Instead, NR has adopted a scheme where up to two hybrid-ARQ acknowledgment bits are punctured, while for a larger number of bits rate matching of the uplink data is used. To avoid the aforementioned error cases, the uplink DAI field in the DCI indicates the amount of resources reserved for uplink hybrid ARQ. Thus, regardless of whether the device missed any previous scheduling assignments or not, the amount of resources to use for the uplink hybrid-ARQ feedback is known.

The mapping of the UCI is such that the more critical bits, that is, hybrid-ARQ acknowledgments, are mapped to the first OFDM symbol after the first demodulation reference signal. Less critical bits, that is CSI reports, are mapped to subsequent symbols.

Unlike the data part, which relies on rate adaptation to handle different radio conditions, this cannot be used for the L1/L2 control-signaling part. Power control could, in principle, be used as an alternative, but this would imply rapid power variations in the time domain, which negatively impact the RF properties. Therefore, the transmission power is kept constant over the PUSCH duration and the amount of resource elements allocated to L1/L2 control signaling—that is, the code rate of the control signaling—is varied. In addition to a semistatic value controlling the amount of PUSCH resources used for UCI, it is also possible to signal this fraction as part of the DCI should a tight control be needed.

Multi-Antenna Transmission 11

Multi-antenna transmission is a key component of NR, especially at higher frequencies. This chapter gives a background to multi-antenna transmission in general, followed by a detailed description on NR multi-antenna precoding.

11.1 INTRODUCTION

The use of multiple antennas for transmission and/or reception can provide substantial benefits in a mobile-communication system.

Multiple antennas at the transmitter and/or receiver side can be used to provide diversity against fading by utilizing the fact that the channels experienced by different antennas may be at least partly uncorrelated, either due to sufficient inter-antenna distance or due to different polarization between the antennas.

Furthermore, by carefully adjusting the phase, and possibly also the amplitude, of each antenna element, multiple antennas at the transmitter side can be used to provide directivity, that is, to focus the overall transmitted power in a certain direction (beam forming) or, in the more general case, to specific locations in space. Such directivity can increase the achievable data rates and range due to higher power reaching the target receiver. Directivity will also reduce the interference to other links, thereby improving the overall spectrum efficiency.

Similarly, multiple receive antennas can be used to provide *receiver-side directivity*, focusing the reception in the direction of a target signal, while suppressing interference arriving from other directions.

Finally, the presence of multiple antennas at both the transmitter and the receiver sides can be used to enable *spatial multiplexing*, that is, transmission of multiple "layers" in parallel using the same time/frequency resources.

In LTE, multi-antenna transmission/reception for diversity, directivity, and spatial multiplexing is a key tool to enable high data rates and high system efficiency. However, multi-antenna transmission/reception is an even more critical component for NR due to the possibility for deployment at much higher frequencies compared to LTE.

There is a well-established and to a large extent correct assumption that radio communication at higher frequencies is associated with higher propagation loss and correspondingly reduced communication range. However, at least part of this is due to an assumption that the dimensions of the receiver antenna scale with the

5G NR: The Next Generation Wireless Access Technology. DOI: https://doi.org/10.1016/B978-0-12-814323-0.00011-9

wavelength, that is, with the inverse of the carrier frequency. As an example, a tenfold increase in the carrier frequency, corresponding to a tenfold reduction in the wave length, is assumed to imply a corresponding tenfold reduction in the physical dimensions of the receiver antenna or a factor of 100 reduction in the physical antenna area. This corresponds to a 20 dB reduction in the energy captured by the antenna.

If the receiver antenna size would instead be kept unchanged as the carrier frequency increases, the reduction in captured energy could be avoided. However, this would imply that the antenna size would increase relative to the wave length, something that inherently increases the directivity of the antenna.[1] The gain with the larger antenna size can thus only be realized if the receive antenna is well directed towards the target signal.

By also keeping the size of the transmitter-side antenna unchanged, in practice increasing the transmit-antenna directivity, the link budget at higher frequencies can be further improved. Assuming line-of-sight propagation and ignoring other losses, the overall link budget would then actually improve for higher frequencies. In practice there are many other factors that negatively impact the overall propagation losses at higher frequencies such as higher atmospheric attenuation and less diffraction leading to degraded non-line-of-sight propagation. Still, the gain from higher antenna directivity at higher frequencies is widely utilized in point-to-point radio links where the use of highly directional antennas at both the transmitter and receiver sides, in combination with line-of-sight links, allows for relatively long-range communication despite operation at very high frequencies.

In a mobile-communication system with devices located in many different directions relative to the base station and the devices themselves having an essentially random rotational direction, the use of fixed highly directional antennas is obviously not applicable. However, a similar effect, that is, an extension of the overall receive antenna area enabling higher-directivity transmission, can also be achieved by means of an antenna panel consisting of many small antenna elements. In this case, the dimension of each antenna element, as well as the distance between antenna elements, is proportional to the wave length. As the frequency increases, the size of each antenna element, as well as their mutual distances, is thus reduced. However, assuming a constant size of the overall antenna configuration, this can be compensated for by increasing the number of antenna elements. Fig. 11.1 shows an example of such an antenna panel consisting of 64 dual-polarized antenna elements and targeting the 28 GHz band. The AAA battery is included in the picture as an indication of the overall size of the antenna panel.

The benefit of such an antenna panel with a large number of small antenna elements, compared to a single large antenna, is that the direction of the transmitter beam can be adjusted by separately adjusting the phase of the signals applied to each antenna element. The same effect can be achieved when a multi-antenna

---

[1]The directivity of an antenna is roughly proportional to the physical antenna area normalized with the square of the wave length.

FIGURE 11.1

Rectangular antenna panel with 64 dual-polarized antenna elements.

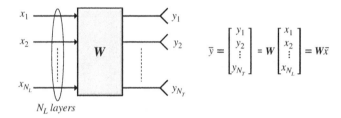

$$\bar{y} = \begin{bmatrix} y_1 \\ y_2 \\ \vdots \\ y_{N_T} \end{bmatrix} = W \begin{bmatrix} x_1 \\ x_2 \\ \vdots \\ x_{N_L} \end{bmatrix} = W\bar{x}$$

FIGURE 11.2

General model of multi-antenna transmission mapping N_L layers to N_T antennas.

panel, such as the one illustrated in Fig. 11.1, is used on the receiver side, that is, the receiver beam direction can be adjusted by separately adjusting the phases of the signals received at each antenna element.

In general, any linear multi-antenna transmission scheme can be modeled according to Fig. 11.2 with N_L layers, captured by the vector $\bar{x}$, being mapped to N_T transmit antennas (the vector $\bar{y}$) by means of multiplication with a matrix W of size $N_T \times N_L$.

The general model of Fig. 11.2 applies most cases of multi-antenna transmission. However, depending on implementation there will be various degrees of constraints that will impact the actual capabilities of the multi-antenna transmission.

One such implementation aspect relates to where, within the overall physical transmitter chain, the multi-antenna processing, that is, the matrix W of Fig. 11.3, is applied. On a high level one can distinguish between two cases:

- The multi-antenna processing is applied within the analog part of the transmitter chain, that is, after digital-to-analog conversion (left part of Fig. 11.3);
- The multi-antenna processing is applied within the digital part of the transmitter chain, that is, before digital-to-analog conversion (right part of Fig. 11.3).

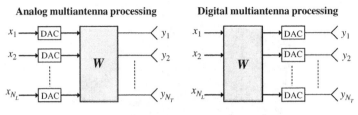

FIGURE 11.3

Analog vs digital multi-antenna processing.

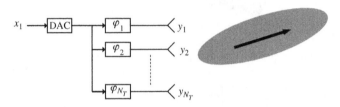

FIGURE 11.4

Analog multi-antenna processing providing beam forming.

The main drawback of digital processing according to the right part of Fig. 11.3 is the implementation complexity, especially the need for one digital-to-analog converter per antenna element. In the case of operation at higher frequencies with a large number of closely spaced antenna elements, analog multi-antenna processing according to the left part of Fig. 11.3 will therefore be the most common case, at least in the short- and medium-term perspectives. In this case, the multi-antenna transmission will typically be limited to per-antenna phase shifts providing beam forming (see Fig. 11.4).

It should be noted that this may not be a severe limitation as operation at higher frequencies is typically more often power-limited than bandwidth-limited, making beam forming more important than, for example, high-order spatial multiplexing. The opposite is often true for lower frequency bands where the spectrum is a more sparse resource with less possibility for wide transmission bandwidths.

Analog processing typically also implies that any beam forming is carried out on a per-carrier basis. For the downlink transmission direction, this implies that it is not possible to frequency multiplex beam-formed transmissions to devices located in different directions relative to the base station. In other words, beam-formed transmissions to different devices located in different directions must be separated in time as illustrated in Fig. 11.5.

In other cases, especially in the case of a smaller number of antenna elements at lower frequencies, multi-antenna processing can be applied in the digital domain according to the right part of Fig. 11.3. This enables much higher

FIGURE 11.5

Time-domain (non-simultaneous) beam forming in multiple directions.

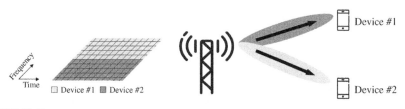

FIGURE 11.6

Simultaneous (frequency-multiplexed) beam forming in multiple directions.

flexibility in the multi-antenna processing with a possibility for high-order spatial multiplexing and with the transmission matrix W being a general $N_T \times N_L$ matrix where each element may include both a phase shift and a scale factor. Digital processing also allows for independent multi-antenna processing for different signals within the same carrier, enabling simultaneous beam-formed transmission to multiple devices located in different directions relative to the base station also by means of frequency multiplexing as illustrated Fig. 11.6.

In the case of digital processing, or more generally in the case where the antenna weights can be flexibly controlled, the transmission matrix W is often refererred to as a *precoder matrix* and the multi-antenna processing is often referred to as *multi-antenna precoding*.

The difference in capabilities between analog and digital multi-antenna processing also applies to the receiver side. In the case of analog processing, the multi-antenna processing is applied in the analog domain before analog-to-digital conversion. In practice, the multi-antenna processing is then limited to receiver-side beam forming where the receiver beam can only be directed in one direction at a time. Reception from two different directions must then take place at different time instances.

Digital implementation, on the other hand, provides full flexibility, supporting reception of multiple layers in parallel and enabling simultaneous beam-formed reception of multiple signals arriving from different directions.

Similar to the transmitter side, the drawback of digital multi-antenna processing on the receiver side is in terms of complexity, especially the need for one analog-to-digital converter per antenna element.

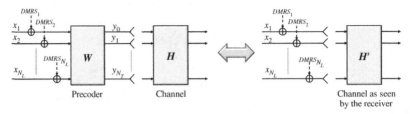

FIGURE 11.7

DMRS precoded jointly with data implying that any multi-antenna precoding is transparent to the receiver.

For the remainder of this chapter we will focus on multi-antenna precoding, that is, multi-antenna transmission with full control over the precoder matrix. The limitations of analog processing and how those imitations are impacting the NR design are discussed in Chapter 12.

One important aspect of multi-antenna precoding is whether or not the precoding is also applied to the demodulation reference signals (DMRSs) used to support coherent demodulation of the precoded signal.

If the DMRSs are not precoded, the receiver needs to be informed about what precoder is used at the transmitter side to enable proper coherent demodulation of the precoded data transmission.

On the other hand, if the reference signals are precoded together with the data, the precoding can, from a receiver point of view, be seen as part of the overall multidimensional channel (see Fig. 11.7). Simply speaking, instead of the "true" $N_R \times N_T$ channel matrix H, the receiver will see a channel H' of size $N_R \times N_L$ that is the concatenation of the channel H with whatever precoding W is applied at the transmitter side. The precoding is thus transparent to the receiver implying that the transmitter can, at least in principle, select an arbitrary precoder matrix and does not need to inform the receiver about the selected precoder.

11.2 DOWNLINK MULTI-ANTENNA PRECODING

All NR downlink physical channels rely on channel-specific DMRSs to support coherent demodulation. Furthermore, a device can assume that the DMRSs are jointly precoded with the data in line with Fig. 11.7. Consequently, any downlink multi-antenna precoding is transparent to the device and the network can, in principle, apply any transmitter-side precoding with no need to inform the device what precoding is applied.[2]

---

[2]The device must still know the number of transmission layers, that is, the number of columns in the precoder matrix applied at the network side.

The specification impact of downlink multi-antenna precoding is therefore mainly related to the measurements and reporting done by the device to support network selection of precoder for downlink PDSCH transmission. These precoder-related measurements and reporting are part of the more general CSI reporting framework based on report configurations as described in Section 8.2. As described there, a CSI report may consist of one or several of the following quantities:

- A *Rank Indicator* (RI), indicating what the device believes is a suitable transmission rank, that is, a suitable number of transmission layers N_L for the downlink transmission;
- A *Precoder-Matrix Indicator* (PMI), indicating what the device believes is a suitable precoder matrix, given the selected rank;
- A *Channel-Quality Indicator* (CQI), in practice indicating what the device believes is a suitable channel-coding rate and modulation scheme, given the selected precoder matrix.

As mentioned above, the PMI reported by a device indicates what the device believes is a suitable precoder matrix to use for downlink transmission to the device. Each possible value of the PMI thus corresponds to one specific precoder matrix. The set of possible PMI values thus corresponds to a set of different precoder matrices, referred to as the *precoder codebook*, that the device can select between when reporting PMI. Note that the device selects PMI based on a certain number of antenna ports N_T, given by the number of antenna ports of the configured CSI-RS associated with the report configuration, and the selected rank N_L. There is thus at least one codebook for each valid combination of N_T and N_L.

It is important to understand that the precoder codebooks for downlink multi-antenna precoding are only used in the context of PMI reporting and do not impose any restrictions on what precoder matrix is eventually used by the network for downlink transmission to the reporting device. The network can use whatever precoder it wants and the precoder selected by the network does not have to be part of any defined codebook.

In many cases it obviously makes sense for the network to use the precoder indicated by the reported PMI. However, in other cases the network may have additional input that speaks in favor of a different precoder. As an example, multi-antenna precoding can be used to enable simultaneous downlink transmission to multiple devices using the same time/frequency resources, so-called *multi-user MIMO* (MU-MIMO). The basic principle of MU-MIMO based on multi-antenna precoding is to choose precoding matrices that not only focus the energy towards the target device but also limit interference to other simultaneously scheduled devices. In this case, the selection of precoding for transmission to a specific device should not only take into account the PMI reported by that device (which only reflects the channel experienced by that device). Rather, the selection of precoding for transmission to a specific device should, in the general case, take into account the PMI reported by all simultaneously scheduled devices.

To conclude on suitable precoding in the MU-MIMO scenario typically also requires more detailed knowledge of the channel experienced by each device, compared to precoding in the case of transmission to a single device. For this reason, NR defines two types of CSI that differ in the structure and size of the precoder codebooks, *Type I CSI* and *Type II CSI*.

- Type I CSI primarily targets scenarios where a single user is scheduled within a given time/frequency resource (no MU-MIMO), potentially with transmission of a relatively large number of layers in parallel (high-order spatial multiplexing);
- Type II CSI primarily targets MU-MIMO scenarios with multiple devices being scheduled simultaneously within the same time/frequency resource but with only a limited number of spatial layers (maximum of two layers) per scheduled device.

The codebooks for Type I CSI are relatively simple and primarily aim at focusing the transmitted energy at the target receiver. Interference between the potentially large number of parallel layers is assumed to be handled primarily by means of receiver processing utilizing multiple receive antennas.

The codebooks for Type II CSI are significantly more extensive, allowing for the PMI to provide channel information with much higher spatial granularity. The more extensive channel information allows the network to select a downlink precoder that not only focuses the transmitted energy at the target device but also limits the interference to other devices scheduled in parallel on the same time/frequency resource. The higher spatial granularity of the PMI feedback comes at the cost of significantly higher signaling overhead. While a PMI report for Type I CSI will consist of at most a few tens of bits, a PMI report for Type II CSI may consist of several hundred bits. Type II CSI is therefore primarily applicable for low-mobility scenarios where the feedback periodicity in time can be reduced.

Below we will give an overview of the different types of CSI. For a more detailed description, for example, see [64].

11.2.1 TYPE I CSI

There are two subtypes of Type I CSI, referred to as *Type I single-panel CSI* and *Type I multi-panel CSI*, corresponding to different codebooks. As the names suggest, the codebooks have been designed assuming different antenna configurations on the network/transmitter side.

Note that an assumption of a specific antenna configuration when designing a codebook does not mean that the codebook cannot be used in deployments based on a different antenna configuration. When a device, based on downlink measurements, selects a precoder matrix from a codebook, it does not make any assumptions regarding the antenna configuration at the network side but simply selects what it believes is the most suitable precoder in the codebook, given the estimated channel conditions.

11.2.1.1 Single-Panel CSI

As the name suggests, the codebooks for Type I single-panel CSI are designed assuming a single antenna panel with $N_1 \times N_2$ cross-polarized antenna elements. An example is illustrated in Fig. 11.8 for the case of $(N_1, N_2) = (4, 2)$, that is, a 16-port antenna.[3]

In general, the precoder matrices W in the codebooks for Type I single-panel CSI can be expressed as the product of two matrices W_1 and W_2 with information about the selected W_1 and W_2 reported separately as different parts of the overall PMI.

The matrix W_1 is assumed to capture long-term frequency-independent characteristics of the channel. A single W_1 is therefore selected and reported for the entire reporting bandwidth (wideband feedback).

In contrast, the matrix W_2 is assumed to capture more short-term and potentially frequency-dependent characteristics of the channel. The matrix can therefore be selected and reported on a subband basis where a subband covers a fraction of the overall reporting bandwidth. Alternatively, the device may not report W_2 at all, in which case the device, when subsequently selecting CQI, should assume that the network randomly selects W_2 on a per PRG (Physical Resource Block Group, see Section 9.8) basis. Note that this does not impose any restrictions on the actual precoding applied at the network side but is only about assumptions made by the device when selecting CQI.

On a high level, the matrix W_1 can be seen as defining a beam or, in some cases a set of neighbor beams, pointing in a specific direction. More specifically, the matrix W_1 can be written as

$$W_1 = \begin{bmatrix} B & 0 \\ 0 & B \end{bmatrix}$$

where each column of the matrix B defines a beam and the 2×2 block structure is due to the two polarizations. Note that, as the matrix W_1 is assumed to only capture long-term frequency-independent channel characteristics, the same beam direction can be assumed to fit both polarization directions.

FIGURE 11.8

Example of assumed antenna structure for Type I single-panel CSI with $(N_1, N_2) = (4, 2)$.

---

[3]Note that there are two antenna ports per cross-polarized antenna element.

Selecting the matrix W_1 or, equivalently, B can thus be seen as selecting a specific beam direction from a large set of possible beam directions defined by the full set of W_1 matrices within the codebook.[4]

In the case of rank 1 or rank 2 transmission, either a single beam or four neighbor beams are defined by the matrix W_1. In the case of four neighboring beams, corresponding to four columns in B, the matrix W_2 then selects the exact beam to be used for the transmission. As W_2 can be reported on a subband basis, it is thus possible to fine-tune the beam direction per subband. In addition, W_2 provides co-phasing between the two polarizations. In the case when W_1 only defines a single beam, corresponding to B being a single-coumn matrix, the matrix W_2 only provides co-phasing between the two polarizations.

For transmission ranks R larger than 2, the matrix W_1 defines N orthogonal beams where $N = \lceil R/2 \rceil$. The N beams, together with the two polarization directions in each beam are then used for transmission of the R layers, with the matrix W_2 only providing cophasing between the two polarizations. Up to eight layers can be transmitted to the same device.

11.2.1.2 Multipanel CSI

In contrast to single-panel CSI, codebooks for Type I multi-panel CSI are designed assuming the joint use of multiple antenna panels at the network side and takes into account that it may be difficult to ensure coherence between transmissions from different panels. More specifically, the design of the multi-panel codebooks assumes an antenna configuration with two or four two-dimensional panels, each with $N_1 \times N_2$ cross-polarized antenna elements. An example of such a multi-panel antenna configuration is illustrated in Fig. 11.9 for the case of four antenna panels and $(N_1, N_2) = (4, 1)$, that is, a 32-port antenna.

The basic principle of Type 1 multi-panel CSI is the same as that of Type 1 single-panel CSI, except that the matrix W_1 defines one beam per polarization

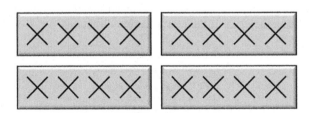

FIGURE 11.9

Example of assumed antenna structure for Type I multi-panel CSI; 32-port antenna with four antenna panels and $(N_1, N_2) = (4, 1)$.

---

[4]Note that even if the matrix W_1 defines a set of beams, these beams are neighbor beams that point in *essentially* the same direction.

and panel. The matrix W_2 then provides per-subband co-phasing between polarizations as well as panels.

The Type I multi-panel CSI supports spatial multiplexing with up to four layers.

11.2.2 **TYPE II CSI**

As already mentioned, Type II CSI provides channel information with significantly higher spatial granularity compared to Type I CSI. Similar to Type I CSI, Type II CSI is based on wideband selection and reporting of beams from a large set of beams. However, while Type I CSI in the end selects and reports a single beam, Type II CSI may select and report up to four orthogonal beams. For each selected beam and each of the two polarizations, the reported PMI then provides an amplitude value (partly wideband and partly subband) and a phase value (subband). In the end, this provides a much more detailed model of the channel, capturing the main rays and their respective amplitude and phase.

At the network side, the PMI delivered from multiple devices can then be used to identify a set of devices to which transmission can be done simultaneously on a set of time/frequency resources (MU-MIMO) and what precoder to use for each transmission.

As the Type II CSI is targeting the MU-MIMO scenario, transmission is limited to up to two layers per device.

11.3 **NR UPLINK MULTIANTENNA PRECODING**

NR support uplink (PUSCH) multi-antenna precoding with up to four layers. However, as mentioned earlier, in the case of DFT-based transform precoding (see Chapter 9), only single-layer transmission is supported.

The device can be configured in two different modes for PUSCH multi-antenna precoding, referred to as *codebook-based* transmission and *non-codebook-based* transmission, respectively. The selection between these two transmission modes is at least partly dependent on what can be assumed in terms of uplink/downlink channel reciprocity, that is, to what extent it can be assumed that the detailed uplink channel conditions can be estimated by the device based on downlink measurements.

Like the downlink, any uplink (PUSCH) multi-antenna precoding is also assumed to be applied to the DMRS used for the PUSCH coherent demodulation. Similar to the downlink transmission direction, uplink precoding is thus transparent to the receiver in the sense that receiver-side demodulation can be carried out without knowledge of the exact precoding applied at the transmitter (device) side. Note though that this does not necessarily imply that the device can freely choose the PUSCH precoder. In the case of codebook-based precoding, the scheduling grant includes information about a precoder, similar to the device providing the network with PMI for downlink multiantenna precoding. However, in contrast to the downlink, where the network may or may not use the precoder matrix indicated by the PMI, in the uplink direction the device is assumed to use the precoder provided by the network. As we will see in Section 11.3.2, also in the case of non-codebook-based transmission will the network have an influence on the final choice of uplink precoder.

Another aspect that may put constraints on uplink multi-antenna transmission is to what extent one can assume coherence between different device antennas, that is, to what extent the relative phase between the signals transmitted on two antennas can be well controlled. Coherence is needed in the case of general multi-antenna precoding where antenna-port-specific weight factors, including specific phase shifts, are applied to the signals transmitted on the different antenna ports. Without coherence between the antenna ports the use of such antenna-port-specific weight factors is obviously meaningless as each antenna port would anyway introduce a more or less random relative phase.

The NR specification allows for different device capabilities with regards to such inter-antenna-port coherence, referred to as *full coherence, partial coherence*, and *no coherence*, respectively.

In the case of full coherence, it can be assumed that the device can control the relative phase between any of the up to four ports that are to be used for transmission.

In the case of partial coherence, the device is capable of *pairwise* coherence, that is, the device can control the relative phase within pairs of ports. However, there is no guarantee of coherence, that is, a controllable phase, between the pairs.

Finally, in the case of no coherence there is no guarantee of coherence between any pair of the device antenna ports.

11.3.1 CODEBOOK-BASED TRANSMISSION

The basic principle of codebook-based transmission is that the network decides on an uplink transmission rank, that is, the number of layers to be transmitted, and a corresponding precoder matrix to use for the transmission. The network informs the device about the selected transmission rank and precoder matrix as part of the uplink scheduling grant. At the device side, the precoder matrix is then applied for the scheduled PUSCH transmission, mapping the indicated number of layers to the antenna ports.

To select a suitable rank and a corresponding precoder matrix, the network needs estimates of the channels between the device antenna ports and the corresponding network receive antennas. To enable this, a device configured for codebook-based PUSCH would typically be configured for transmission of at least one multi-port SRS. Based on measurements on the configured SRS, the network can sound the channel and determine a suitable rank and precoder matrix.

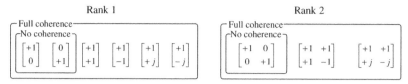

FIGURE 11.10

Uplink codebooks for the case of two antenna ports.

The network cannot select an arbitrary precoder. Rather, for a given combination of number of antenna ports N_T ($N_T = 2$ or $N_T = 4$) and transmission rank N_L ($N_L \leq N_T$), the network selects the precoder matrix from a limited set of available precoders (the "uplink codebook").

As an example, Fig. 11.10 illustrates the available precoder matrices, that is, the code books for the case of two antenna ports.

When selecting the precoder matrix, the network needs to consider the device capability in terms of antenna-port coherence (see above). For devices not supporting coherence, only the first two precoder matrixes can therefore be used in the case of single-rank transmission.

It can be noted that restricting the codebook selection to these two matrices is equivalent to selecting either the first or second antenna port for transmission. In the case of such *antenna selection*, a well-controlled phase, that is, coherence between the antenna ports, is not required. On the other hand, the remaining precoder vectors imply linear combination of the signals on the different antenna ports, which requires coherence between the antenna ports.

In the case of rank-2 transmission ($N_L = 2$) only the first matrix, which does not imply any coupling between the antenna ports, can be selected for devices that do not support coherence.

To further illustrate the impact of no, partial, and full coherence, Fig. 11.11 illustrates the full set of rank-1 precoder matrices for the case of four antenna ports. Once again, the matrices corresponding to no coherence are limited to antenna-port selection. The extended set of matrices corresponding to partial coherence allows for linear combination within pairs of antenna ports with selection between the pairs. Finally, full coherence allows for a linear combination over all four antenna ports.

The above-described NR codebook-based transmission for PUSCH is essentially the same as the corresponding codebook-based transmission for LTE except that NR supports somewhat more extensive codebooks. Another more fundamental extension of NR codebook-based PUSCH transmission, compared to LTE, is that a device can be configured to transmit *multiple* multi-port SRS.[5] In the case of such *multi-SRS transmission*, the network feedback is extended with a one-bit

---

[5]Release 15 is limited to two SRSs.

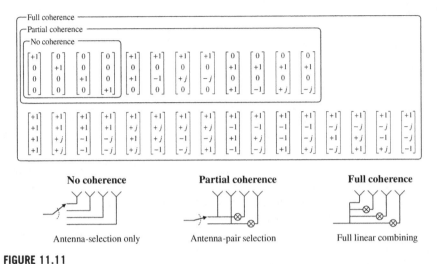

FIGURE 11.11

Single-layer uplink codebooks for the case of four antenna ports.

SRS resource indicator (SRI) indicating one of the configured SRSs. The device should then use the precoder provided in the scheduling grant and map the output of the precoding to the antenna ports corresponding to the SRS indicated in the SRI. In terms of the spatial filter F discussed in Chapter 8, the different SRSs would typically be transmitted using different spatial filters. The device should then transmit the precoder signal using the same spatial filter as used for the SRS indicated by the SRI.

One way to visualize the use of multiple SRS for codebook-based PUSCH transmission is to assume that the device transmits the multi-port SRS within separate, relatively large beams (see Fig. 11.12). These beams may, for example, correspond to different device antenna panels with different directions, where each panel includes a set of antenna elements, corresponding to the antenna ports of each multi-port SRS. The SRI received from the network then determines what beam to use for the transmission while the precoder information (number of layers and precoder) determines how the transmission is to be done within the selected beam. As an example, in the case of full-rank transmission the device will do full-rank transmission within the beam corresponding to the SRS selected by the network and signaled by means of SRI (upper part of Fig. 11.12). At the other extreme, in the case of single-rank transmission the precoding will in practice create additional beam forming within the wider beam indicated by the SRI (lower part of Fig. 11.12).

Codebook-based precoding is typically used when uplink/downlink reciprocity does not hold, that is, when uplink measurements are needed in order to determine a suitable uplink precoding.

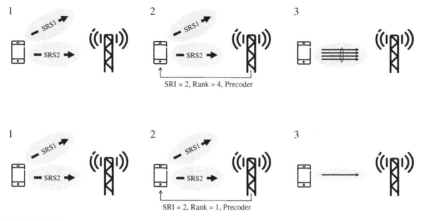

FIGURE 11.12

Codebook-based transmission based on multiple SRS. Full-rank transmission (upper part) and single-rank transmission (lower part).

11.3.2 NON-CODEBOOK-BASED PRECODING

In contrast to codebook-based precoding, which is based on network measurements and selection of uplink precoder, non-codebook-based precoding is based on device measurements and precoder indications to the network. The basic principle of uplink non-codebook-based precoding is illustrated in Fig. 11.13, with further explanation below.

Based on downlink measurements, in practice measurements on a configured CSI-RS, the device selects what it believes is a suitable uplink multi-layer precoder. Non-codebook-based precoding is thus based on an assumption of channel reciprocity, that is, that the device can acquire detailed knowledge of the uplink channel based on downlink measurements. Note that there are no restrictions on the device selection of precoder, thus the term "non-codebook-based."

Each column of a precoder matrix W can be seen as defining a digital "beam" for the corresponding layer. The device selection of precoder for N_L layers can thus be seen as the selection of N_L different beam directions where each beam corresponds to one possible layer.

In principle, PUSCH transmission could be done directly as transmission of N_L layers based on the device-selected precoding. However, device selection of a precoder based on downlink measurements may not necessarily be the best precoder from a network point of view. Thus, the NR non-codebook-based precoding includes an additional step where the network can modify the device-selected precoder, in practice remove some "beams," or equivalently some columns, from the selected precoder.

To enable this, the device applies the selected precoder to a set of configured SRSs, with one SRS transmitted on each layer or "beam" defined by the precoder

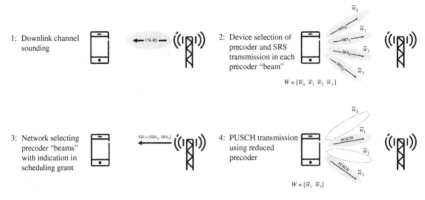

FIGURE 11.13

Non-codebook-based precoding.

(step 2 in Fig. 11.13). Based on measurements on the received SRS, the network can then decide to modify the device-selected precoder for each scheduled PUSCH transmission. This is done by indicating a subset of the configured SRSs within the SRS resource indicator (SRI) included in the scheduling grant (step 3).[6] The device then carries out the scheduled PUSCH transmission (step 4) using a reduced precoder matrix where only the columns corresponding to the SRSs indicated within the SRI are included. Note that the SRI then also implicitly defines the number of layers to be transmitted.

It should be noted that the device indication of precoder selection (step 2 in Fig. 11.13) is not done for each scheduled transmission. The uplink SRS transmission indicating device precoder selection can take place periodically (periodic or semipersistent SRS) or on demand (aperiodic SRS). In contrast, the network indication of precoder, that is in practice the network indication of the subset of beams of the device precoder, is then done for each scheduled PUSCH transmission.

---

[6]For a device configured for non-codebook-based precoding the SRI may thus indicate multiple SRSs, rather than a single SRS which is the case for codebook-based precoding (see Section 11.3.1).

Beam Management

Chapter 11 discussed multi-antenna transmission in general and then focused on multi-antenna precoding. A general assumption for the discussion on multi-antenna precoding was the possibility for detailed control, including both phase adjustment and amplitude scaling, of the different antenna elements. In practice this requires that multi-antenna processing at the transmitter side is carried out in the digital domain before digital-to-analog conversion. Likewise, the receiver multi-antenna processing must be carried out *after* analog-to-digital conversion.

However, in the case of operation at higher frequencies with a large number of closely space antenna elements, the antenna processing will rather be carried out in the analog domain with focus on beam-forming. As analog antenna processing will be carried out on a carrier basis, this also implies that beam-formed transmission can only be done in one direction at a time. Downlink transmissions to different devices located in different directions relative to the base station must therefore be separated in time. Likewise, in the case of analog-based receiver-side beam-forming, the receive beam can only focus in one direction at a time.

The ultimate task of beam management is, under these conditions, to establish and retain a suitable *beam pair*, that is, a transmitter-side beam direction and a corresponding receiver-side beam direction that jointly provide good connectivity.

As illustrated in Fig. 12.1, the best beam pair may not necessarily correspond to transmitter and receiver beams that are physically pointing directly towards each other. Due to obstacles in the surrounding environment, such a "direct" path between the transmitter and receiver may be blocked and a reflected path may provide better connectivity, as illustrated in the right-hand part of Fig. 12.1. This is especially true for operation in higher-frequency bands with less "around-the-corner" dispersion. The beam-management functionality must be able to handle such a situation and establish and retain a suitable beam pair also in this case.

Fig. 12.1 illustrates the case of beam forming in the downlink direction, with beam-based transmission at the network side and beam-based reception at the device side. However, beam forming is at least as relevant for the uplink transmission direction with beam-based transmission at the device side and corresponding beam-based reception at the network side.

In many cases, a suitable transmitter/receiver beam pair for the downlink transmission direction will also be a suitable beam pair for the uplink transmission direction and vice versa. In 3GPP this is referred to as (downlink/uplink) *beam correspondence*. In the case of beam correspondence, it is sufficient to

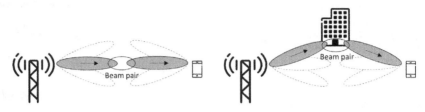

FIGURE 12.1

Illustration of beam pairs in the downlink direction. Direct (left) and via reflection (right).

explicitly determine a suitable beam pair in one of the transmission directions. The same pair can then be used also in the opposite transmission direction.

As beam management is not intended to track fast and frequency-selective channel variations, beam correspondence does not require that downlink and uplink transmission take place on the same carrier frequency. The concept of beam correspondence is thus applicable also for FDD operation in paired spectrum.

In general, beam management can be divided into different parts:

- Initial *beam establishment*;
- *Beam adjustment*, primarily to compensate for movements and rotations of the mobile device, but also for gradual changes in the environment;
- *Beam recovery* to handle the situation when rapid changes in the environment disrupt the current beam pair.

12.1 INITIAL BEAM ESTABLISHMENT

Initial beam establishment includes the procedures and functions by which a beam pair is initially established in the downlink and uplink transmission directions, for example, when a connection is established. As will be described in more detail in Chapter 16, during initial cell search a device will acquire a so-called *SS block* transmitted from a cell, with the possibility for multiple SS blocks being transmitted in sequence within different downlink beams. By associating each such SS block, in practice the different downlink beams, with a corresponding random-access occasion and preamble (see Section 16.2.1.5), the subsequent uplink random-access transmission can be used by the network to identify the downlink beam acquired by the device, thereby establishing an initial beam pair.

When communication continues after connection set up the device can assume that network transmissions to the device will be done using the same spatial filter, in practice the same transmitter beam, as used for the acquired SS block. Consequently, the device can assume that the receiver beam used to acquire the SS block will be a suitable beam also for the reception of subsequent downlink transmissions. Likewise, subsequent uplink transmissions should be done using the same spatial filter (the same beam) as used for the random-access transmission, implying that the network can assume that the uplink receiver beam established at initial access will remain valid.

12.2 BEAM ADJUSTMENT

Once an initial beam pair has been established, there is a need to regularly reevaluate the selection of transmitter-side and receiver-side beam directions due to movements and rotations of the mobile device. Furthermore, even for stationary devices, movements of other objects in the environment may block or unblock different beam pairs, implying a possible need to reevaluate the selected beam directions. This *beam adjustment* may also include refining the beam shape, for example making the beam more narrow compared to a relatively wider beam used for initial beam establishment.

In the general case, beam-forming is about beam pairs consisting of transmitter-side beam-forming and receiver-side beam-forming. Hence, beam adjustment can be divided into two separate procedures:

- Reevaluation and possible adjustment of the transmitter-side beam direction given the current receiver-side beam direction;
- Reevaluation and possible adjustment of the receiver-side beam direction given the current transmitter-side beam direction.

As described above, in the general case beam forming, including beam adjustment, needs to be carried out for both the downlink and uplink transmission directions. However, as also discussed, if downlink/uplink beam correspondence can be assumed, explicit beam adjustment only has to be carried out in one of the directions, for example, in the downlink direction. It can then be assumed that the adjusted downlink beam pair is appropriate also for the opposite transmission direction.

12.2.1 DOWNLINK TRANSMITTER-SIDE BEAM ADJUSTMENT

Downlink transmitter-side beam adjustment aims at refining the network transmit beam, given the receiver beam currently used at the device side. To enable this, the device can measure on a set of reference signals, corresponding to different downlink beams (see Fig. 12.2). Assuming analog beam forming, transmissions within the different downlink beams must be done in sequence, that is, by means of a beam sweep.

The result of the measurements is then reported to the network which, based on the reporting, may decide to adjust the current beam. Note that this adjustment may not necessarily imply the selection of one of the beams that the device has measured on. The network could, for example, decide to transmit using a beam direction between two of the reported beams.

Also note that, during measurements done for transmitter-side beam adjustment, the device receiver beam should be kept fixed in order for the measurements to capture the quality of the different transmitter beams *given the current receive beam*.

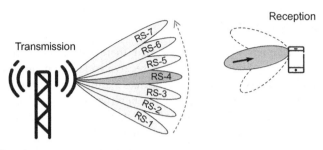

FIGURE 12.2

Downlink transmitter-side beam adjustment.

To enable measurements and reporting on a set of beams as outlined in Fig. 12.2, the reporting framework based on report configurations (see Section 8.2) can be used. More specifically, the measurement/reporting should be described by a report configuration having L1-RSRP as the quantity to be reported.

The set of reference signals to measure on, corresponding to the set of beams, should be included in the NZP-CSI-RS resource set associated with the report configuration. As described in Section 8.1.6, such a resource set may either include a set of configured CSI-RS or a set of SS blocks. Measurements for beam management can thus be carried out on either CSI-RS or SS block. In the case of L1-RSRP measurements based on CSI-RS, the CSI-RS should be limited to single-port or dual-port CSI-RS. In the latter case, the reported L1-RSRP should be a linear average of the L1-RSRP measured on each port.

The device can report measurements corresponding to up to four reference signals (CSI-RS or SS blocks), in practice up to four beams, in a single reporting instance. Each such report would include:

- Indications of the up to four reference signals, in practice beams, that this specific report relates to;
- The measured L1-RSRP for the strongest beam;
- For the remaining up to three beams: The difference between the measured L1-RSRP and the measured L1-RSRP of the best beam.

12.2.2 DOWNLINK RECEIVER-SIDE BEAM ADJUSTMENT

Receiver-side beam adjustment aims at finding the best receive beam, given the current transmit beam. To enable this, the device should once again be configured with a set of downlink reference signals that, in this case, are transmitted within *the same* network-side beam (the current serving beam). As outlined in Fig. 12.3, the device can then do a receiver-side beam sweep to measure on the configured reference signals in sequence over a set of receiver beams. Based on these measurements the device may adjust its current receiver beam.

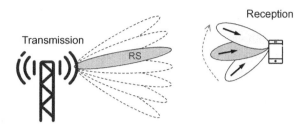

FIGURE 12.3

Downlink receiver-side beam adjustment.

Downlink receiver-side beam adjustment can be based on similar report con-figurations as for transmitter-side beam adjustment. However, as the receiver-side beam adjustment is done internally within the device, there is no report quantity associated with receiver-side beam adjustment. According to Section 8.2, the report quantity should thus be set to "None."

To allow for analog beam-forming at the receiver side, the different reference signals within the resource set should be transmitted in different symbols, allow-ing for the receiver-side beam to sweep over the set of reference signals. At the same time, the device should be allowed to assume that the different reference signals in the resource set are transmitted using the same spatial filter, in practice the same transmit beam. In general, a configured resource set includes a *"repeti-tion"* flag that indicates whether or not a device can assume that all reference sig-nals within the resource set are transmitted using the same spatial filter. For a resource set to be used for downlink receiver side beam adjustment, the repetition flag should thus be set.

12.2.3 UPLINK BEAM ADJUSTMENT

Uplink beam adjustment serves the same purpose as downlink beam adjustment, that is, to retain a suitable beam pair which, in the case of uplink beam adjust-ment, implies a suitable transmitter beam at the device side and a corresponding suitable receiver beam at the network side.

As discussed above, if beam correspondence can be assumed and if a suitable downlink beam pair has been established and retained, explicit uplink beam management is not needed. Rather, a suitable beam pair for the downlink transmis-sion direction can be assumed to be suitable also for the uplink direction. Note that the opposite would also be true, that is, if a suitable beam pair is established and retained for the uplink direction, the same beam pair could also be used in the down-link direction without the need for explicit downlink beam management.

If explicit uplink beam adjustment is needed it can be done in essentially the same way as for downlink beam adjustment with the main difference being that measurements are done by the network based on configured SRS, rather than CSI-RS or SS block.

12.2.4 BEAM INDICATION AND TCI

Downlink beam-forming can be done transparent to the device, that is, the device does not need to know what beam is used at the transmitter.

However, NR also supports beam indication. In practice this implies informing the device that a certain PDSCH and/or PDCCH transmission uses the same transmission beam as a configured reference signal (CSI-RS or SS block). More formally, it implies informing the device that a certain PDSCH and/or PDCCH is transmitted using the same spatial filter as the configured reference signal.

In more detail, beam indication is based on the configuration and downlink signaling of so-called *Transmission Configuration Indication* (TCI) states. Each TCI state includes, among other things, information about a reference signal (a CSI-RS or an SS block). By associating a certain downlink transmission (PDCCH or PDSCH) with a certain TCI, the network informs the device that it can assume that the downlink transmission is done using the same spatial filter as the reference signal associated with that TCI.

A device can be configured with up to 64 *candidate TCI states*. For beam indication for PDCCH, a subset of the M configured candidate states is assigned by RRC signaling to each configured CORESET. By means of MAC signaling, the network can then more dynamically indicate a specific TCI state, within the per-CORESET-configured subset, to be valid. When monitoring for PDCCH within a certain CORESET, the device can assume that the PDCCH transmission uses the same spatial filter as the reference signal associated with the MAC-indicated TCI. In other words, if the device has earlier determined a suitable receiver-side beam direction for reception of the reference signal, the device can assume that the same beam direction is suitable for reception of the PDCCH.

For PDSCH beam indication, there are two alternatives depending on the scheduling offset, that is, depending on the transmission timing of the PDSCH relative to the corresponding PDCCH carrying scheduling information for the PDSCH.

If this scheduling offset is larger than N symbols, the DCI of the scheduling assignment may explicitly indicate the TCI state for the PDSCH transmission.[1] To enable this, the device is first configured with a set of up to eight TCI states from the originally configured set of candidate TCI states. A three-bit indicator within the DCI then indicates the exact TCI state valid for the scheduled PDSCH transmission.

If the scheduling offset is smaller or equal to N symbols, the device should instead assume that the PDSCH transmission is QCL with the corresponding PDCCH transmission. In other words, the TCI state for the PDCCH state indicated by MAC signaling should be assumed to be valid also for the corresponding scheduled PDSCH transmission.

The reason for limiting the fully dynamic TCI selection based on DCI signaling to situations when the scheduling offset is larger than a certain value is simply that, for shorter scheduling offsets, there will not be sufficient time for the device to decode the TCI information within the DCI and adjust the receiver beam accordingly before the PDSCH is to be received.

---

[1]The exact value of N is still under discussion in 3GPP.

12.3 BEAM RECOVERY

In some cases, movements in the environment or other events, may lead to a currently established beam pair being rapidly blocked without sufficient time for the regular beam adjustment to adapt. The NR specification includes specific procedures to handle such *beam-failure* events, also referred to as *beam (failure) recovery.*

In many respects, beam failure is similar to the concept of *radio-link failure* (RLF) already defined for current radio-access technologies such as LTE and one could in principle utilize already-established RLF-recovery procedures to recover also from beam-failure events. However, there are reasons to introduce additional procedures specifically targeting beam failure.

- Especially in the case of narrow beams, beam failure, that is, loss of connectivity due to a rapid degradation of established beam pairs, can be expected to occur more frequently compared to RLF, which typically corresponds to a device moving out of coverage from the currently serving cell;
- RLF typically implies loss of coverage to the currently serving cell in which case connectivity must be re-established to a new cell, perhaps even on a new carrier. After beam failure, connectivity can often be reestablished by means of a new beam pair within the current cell. As a consequence, recovery from beam failure can often be achieved by means of lower-layer functionality, allowing for faster recovery compared to the higher-layer mechanisms used to recover from RLF.

In general, beam failure/recovery consists of the following steps:

- *Beam-failure detection*, that is, the device detecting that a beam failure has occurred;
- *Candidate-beam identification*, that is, the device trying to identify a new beam or, more exactly, a new beam pair by means of which connectivity may be restored;
- *Recovery-request transmission*, that is, the device transmitting a beam-recovery request to the network;
- Network response to the beam-recovery request.

12.3.1 BEAM-FAILURE DETECTION

Fundamentally, a beam failure is assumed to have happened when the error probability for the downlink control channel (PDCCH) exceeds a certain value. However, similar to radio-link failure, rather than actually measuring the PDCCH error probability, the device declares a beam failure based on measurements of the quality of some reference signal. This is often expressed as measuring a *hypothetical error rate*. More specifically, the device should declare beam failure based on measured L1-RSRP of a periodic CSI-RS or an SS block that is spatially QCL with the PDCCH. By default, the device should declare beam failure based on measurement on the reference signal (CSI-RS or SS block) associated with the PDCCH TCI state. However, there is also a possibility to explicitly configure a different CSI-RS on which to measure for beam-failure detection.

Each time instant the measured L1-RSRP is below a configured value is defined as a *beam-failure instance*. If the number of consecutive beam-failure instances exceeds a configured value, the device declares a beam failure and initiates the *beam-failure-recovery* procedure.

12.3.2 NEW-CANDIDATE-BEAM IDENTIFICATION

As a first step of the beam-recovery procedure, the device tries to find a new beam pair on which connectivity can be restored. To enable this, the device is configured with a resource set consisting of a set of CSI-RS, or alternatively a set of SS blocks. In practice, each of these reference signals is transmitted within a specific downlink beam. The resource set thus corresponds to a set of *candidate beams*.

Similar to normal beam establishment, the device measures the L1-RSRP on the reference signals corresponding to the set of candidate beams. If the L1-RSRP exceeds a certain configured target, the reference signal is assumed to correspond to a beam by means of which connectivity may be restored. It should be noted that, when doing this, the device has to consider different receiver-side beam directions when applicable, that is, what the device determines is, in practice, a candidate beam pair.

12.3.3 DEVICE RECOVERY REQUEST AND NETWORK RESPONSE

If a beam failure has been declared and a new candidate beam pair has been identified, the device carries out a *beam-recovery request*. The aim of the recovery request is to inform the network that the device has detected a beam failure. The recovery request may also include information about the candidate beam identified by the device.

The beam-recovery request is in essence a two-step contention-free random-access request consisting of preamble transmission and random-access response.[2] Each reference signal corresponding to the different candidate beams is associated with a specific preamble configuration (RACH occasion and preamble sequence, see Chapter 16). Given the identified beam, the preamble transmission should be carried out using the associated preamble configuration. Furthermore, the preamble should be transmitted within the uplink beam that coincides with the identified downlink beam.

---

[2]See Section 16.2 for more details on the NR random-access procedure including preamble structure.

It should be noted that each candidate beam may not necessarily be associated with a unique preamble configuration. There are different alternatives:

- Each candidate beam is associated with a unique preamble configuration. In this case, the network can directly identify the identified downlink beam from the received preamble;
- The candidate beams are divided into groups where all beams within the same group correspond to the same preamble configuration, while beams of different groups correspond to different preamble configurations. In this case, the received preamble only indicates the group to which the identified downlink beam belongs;
- All candidate beams are associated with the same preamble configuration. In this case, the preamble reception only indicates that beam failure has occurred and that the device requests a beam-failure recovery.

Under the assumption that the candidate beams are originating from the same site it can also be assumed that the random-access transmission is well time-aligned when arriving at the receiver. However, there may be substantial differences in the overall path loss for different candidate beam pairs. The configuration of the beam-recovery-request transmission thus includes parameters for power ramping (see Section 16.2).

Once a device has carried out a beam-recovery request it monitors downlink for a network response. When doing so, the device may assume that the network, when responding to the request, is transmitting PDCCH QCL with the RS associated with the candidate beam included in the request.

The monitoring for the recovery-request response starts four slots after the transmission of the recover request. If no response is received within a window of a configurable size, the device retransmits the recovery response according to the configured power-ramping parameters.

Retransmission Protocols

13

Transmissions over wireless channels are subject to errors, for example, due to variations in the received signal quality. To some degree, such variations can be counteracted through link adaptation as will be discussed in Chapter 14. However, receiver noise and unpredictable interference variations cannot be counteracted. Therefore, virtually all wireless communication systems employ some form of *Forward Error Correction* (FEC), adding redundancy to the transmitted signal allowing the receiver to correct errors and tracing its roots to the pioneering work of Shannon [69]. In NR, LDPC coding is used for error correction as discussed in Section 9.2.

Despite the error-correcting code, there will be data units received in error, for example, due to a too high noise or interference level. *Hybrid Automatic Repeat Request* (HARQ), first proposed by Wozencraft and Horstein [72] and relying on a combination of error-correcting coding and retransmission of erroneous data units, is therefore commonly used in many modern communication systems. Data units in error despite the error correcting coding are detected by the receiver, which requests a retransmission from the transmitter.

In NR, three different protocol layers all offer retransmission functionality—MAC, RLC, and PDCP—as already mentioned in the introductory overview in Chapter 6. The reasons for having a multilevel retransmission structure can be found in the trade-off between fast and reliable feedback of the status reports. The hybrid-ARQ mechanism in the MAC layer targets very fast retransmissions and, consequently, feedback on success or failure of the downlink transmission is provided to the gNB after each received transport block (for uplink transmission no explicit feedback needs to be transmitted as the receiver and scheduler are in the same node). Although it is in principle possible to attain a very low error probability of the hybrid-ARQ feedback, it comes at a cost in transmission resources such as power. In many cases, a feedback error rate of 0.1−1% is reasonable, which results in a hybrid-ARQ residual error rate of a similar order. In many cases this residual error rate is sufficiently low, but there are cases when this is not the case. One obvious case is services requiring ultra-reliable delivery of data combined with low latency. In such cases, either the feedback error rate needs to be decreased and the increased cost in feedback signaling has to be accepted, or additional retransmissions can be performed without relying on feedback signaling, which comes at a decreased spectral efficiency.

A low error rate is not only of interest for URLLC type of services, but is also important from a data-rate perspective. High data rates with TCP may require virtually error-free delivery of packets to the TCP layer. As an example, for sustainable data rates exceeding 100 Mbit/s, a packet-loss probability less than 10^{-5} is required [65]. The reason is that TCP assumes packet errors to be due to congestion in the network. Any packet error therefore triggers the TCP congestion-avoidance mechanism with a corresponding decrease in data rate.

Compared to the hybrid-ARQ acknowledgments, the RLC status reports are transmitted relatively infrequently and thus the cost of obtaining a reliability of 10^{-5} or lower is relatively small. Hence, the combination of hybrid-ARQ and RLC attains a good combination of small round-trip time and a modest feedback overhead where the two components complement each other—fast retransmissions due to the hybrid-ARQ mechanism and reliable packet delivery due to the RLC.

The PDCP protocol is also capable of handling retransmissions, as well as ensuring in-sequence delivery. PDCP-level retransmissions are mainly used in the case of inter-gNB handover as the lower protocols in this case are flushed. Not-yet-acknowledged PDCP PDUs can be forwarded to the new gNB and transmitted to the device. In the case that some of these were already received by the device, the PDCP duplicate detection mechanism will discard the duplicates. The PDCP protocol can also be used to obtain selection diversity by transmitting the same PDUs on multiple carriers. The PDCP in the receiving end will in this case remove any duplicates in case the same information was received successfully on multiple carriers.

In the following sections, the principles behind the hybrid-ARQ, RLC, and PDCP protocols will be discussed in more detail. Note that these protocols are present also in LTE where they to a large extent provide the same functionality. However, the NR versions are enhanced to significantly reduce the delays.

13.1 HYBRID-ARQ WITH SOFT COMBINING

The hybrid-ARQ protocol is the primary way of handling retransmissions in NR. In case of an erroneously received packet, a retransmission is requested. However, despite it not being possible to decode the packet, the received signal still contains information, which is lost by discarding erroneously received packets. This shortcoming is addressed by *hybrid-ARQ with soft combining*. In hybrid-ARQ with soft combining, the erroneously received packet is stored in a buffer memory and later combined with the retransmission to obtain a single, combined packet that is more reliable than its constituents. Decoding of the error-correction code operates on the combined signal.

Although the protocol itself primarily resides in the MAC layer, there is also physical layer functionality involved in the form of soft combining.

Retransmissions of codeblock groups, that is, retransmission of a part of the transport block, are handled by the physical layer from a specification perspective, although it could equally well have been described as part of the MAC layer.

The basis for the NR hybrid-ARQ mechanism is, similarly to LTE, a structure with multiple stop-and-wait protocols, each operating on a single transport block. In a stop-and-wait protocol, the transmitter stops and waits for an acknowledgment after each transmitted transport block. This is a simple scheme; the only feedback required is a single bit indicating positive or negative acknowledgment of the transport block. However, since the transmitter stops after each transmission, the throughput is also low. Therefore, multiple stop-and-wait processes operating in parallel are used such that, while waiting for acknowledgment from one process, the transmitter can transmit data to another hybrid-ARQ process. This is illustrated in Fig. 13.1; while processing the data received in the first hybrid-ARQ process the receiver can continue to receive using the second process, etc. This structure, multiple hybrid-ARQ processes operating in parallel to form one hybrid-ARQ entity, combines the simplicity of a stop-and-wait protocol while still allowing continuous transmission of data, and is used in LTE as well as NR.

There is one hybrid-ARQ entity per carrier the receiver is connected to. Spatial multiplexing of more than four layers to a single device in the downlink, where two transport blocks can be transmitted in parallel on the same transport channel as described in Section 9.1, is supported by one hybrid-ARQ entity having two sets of hybrid-ARQ processes with independent hybrid-ARQ acknowledgments.

NR uses an *asynchronous* hybrid-ARQ protocol in both downlink and uplink, that is, the hybrid-ARQ process which the downlink or uplink transmission relates to is explicitly signaled as part of the downlink control information (DCI). LTE uses the same scheme for the downlink but not for the uplink, where LTE uses a synchronous protocol (although later LTE releases added support for an

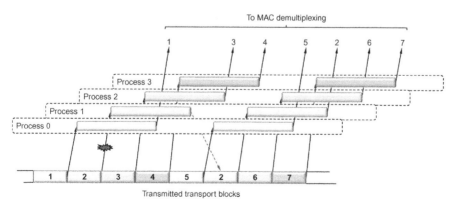

FIGURE 13.1

Multiple hybrid-ARQ processes.

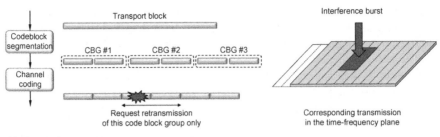

FIGURE 13.2

Codeblock-group retransmission.

asynchronous protocol as well). There are several reasons why NR adopted an asynchronous protocol in both directions. One reason is that synchronous hybrid-ARQ operation does not allow dynamic TDD. Another reason is that operation in unlicensed spectra, to be introduced in later NR releases, is more efficient with asynchronous operation as it cannot be guaranteed that the radio resources are available at the time for a synchronous retransmission. Thus, NR settled for an asynchronous scheme in both uplink and downlink with up to 16 processes. Having a larger maximum number of hybrid-ARQ processes than in LTE[1] is motivated by the possibility for remote radio heads, which incurs a certain front-haul delay, together with the shorter slot durations at high frequencies. It is important though, that the larger number of maximum hybrid-ARQ processes does not imply a longer roundtrip time as not all processes need to be used, it is only an upper limit of the number of processes possible to address.

Large transport block sizes are segmented into multiple codeblocks prior to coding, each with its own 24-bit CRC (in addition to the overall transport-block CRC). This was discussed already in Section 9.2 and the reason is primarily complexity; the size of a codeblock is large enough to give good performance while still having a reasonable decoding complexity. Since each codeblock has its own CRC, errors can be detected on individual codeblocks as well as on the overall transport block. A relevant question is if retransmission should be limited to transport blocks or whether there are benefits of retransmitting only the codeblocks that are erroneously received. For the very large transport block sizes used to support data rates of several gigabits per second, there can be hundreds of codeblocks in a transport block. If only one or a few of them are in error, retransmitting the whole transport block results in a low spectral efficiency compared to retransmitting only the erroneous codeblocks. One example where only some codeblocks are in error is a situation with bursty interference where some OFDM symbols are hit more severely than others, as illustrated in Fig. 13.2, for example, due to one downlink transmission preempting another as discussed in Section 14.1.2.

---

[1] In LTE, eight processes are used for FDD and up to 15 processes for TDD, depending on the uplink−downlink configuration.

To correctly receive the transport block for the example above, it is sufficient to retransmit the erroneous codeblocks. At the same time, the control signaling overhead would be too large if individual codeblocks can be addressed by the hybrid-ARQ mechanism. Therefore, so-called *codeblock groups* (CBGs) are defined. If per-CBG retransmission is configured, feedback is provided per CBG instead of per transport block and only the erroneously received codeblock groups are retransmitted, which consumes less resources than retransmitting the whole transport block. Two, four, six, or eight codeblock groups can be configured with the number of codeblocks per codeblock group varying as a function of the total number of codeblocks in the initial transmission. Note that the codeblock group a codeblock belongs to is determined from the initial transmission and does not change between the transmission attempts. This is to avoid error cases which could arise if the codeblocks were repartitioned between two retransmissions.

The CBG retransmissions are handled as part of the physical layer from a specification perspective. There is no fundamental technical reason for this but rather a way to reduce the specification impact from CBG-level retransmissions. A consequence of this is that it is not possible, in the same hybrid-ARQ process, to mix transmission of new CBGs belonging to another transport block with retransmissions of CBGs belonging to the incorrectly received transport block.

13.1.1 SOFT COMBINING

An important part of the hybrid-ARQ mechanism is the use of *soft combining*, which implies that the receiver combines the received signal from multiple transmission attempts. By definition, a hybrid-ARQ retransmission must represent the same set of information bits as the original transmission. However, the set of coded bits transmitted in each retransmission may be selected differently as long as they represent the same set of information bits. Depending on whether the retransmitted bits are required to be identical to the original transmission or not, the soft combining scheme is often referred to as *Chase combining*, first proposed in Ref. [22], or *Incremental Redundancy* (IR), which is used in NR. With incremental redundancy, each retransmission does not have to be identical to the original transmission. Instead, *multiple sets* of coded bits are generated, each representing the same set of information bits [67,71]. The rate-matching functionality of NR, described in Section 9.3, is used to generate different sets of coded bits as a function of the redundancy version as illustrated in Fig. 13.3.

In addition to a gain in accumulated received E_b/N_0, incremental redundancy also results in a coding gain for each retransmission (until the mother code rate is reached). The gain with incremental redundancy compared to pure energy accumulation (Chase combining) is larger for high initial code rates [24]. Furthermore, as shown in Ref. [33], the performance gain of incremental redundancy compared to Chase combining can also depend on the relative power difference between the transmission attempts.

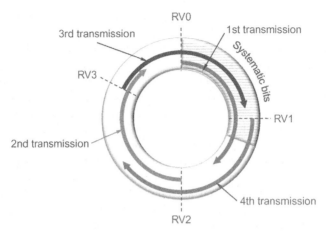

FIGURE 13.3

Example of incremental redundancy.

In the discussion so far, it has been assumed that the receiver has received all the previously transmitted redundancy versions. If all redundancy versions provide the same amount of information about the data packet, the order of the redundancy versions is not critical. However, for some code structures, not all redundancy versions are of equal importance. This is the case for the LDPC codes used in NR; the systematic bits are of higher importance than the parity bits. Hence, the initial transmission should at least include all the systematic bits and some parity bits. In the retransmission(s), parity bits not in the initial transmission can be included. This is the background to why systematic bits are inserted first in the circular buffer in Section 9.3. The starting points in the circular buffer are defined such that both RV0 and RV3 are self-decodable, that is, includes the systematic bits under typical scenarios. This is also the reason RV3 is located after nine o'clock in Fig. 13.3, as this allows more of the systematic bits to be included in the transmission. With the default order of the redundancy versions 0, 2, 3, 1, every second retransmission is typically self-decodable.

Hybrid ARQ with soft combining, regardless of whether Chase or incremental redundancy is used, leads to an implicit reduction of the data rate by means of retransmissions and can thus be seen as implicit link adaptation. However, in contrast to link adaptation based on explicit estimates of the instantaneous channel conditions, hybrid-ARQ with soft combining implicitly adjusts the coding rate based on the result of the decoding. In terms of overall throughput this kind of implicit link adaptation can be superior to explicit link adaptation, as additional redundancy is only added *when needed*—that is, when previous higher-rate transmissions were not possible to decode correctly. Furthermore, as it does not try to predict any channel variations, it works equally well, regardless of the speed at which the terminal is moving. Since implicit link adaptation can provide a gain in system throughput, a valid question is why explicit link adaptation is necessary at all. One major reason for having explicit link adaptation is the reduced delay. Although relying on implicit link adaptation alone is sufficient from a system throughput perspective, the end-user service quality may not be acceptable from a delay perspective.

For proper operation of soft combining, the receiver needs to know when to perform soft combining prior to decoding and when to clear the soft buffer—that is, the receiver needs to differentiate between the reception of an initial transmission (prior to which the soft buffer should be cleared) and the reception of a retransmission. Similarly, the transmitter must know whether to retransmit erroneously received data or to transmit new data. This is handled by the *new-data indicator* as discussed further below for downlink and uplink hybrid-ARQ, respectively.

13.1.2 DOWNLINK HYBRID-ARQ

In the downlink, retransmissions are scheduled in the same way as new data—that is, they may occur at any time and at an arbitrary frequency location within the downlink cell bandwidth. The scheduling assignment contains the necessary hybrid-ARQ-related control signaling—hybrid-ARQ process number, new-data indicator, CBGTI, and CBGFI in case per-CBG retransmission is configured, as well as information to handle the transmission of the acknowledgment in the uplink such as timing and resource indication information.

Upon receiving a scheduling assignment in the DCI, the receiver tries to decode the transport block, possibly after soft combining with previous attempts as described above. Since transmissions and retransmissions are scheduled using the same framework in general, the device needs to know whether the transmission is a new transmission, in which case the soft buffer should be flushed, or a retransmission, in which case soft combining should be performed. Therefore, an explicit *new-data indicator* is included for the scheduled transport block as part of the scheduling information transmitted in the downlink. The new-data indicator is toggled for a new transport block—that is, it is essentially a single-bit sequence number. Upon reception of a downlink scheduling assignment, the device checks the new-data indicator to determine whether the current transmission should be soft combined with the received data currently in the soft buffer for the hybrid-ARQ process in question, or if the soft buffer should be cleared.

The new-data indicator operates on the transport-block level. However, if per-CBG retransmissions are configured, the device needs to know which CBGs are retransmitted and whether the corresponding soft buffer should be flushed or not. This is handled through two additional information fields present in the DCI in case per-CBG retransmission is configured, the *CBG Transmit Indicator* (CBGTI) and the *CBG Flush Indicator* (CBGFI). The CBGTI is a bitmap indicating whether a certain CBG is present in the downlink transmission or not (see

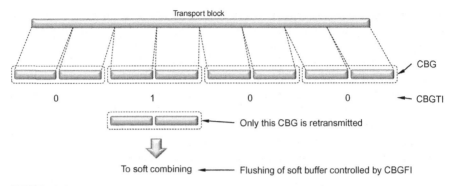

Transport block

CBG

0   1   0   0   ← CBGTI

Only this CBG is retransmitted

To soft combining ◄────── Flushing of soft buffer controlled by CBGFI

FIGURE 13.4

Illustration of per-CBG retransmission.

Fig. 13.4). The CBGFI is a single bit, indicating whether the CBGs indicated by the CBGTI should be flushed or whether soft combining should be performed.

The result of the decoding operation—a positive acknowledgment in the case of a successful decoding and a negative acknowledgment in the case of unsuccessful decoding—is fed back to the gNB as part of the uplink control information. If CBG retransmissions are configured, a bitmap with one bit per CBG is fed back instead of a single bit representing the whole transport block.

13.1.3 UPLINK HYBRID-ARQ

The uplink uses the same asynchronous hybrid-ARQ protocol as the downlink. The necessary hybrid-ARQ-related information—hybrid-ARQ process number, new-data indicator, and, if per-CBG retransmission is configured, the CBGTI—is included in the scheduling grant.

To differentiate between new transmissions and retransmissions of data, the new-data indicator is used. Toggling the new-data indicator requests transmission of a new transport block, otherwise the previous transport block for this hybrid-ARQ process should be retransmitted (in which case the gNB can perform soft combining). The CBGTI is used in a similar way as in the downlink, namely to indicate the codeblock groups to retransmit in the case of per-CBG retransmission. Note that no CBGFI is needed in the uplink as the soft buffer is located in the gNB which can decide whether to flush the buffer or not based on the scheduling decisions.

13.1.4 TIMING OF UPLINK ACKNOWLEDGMENTS

In LTE, the time from downlink data reception to transmission of the acknowledgment is fixed in the specifications. This is possible for full-duplex transmission, for example, FDD, in which case the acknowledgment is transmitted almost

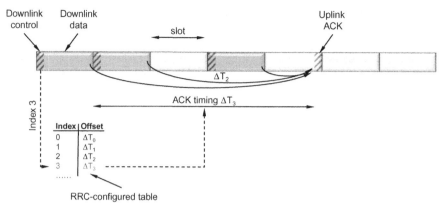

FIGURE 13.5

Determining the acknowledgment timing.

3 ms after the end of data reception in LTE.[2] A similar approach can be used if the uplink–downlink allocation is semistatically configured in the case of half-duplex operation, for example, semistatic TDD as in LTE. Unfortunately, this type of scheme with predefined timing instants for the acknowledgments does not blend well with dynamic TDD, one of the cornerstones of NR, as an uplink opportunity cannot be guaranteed a fixed time after the downlink transmission due to the uplink–downlink direction being dynamically controlled by the scheduler. Coexistence with other TDD deployments in the same frequency band may also impose restrictions when it is desirable, or possible, to transmit in the uplink. Furthermore, even if it would be possible, it may not be desirable to change the transmission direction from downlink to uplink in each slot as this would increase the switching overhead. Consequently, a more flexible scheme capable of dynamically controlling when the acknowledgment is transmitted is adopted in NR.

The hybrid-ARQ timing field in the downlink DCI is used to control the transmission timing of the acknowledgment in the uplink. This three-bit field is used as an index into an RRC-configured table providing information on when the hybrid-ARQ acknowledgment should be transmitted relative to the reception of the PDSCH (see Fig. 13.5). In this particular example, three slots are scheduled in the downlink before an acknowledgment is transmitted in the uplink. In each downlink assignment, different acknowledgment timing indices have been used, which in combination with the RRC-configured table result in all three slots being acknowledged at the same time (multiplexing of these acknowledgments in the same slot is discussed below).

---

[2]The time depends on the timing advance value. For the largest possible timing advance, the time is 2.3 ms in LTE.

Table 13.1 Minimum Processing Time (PDSCH Mapping Type A, Feedback on PUCCH)

DM-RS Configuration	Device Capability	Subcarrier Spacing				LTE Rel 8
		15 kHz	**30 kHz**	**60 kHz**	**120 kHz**	
Front-loaded	Baseline	0.57 ms	0.36 ms	0.30 ms	0.18 ms	2.3 ms
	Aggressive	0.18–0.29 ms	0.08–0.17 ms			
Additional	Baseline	0.92 ms	0.46 ms	0.36 ms	0.21 ms	
	Aggressive	0.85 ms	0.4 ms			

Furthermore, NR is designed with very low latency in mind and is therefore capable of transmitting the acknowledgment much sooner after the end of the downlink data reception than the corresponding LTE timing relation. All devices support the baseline processing times listed in Table 13.1, with even faster processing optionally supported by some devices. The capability is reported per subcarrier spacing. One part of the processing time is constant in symbols across different subcarrier spacing, that is, the time in microseconds scales with the subcarrier spacing, but there is also a part of the processing time fixed in microseconds and independent of the subcarrier spacing. Hence, the processing times listed in the table are not directly proportional to the subcarrier spacing although there is a dependency. There is also a dependency on the reference signal configuration; if the device is configured with additional reference signal occasions later in the slot, the device cannot start the processing until at least some of these reference signals have been received and the overall processing time is longer. Nevertheless, the processing is much faster than the corresponding LTE case as a result of stressing the importance of low latency in the NR design.

For proper transmission of the acknowledgment it is not sufficient for the device to know *when* to transmit, which is obtained from the timing field discussed above, but also *where* in the resource domain (frequency resources and, for some PUCCH formats, the code domain). In the original LTE design, this is primarily obtained from the location of the PDCCH scheduling the transmission. For NR with its flexibility in the transmission timing of the acknowledgment, such a scheme is not sufficient. In the case that two devices are instructed to transmit their acknowledgment at the same time even if they were scheduled at different time instants, it is necessary to provide the devices with separate resources. This is handled through the *PUCCH resource indicator*, which is a three-bit index selecting one of eight RRC-configured resource sets as described in Section 10.2.7.

13.1.5 MULTIPLEXING OF HYBRID-ARQ ACKNOWLEDGMENTS

In the previous section, the timing of the hybrid-ARQ acknowledgments in the example was such that multiple transport blocks need to be acknowledged at the same time. Other examples where multiple acknowledgments need to be transmitted in the uplink at the same time are carrier aggregation and per-CBG retransmissions. NR therefore supports multiplexing of acknowledgments for multiple transport blocks received by a device into one multi-bit acknowledgment message. The multiple bits can be multiplexed using either a semi-static codebook or a dynamic codebook with RRC configuration selecting between the two.

The semi-static codebook can be viewed as a matrix consisting of a time-domain dimension and a component-carrier (or CBG or MIMO layer) dimension, both of which are semi-statically configured. The size in the time domain is given by the maximum and minimum hybrid-ARQ acknowledgment timings configured in Table 13.1, and the size in the carrier domain is given by the number of simultaneous transport blocks (or CBGs) across all component carriers. An example is provided in Fig. 13.6, where the acknowledgment timings are one, two, three, and four, respectively, and three carriers, one with two transport blocks, one with one transport block, and one with four CBGs, are configured. Since the codebook size is fixed, the number of bits to transmit in a hybrid-ARQ report is known ($4\cdot7 = 28$ bits in the example in Fig. 13.6) and the appropriate format for the uplink control signaling can be selected. Each entry in the matrix represents the

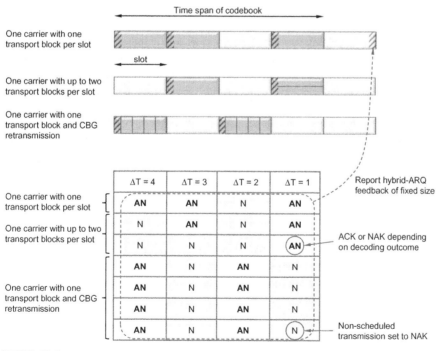

FIGURE 13.6

Example of semistatic hybrid-ARQ acknowledgment codebook.

decoding outcome, positive or negative acknowledgment, of the corresponding transmission. Not all transmission opportunities possible with the codebook are used in this example and for entries in the matrix without a corresponding transmission, a negative acknowledgment is transmitted. This provides robustness; in the case of missed downlink assignment a negative acknowledgment is provided to the gNB, which can retransmit the missing transport block (or CBG).

One drawback with the semistatic codebook is the potentially large size of a hybrid-ARQ report. For a small number of component carriers and no CBG retransmissions, this is less of a problem, but if a large number of carriers and codeblock groups are configured out of which only a small number is simultaneously used, this may become more of an issue.

To address the drawback of a potentially large semi-static codebook size in some scenarios, NR also supports a dynamic codebook. In fact, this is the default codebook used unless the system is configured otherwise. With a dynamic codebook, only the acknowledgment information for the *scheduled* carriers[3] is included in the report, instead of all carriers, scheduled or not, as is the case with a semi-static codebook. Hence, the size of the codebook (the matrix in Fig. 13.6) is dynamically varying as a function of the number of scheduled carriers. In essence, only the bold entries in the example in Fig. 13.6 would be included in the hybrid-ARQ report and the non-bold entries with a gray background (which correspond to non-scheduled carriers) would be omitted. This reduces the size of the acknowledgment message.

A dynamic codebook would be straightforward if there were no errors in the downlink control signaling. However, in the presence of an error in the downlink control signaling, the device and gNB may have different understanding on the number of scheduled carriers, which would lead to an incorrect codebook size and possibly corrupt the feedback report for all carriers, and not only for the ones for which the downlink controls signaling was missed. Assume, as an example, that the device was scheduled for downlink transmission in two subsequent slots but missed the PDCCH and hence scheduling assignment for the first slot. In response the device will transmit an acknowledgment for the second slot only, while the gNB tries to receive acknowledgments for two slots, leading to a mismatch.

To handle these error cases, NR uses the *downlink assignment index* (DAI) included in the DCI containing the downlink assignment. The DAI field is further split into two parts, a counter DAI (cDAI) and, in the case of carrier aggregation, a total DAI (tDAI). The counter DAI included in the DCI indicates the number of scheduled downlink transmissions up to the point the DCI was received in a carrier first, time second manner. The total DAI included in the DCI indicates the total number of downlink transmissions across all carriers up to this point in time,

---

[3]The description here uses the term "carrier" but the same principle is equally applicable to per-CBG retransmission or multiple transport blocks in the case of MIMO and "transmission instant" is a more generic term, albeit the description would be harder to read.

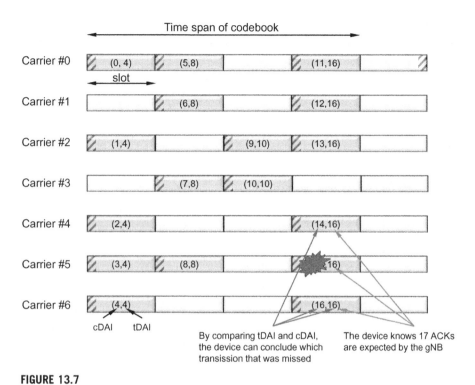

FIGURE 13.7

Example of dynamic hybrid-ARQ acknowledgment codebook.

that is, the highest cDAI at the current point in time (see Fig. 13.7 for an example). The counter DAI and total DAI are represented with decimal numbers with no limitation; in practice two bits are used for each and the numbering will wrap around, that is, what is signaled is the numbers in the figure modulo four. As seen in this example, the dynamic codebook needs to account for 17 acknowledgments (numbered 0−16). This can be compared with the semistatic codebook which would require 28 entries regardless of the number of transmissions.

Furthermore, in this example, one transmission on component carrier five is lost. Without the DAI mechanism, this would result in misaligned codebooks between the device and the gNB. However, as long as the device receives at least one component carrier, it knows the value of the total DAI and hence the size of the codebook at this point in time. Furthermore, by checking the values received for the counter DAI, it can conclude which component carrier was missed and that a negative acknowledgment should be assumed in the codebook for this position.

In the case that CBG retransmission is configured for some of the carriers, the dynamic codebook is split into two parts, one for the non-CBG carriers and one for the CBG carriers. Each codebook is handled according to the principles outlined above. The reason for the split is that for the CBG carriers, the device needs to generate feedback for each of these carriers according to the largest CBG configuration.

13.2 **RLC**

The *radio-link control* (RLC) protocol takes data in the form of RLC SDUs from PDCP and delivers them to the corresponding RLC entity in the receiver by using functionality in MAC and physical layers. The relation between RLC and MAC, including multiplexing of multiple logical channels into a single transport channel, is illustrated in Fig. 13.8.

There is one RLC entity per logical channel configured for a device with the RLC entity being responsible for one or more of:

- Segmentation of RLC SDUs;
- Duplicate removal; and
- RLC retransmission.

Unlike LTE, there is no support for concatenation or in-sequence delivery in the RLC protocol. This is a deliberate choice done to reduce the overall latency as discussed further in the following sections. It has also impacted the header design. Also, note that the fact that there is one RLC entity per logical channel and one hybrid-ARQ entity per cell (component carrier) implies that RLC retransmissions can occur on a different cell (component carrier) than the original transmission. This is not the case for the hybrid-ARQ protocol where retransmissions are bound to the same component carrier as the original transmission.

Different services have different requirements; for some services (for example, transfer of a large file), error-free delivery of data is important, whereas for other applications (for example, streaming services), a small amount of missing packets is not a problem. The RLC can therefore operate in three different modes, depending on the requirements from the application:

- *Transparent mode* (TM), where the RLC is completely transparent and is essentially bypassed. No retransmissions, no duplicate detection, and no segmentation/reassembly take place. This configuration is used for control-plane broadcast channels such as BCCH, CCCH, and PCCH, where the information should reach multiple users. The size of these messages is selected such that all intended devices are reached with a high probability and hence there is neither need for segmentation to handle varying channel conditions, nor retransmissions to provide error-free data transmission. Furthermore, retransmissions are not feasible for these channels as there is no possibility for the device to feedback status reports as no uplink has been established.

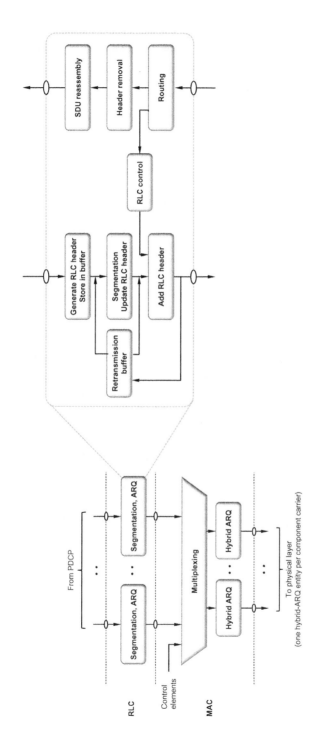

FIGURE 13.8

MAC and RLC.

- *Unacknowledged mode* (UM) supports segmentation but not retransmissions. This mode is used when error-free delivery is not required, for example, voice-over-IP.
- *Acknowledged mode* (AM) is the main mode of operation for the DL-SCH and UL-SCH. Segmentation, duplicate removal, and retransmissions of erroneous data are all supported.

In the following sections, the operation of the RLC protocol is described, focusing on acknowledged mode.

13.2.1 SEQUENCE NUMBERING AND SEGMENTATION

In unacknowledged and acknowledged modes, a sequence number is attached to each incoming SDU using 6 or 12 bits for unacknowledged mode and 12 or 18 bits for acknowledged mode. The sequence number is included in the RLC PDU header in Fig. 13.9. In the case of a non-segmented SDU, the operation is straightforward; the RLC PDU is simply the RLC SDU with a header attached. Note that this allows the RLC PDUs to be generated in advance as the header, in the absence of segmentation, does not depend on the scheduled transport block size. This is beneficial from a latency perspective and the reason the header structure is changed compared to the one used in LTE.

However, depending on the transport-block size after MAC multiplexing, the size of (the last) of the RLC PDUs in a transport block may not match the RLC SDU size. To handle this, an SDU can be segmented into multiple segments. If no segmentation takes place, padding would need to be used instead, leading to degraded spectral efficiency. Hence, dynamically varying the number of RLC PDUs used to fill the transport block, together with segmentation to adjust the size of the last RLC PDU, ensures the transport block is efficiently utilized.

Segmentation is simple; the last preprocessed RLC SDU can be split into two segments, the header of the first segment is updated, and to the second segment a new header is added (which is not time critical as it is not being transmitted in the current transport block). Each SDU segment carries the same sequence number as the original unsegmented SDU and this sequence number is part of the RLC header. To distinguish whether the PDU contains a complete SDU or a segment, a *segmentation information* (SI) field is also part of the RLC header, indicating whether the PDU is a complete SDU, the first segment of the SDU, the last segment of the SDU, or a segment between the first and last segments of the SDU. Furthermore, in the case of a segmented SDU, a 16-bit *segmentation offset* (SO) is included in all segments except the first one and used to indicate which byte of the SDU the segment represents. There is also a *poll bit* (P) in the header used to request a status report for acknowledged mode as described further below, and a *data/control indicator*, indicating whether the RLC PDU contains data to/from a logical channel or control information required for RLC operation.

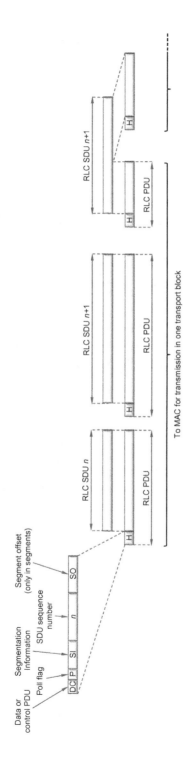

FIGURE 13.9

Generation of RLC PDUs from RLC SDUs (acknowledged mode assumed for the header structure).

The header structure above holds for acknowledged mode. The header for unacknowledged mode is similar but does not include either the poll bit or the data/control indicator. Furthermore, the sequence number is included in the case of segmentation only.

In LTE, the RLC can also perform concatenation of RLC SDUs into a single PDU. However, this functionality is not present in NR in order to reduce latency. If concatenation would be supported, an RLC PDU cannot be assembled until the uplink grant is received as the scheduled transport-block size is not known in advance. Consequently, the uplink grant must be received well in advance to allow sufficient processing time in the device. Without concatenation, the RLC PDUs can be assembled in advance, prior to receiving the uplink grant, and thereby reducing the processing time required between receiving an uplink grant and the actual uplink transmission.

13.2.2 ACKNOWLEDGED MODE AND RLC RETRANSMISSIONS

Retransmission of missing PDUs is one of the main functionalities of the RLC in acknowledged mode. Although most of the errors can be handled by the hybrid-ARQ protocol, there are, as discussed at the beginning of the chapter, benefits of having a second-level retransmission mechanism as a complement. By inspecting the sequence numbers of the received PDUs, missing PDUs can be detected and a retransmission requested from the transmitting side.

RLC acknowledged mode in NR is similar to its counterpart in LTE with one exception—reordering to ensure in-sequence delivery is not supported in NR. Removing in-sequence delivery from the RLC also helps reduce the overall latency as later packets do not have to wait for retransmission of an earlier missing packet before being delivered to higher layers, but can be forwarded immediately. This also leads to reduced buffering requirements positively impacting the amount of memory used for RLC buffering. In LTE, which does support in-sequence delivery from the RLC protocol, an RLC SDU cannot be forwarded to higher layers unless all previous SDUs have been correctly received. A single missing SDU, for example, due to a momentary interference burst, can thus block delivery of subsequent SDUs for quite some time, even if those SDUs would be useful to the application, a property which is clearly not desirable in a system targeting very low latency.

In acknowledged mode, the RLC entity is bidirectional—that is, data may flow in both directions between the two peer entities. This is necessary as the reception of PDUs needs to be acknowledged back to the entity that transmitted those PDUs. Information about missing PDUs is provided by the receiving end to the transmitting end in the form of so-called *status reports*. Status reports can either be transmitted autonomously by the receiver or requested by the transmitter. To keep track of the PDUs in transit, the sequence number in the header is used.

Both RLC entities maintain two windows in acknowledged mode, the transmission and reception windows, respectively. Only PDUs in the transmission window are eligible for transmission; PDUs with sequence number below the start of the window have already been acknowledged by the receiving RLC. Similarly, the receiver only accepts PDUs with sequence numbers within the reception window. The receiver also discards any duplicate PDUs as only one copy of each SDU should be delivered to higher layers.

The operation of the RLC with respect to retransmissions is perhaps best understood by the simple example in Fig. 13.10, where two RLC entities are illustrated, one in the transmitting node and one in the receiving node. When operating in acknowledged mode, as assumed below, each RLC entity has both transmitter and receiver functionality, but in this example only one of the directions is discussed as the other direction is identical. In the example, PDUs numbered from n to $n + 4$ are awaiting transmission in the transmission buffer. At time t_0, PDUs with sequence number up to and including n have been transmitted and correctly received, but only PDUs up to and including $n - 1$ have been acknowledged by the receiver. As seen in the figure, the transmission window starts from n, the first not-yet-acknowledged PDU, while the reception window starts from $n + 1$, the next PDU expected to be received. Upon reception of a PDU n, the SDU is reassembled and delivered to higher layers, that is, the PDCP. For a PDU containing a complete SDU, reassembly is simply header removal, but in the case of a segmented SDU, the SDU cannot be delivered until PDUs carrying all the segments have been received.

The transmission of PDUs continues and, at time t_1, PDUs $n + 1$ and $n + 2$ have been transmitted but, at the receiving end, only PDU $n + 2$ has arrived. As soon as a complete SDU is received, it is delivered to higher layers, hence PDU $n + 2$ is forwarded to the PDCP layer without waiting for the missing PDU $n + 1$. One reason PDU $n + 1$ is missing could be that it is under retransmission by the hybrid-ARQ protocol and therefore has not yet been delivered from the hybrid-ARQ to the RLC. The transmission window remains unchanged compared to the previous figure, as none of the PDUs n and higher have been acknowledged by the receiver. Hence, any of these PDUs may need to be retransmitted as the transmitter is not aware of whether they have been received correctly or not.

The reception window is not updated when PDU $n + 2$ arrives, the reason being the missing PDU $n + 1$. Instead the receiver starts a timer, the *t-Reassembly* timer. If the missing PDU $n + 1$ is not received before the timer expires, a retransmission is requested. Fortunately, in this example, the missing PDU arrives from the hybrid-ARQ protocol at time t_2, before the timer expires. The reception window is advanced and the reassembly timer is stopped as the missing PDU has arrived. PDU $n + 1$ is delivered for reassembly into SDU $n + 1$.

Duplicate detection is also the responsibility of the RLC, using the same sequence number as used for retransmission handling. If PDU $n + 2$ arrives again (and is within the reception window), despite it having already been received, it is discarded.

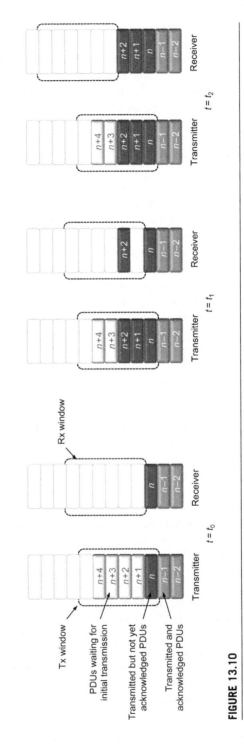

FIGURE 13.10

SDU delivery in acknowledged mode.

The transmission continues with PDUs $n+3$, $n+4$, and $n+5$, as shown in Fig. 13.11. At time t_3, PDUs up to $n+5$ have been transmitted. Only PDU $n+5$ has arrived and PDUs $n+3$ and $n+4$ are missing. Similar to the case above, this causes the reassembly timer to start. However, in this example no PDUs arrive prior to the expiration of the timer. The expiration of the timer at time t_4 triggers the receiver to send a control PDU containing a status report, indicating the missing PDUs, to its peer entity. Control PDUs have higher priority than data PDUs to avoid the status reports being unnecessarily delayed and negatively impacting the retransmission delay. Upon receipt of the status report at time t_5, the transmitter knows that PDUs up to $n+2$ have been received correctly and the transmission window is advanced. The missing PDUs $n+3$ and $n+4$ are retransmitted and, this time, correctly received.

Finally, at time t_6, all PDUs, including the retransmissions, have been delivered by the transmitter and successfully received. As $n+5$ was the last PDU in the transmission buffer, the transmitter requests a status report from the receiver by setting a flag in the header of the last RLC data PDU. Upon reception of the PDU with the flag set, the receiver will respond by transmitting the requested status report, acknowledging all PDUs up to and including $n+5$. Reception of the status report by the transmitter causes all the PDUs to be declared as correctly received and the transmission window is advanced.

Status reports can, as mentioned earlier, be triggered for multiple reasons. However, to control the amount of status reports and to avoid flooding the return link with an excessive number of status reports, it is possible to use a status prohibit timer. With such a timer, status reports cannot be transmitted more often than once per time interval as determined by the timer.

The example above basically assumed each PDU carrying a non-segmented SDU. Segmented SDUs are handled the same way, but an SDU cannot be delivered to the PDCP protocol until all the segments have been received. Status reports and retransmissions operate on individual segments; only the missing segment of a PDU needs to be retransmitted.

In the case of a retransmission, all RLC PDUs may not fit into the transport block size scheduled for the RLC retransmission. Resegmentation following the same principle as the original segmentation is used in this case.

13.3 **PDCP**

The *Packet Data Convergence Protocol* (PDCP) is responsible for:

- Header compression;
- Ciphering and integrity protection;
- Routing and duplication for split bearers; and
- Retransmission, reordering, and SDU discard.

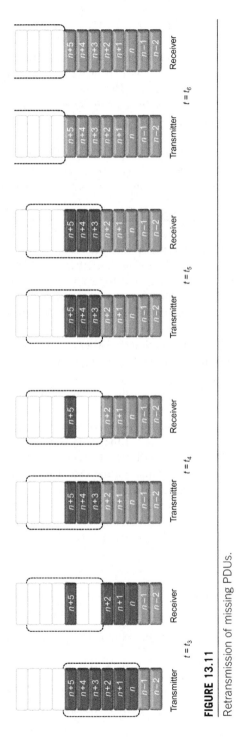

FIGURE 13.11

Retransmission of missing PDUs.

Header compression, with the corresponding decompression functionality at the receiver side, can be configured and serves the purpose of reducing the number of bits transmitted over the radio interface. Especially for small payloads, such as voice-over-IP and TCP acknowledgments, the size of an uncompressed IP header is in the same range as the payload itself, 40 bytes for IP v4 and 60 bytes for IP v6, and can account for around 60% of the total number of bits sent. Compressing this header to a couple of bytes can therefore increase the spectral efficiency by a large amount. The header compression scheme in NR is based on Robust Header Compression (ROHC) [38], a standardized header-compression framework also used for several other mobile-communication technologies, for example, LTE. Multiple compression algorithms, denoted profiles, are defined, each specific to the particular network layer and transport layer protocol combination such as TCP/IP and RTP/UDP/IP. Header compression is developed to compress IP packets. Hence it is applied to the data part only and not the SDAP header (if present).

Integrity protection ensures that the data originate from the correct source and ciphering protects against eavesdropping. PDCP is responsible for both these functions, if configured. Integrity protection and ciphering are used for both the data plane and the control plane and applied to the payload only and not the PDCP control PDUs or SDAP headers.

For dual connectivity and split bearers (see Chapter 6, for a more in-depth discussion on dual connectivity), PDCP can provide routing and duplication functionality. With dual connectivity, some of the radio bearers are handled by the master cell group, while others are handled by the secondary cell group. There is also a possibility to split a bearer across both cell groups. The routing functionality of the PDCP is responsible for routing the data flows for the different bearers to the correct cell groups, as well as handling flow control between the central unit (gNB-CU) and distributed unit (gNB-DU) in the case of a split gNB.

Duplication implies that the same data can be transmitted on two separate logical channels where configuration ensures that the two logical channels are mapped to different carriers. This can be used in combination with carrier aggregation or dual connectivity to provide additional diversity. If multiple carriers are used to transmit the same data, the likelihood that reception of the data on at least one carrier is correct increases. If multiple copies of the same SDU are received, the receiving-side PDCH discards the duplicates. This results in selection diversity which can be essential to providing very high reliability.

Retransmission functionality, including the possibility for reordering to ensure in-sequence delivery, is also part of the PDCP. A relevant question is why the PDCP is capable of retransmissions when there are two other retransmission functions in lower layers, the RLC ARQ and the MAC hybrid-ARQ functions. One reason is inter-gNB handover. Upon handover, undelivered downlink data packets will be forwarded by the PDCP from the old gNB to the new gNB. In this case, a new RLC entity (and hybrid-ARQ entity) is established in the new gNB and the RLC status is lost. The PDCP retransmission functionality ensures that no packets are lost as a result of this handover. In the uplink, the PDCP entity in the device will handle retransmission of all uplink packets not yet delivered to the gNB as the hybrid-ARQ buffers are flushed upon handover.

In-sequence delivery is not ensured by the RLC to reduce the overall latency. In many cases, rapid delivery of the packets is more important than guaranteed in-sequence delivery. However, if in-sequence delivery is important, the PDCP can be configured to provide this.

Retransmission and in-sequence delivery, if configured, is jointly handled in the same protocol, which operates similarly to the RLC ARQ protocol except that no segmentation is supported. A so-called count value is associated with each SDU, where the count is a combination of the PDCP sequence number and the hyper-frame number. The count value is used to identify lost SDUs and request retransmission, as well as reorder received SDUs before delivery to upper layers is reordering is configured. Reordering basically buffers a received SDU and does not forward it to higher layers until all lower-numbered SDUs have been delivered. Referring to Fig. 13.10, this would be similar to not delivering SDU $n + 2$ until $n + 1$ has been successfully received and delivered. There is also a possibility to configure a discard timer for each PDCP SDU; when the timer expires the corresponding SDU is discarded and not transmitted.

Scheduling

14

NR is essentially a scheduled system, implying that the scheduler determines when and to which devices the time, frequency, and spatial resources should be assigned and what transmission parameters, including data rate, to use. Scheduling can be either dynamic or semistatic. Dynamic scheduling is the basic mode-of-operation where the scheduler for each time interval, for example, a slot, determines which devices are to transmit and receive. Since scheduling decisions are taken frequently, it is possible to follow rapid variations in the traffic demand and radio-channel quality, thereby efficiently exploiting the available resources. Semi-static scheduling implies that the transmission parameters are provided to the devices in advance and not on a dynamic basis.

In the following, dynamic downlink and uplink scheduling will be discussed, including bandwidth adaptation, followed by a discussion on non-dynamic scheduling and finally a discussion on discontinuous reception as a way to reduce device power consumption.

14.1 DYNAMIC DOWNLINK SCHEDULING

Fluctuations in the received signal quality due to small-scale as well as large-scale variations in the environment are an inherent part in any wireless communication system. Historically, such variations were seen as a problem, but the development of *channel-dependent scheduling*, where transmissions to an individual device take place when the radio-channel conditions are favorable, allows these variations to be exploited. Given a sufficient number of devices in the cell having data to transfer, there is a high likelihood of at least some devices having favorable channel conditions at each point in time and able to use a correspondingly high data rate. The gain obtained by transmitting to users with favorable radio-link conditions is commonly known as multiuser diversity. The larger the channel variations and the larger the number of users in a cell, the larger the multiuser diversity gain. Channel-dependent scheduling was introduced in the later versions of the 3G standard known as HSPA [21] and is also used in LTE as well as NR.

There is a rich literature in the field of scheduling and how to exploit variations in the time and frequency domains (see, for example, Ref. [28] and the references therein). Lately, there has also been a large interest in various massive multiuser MIMO schemes [55] where a large number of antenna elements are

5G NR: The Next Generation Wireless Access Technology. DOI: https://doi.org/10.1016/B978-0-12-814323-0.00014-4

used to create very narrow "beams," or, expressed differently, isolate the different users in the spatial domain. It can be shown that, under certain conditions, the use of a large number of antennas results in an effect known as "channel hardening." In essence, the rapid fluctuations of the radio-channel quality disappear, simplifying the time–frequency part of the scheduling problem at the cost of a more complicated handling of the spatial domain.

In NR, the *downlink scheduler* is responsible for dynamically controlling the device(s) to transmit to. Each of the scheduled devices is provided with a *scheduling assignment* including information on the set of time–frequency resources upon which the device's DL-SCH[1] is transmitted, the modulation-and-coding scheme, hybrid-ARQ-related information, and multi-antenna parameters as outlined in Chapter 10. In most cases the scheduling assignment is transmitted just before the data on the PDSCH, but the timing information in the scheduling assignment can also schedule in OFDM symbols later in the slot or in later slots. One use for this is bandwidth adaptation as discussed below. Changing the bandwidth part may take some time and hence data transmission may not occur in the same slot as the control signaling was received in.

It is important to understand that NR *does not* standardize the scheduling behavior. Only a set of supporting mechanisms are standardized on top of which a vendor-specific scheduling strategy is implemented. The information needed by the scheduler depends on the specific scheduling strategy implemented, but most schedulers need information about at least:

- Channel conditions at the device, including spatial-domain properties;
- Buffer status of the different data flows; and
- Priorities of the different data flows, including the amount of data pending retransmission.

Additionally, the interference situation in neighboring cells can be useful if some form of interference coordination is implemented.

Information about the channel conditions at the device can be obtained in several ways. In principle, the gNB can use any information available, but typically the CSI reports from the device are used as discussed in Section 8.1. There is a wide range of CSI reports that can be configured where the device reports the channel quality in the time, frequency, and spatial domains. The amount of correlation between the spatial channels to different devices is also of interest to estimate the degree of spatial isolation between two devices in the case they are candidates for being scheduled on the same time–frequency resources using multiuser MIMO. Uplink sounding using SRS transmission can, together with assumptions on channel reciprocity, also be used to assess the downlink channel quality. Various other quantities can be used as well, for example, signal-strength measurements for different beam candidates.

---

[1]In the case of carrier aggregation there is one DL-SCH (or UL-SCH) per component carrier.

The buffer status and traffic priorities are easily obtained in the downlink case as the scheduler and the transmission buffers reside in the same node. Prioritization of different traffic flows is purely implementation-specific, but retransmissions are typically prioritized over transmission of new data, at least for data flows of the same priority. Given that NR is designed to handle a much wider range of traffic types and applications than previous technologies, such as LTE, priority handling in the scheduler can in many cases be even more emphasized than in the past. In addition to selecting data from different data flows, the scheduler also has the possibility to select the transmission duration. For example, for a latency-critical service with its data mapped to a certain logical channel, it may be advantageous to select a transmission duration corresponding to a fraction of a slot, while for another service on another logical channel, a more traditional approach of using the full slot duration for transmission might be a better choice. It may also be the case that, for latency reasons and shortage of resources, an urgent transmission using a small number of transmissions needs to preempt an already ongoing transmission using the full slot. In this case, the preempted transmission is likely to be corrupted and require a retransmission, but this may be acceptable given the very high priority of the low-latency transmission. There are also some mechanisms in NR which can be used to mitigate this, as discussed in Section 14.1.2.

Different downlink schedulers may coordinate their decisions to increase the overall performance, for example, by avoiding transmission on a certain frequency range in one cell to reduce the interference towards another cell. In the case of (dynamic) TDD, the different cells can also coordinate the transmission direction, uplink or downlink, between the cells to avoid detrimental interference situations. Such coordination can take place on different time scales. Typically, the coordination is done at a slower rate than the scheduling decisions in each cell as the requirements on the backhaul connecting different gNBs otherwise would be too high.

In the case of carrier aggregation, the scheduling decisions are taken per carrier and the scheduling assignments are transmitted separately for each carrier, that is, a device scheduled to receive data from multiple carriers simultaneously receives multiple PDCCHs. A PDCCH received can either point to the same carrier, known as self-scheduling, or to another carrier, commonly referred to as cross-carrier scheduling (see Fig. 14.1). In the case of cross-carrier scheduling of a carrier with a different numerology than the one upon which the PDCCH was transmitted, timing offsets in the scheduling assignment, for example, which slot the assignment relates to, are interpreted in the PDSCH numerology (and not the PDCCH numerology).

The scheduling decisions for the different carriers are not taken in isolation. Rather, the scheduling of the different carriers for a given device needs to be coordinated. For example, if a certain piece of data is scheduled for transmission on one carrier, the same piece of data should normally not be scheduled on another carrier as well. However, it is in principle possible to schedule the same

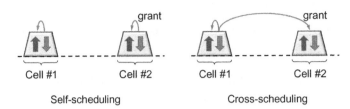

FIGURE 14.1

Self-scheduling and cross-carrier scheduling.

data on multiple carriers. This can be used to increase reliability; with multiple carriers transmitting the same data the likelihood of successful reception on at least one carrier is increased. At the receiver the RLC (or PDCP) layer can be configured to remove duplicates in case the same data are successfully received on multiple carriers. This results in selection diversity.

14.1.1 BANDWIDTH ADAPTATION

NR support a very wide transmission bandwidth, up to several 100 MHz on a single carrier. This is useful for rapid delivery of large payloads but is not needed for smaller payload sizes or for monitoring the downlink control channels when not scheduled. Hence, as mentioned already in Chapter 5 NR supports *receiver-bandwidth adaptation* such that the device can use a narrow bandwidth for monitoring control channels and only open the full bandwidth when a large amount of data is scheduled. This can be seen as discontinuous reception in the frequency domain.

Opening the wideband receiver can be done by using the bandwidth part indicator field in the DCI. If the bandwidth part indicator points to a different bandwidth part than the currently active one, the active bandwidth part is changed (see Fig. 14.2). The time it takes to change the active bandwidth part depends on several factors, for example, if the center frequency changes and the receiver needs to retune or not, but can be in the order of a slot. Once activated, the device uses the new, and wider, bandwidth part for its operation.

Upon completion of the data transfer requiring the wider bandwidth, the same mechanism can be used to revert back to the original bandwidth part. There is also a possibility to configure a timer to handle the bandwidth-part switching instead of explicit signaling. In this case, one of the bandwidth parts is configured as the default bandwidth part. If no default bandwidth part is explicitly configured, the initial bandwidth part obtained from the random-access procedure is used as the default bandwidth part. Upon receiving a DCI indicating a bandwidth part other than the default one, the timer is started. When the timer expires, the device switches back to the default bandwidth part. Typically, the default bandwidth part is narrower and can hence help reducing the device power consumption.

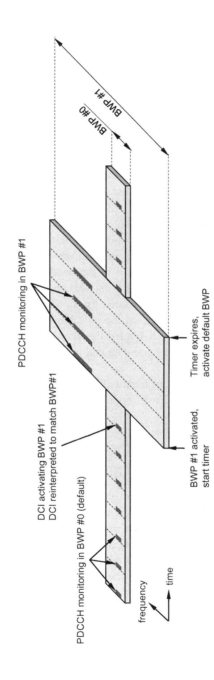

FIGURE 14.2

Illustration of bandwidth adaptation principle.

The introduction of bandwidth adaptation in NR raised several design questions not present in LTE, in particular related to the handling of controls signaling as many transmission parameters are configured per bandwidth part and the DCI payload size therefore may differ between different bandwidth parts. The frequency-domain resource allocation field is an obvious example; the larger the bandwidth part, the larger the number of bits for frequency-domain resource allocation. This is not an issue as long as the downlink data transmission uses the same bandwidth part as the DCI control signaling.[2] However, in the case of bandwidth adaptation this is not true as the bandwidth part indicator in the DCI received in one bandwidth part can point to *another* differently sized bandwidth part for data reception. This raises the issue on how to interpret the DCI if the bandwidth part index points to another bandwidth part than the current one, as the DCI fields in the detected DCI may not match what is needed in the bandwidth part pointed to by the index field.

One possibility to address this would be to blindly monitor for multiple DCI payload sizes, one for each configured bandwidth parts, but unfortunately this would imply a large burden on the device. Instead, an approach where the DCI fields detected are reinterpreted to be useful in the bandwidth part pointed to by the index is used. A simple approach has been selected where the bitfields are padded or truncated to match what is assumed by the bandwidth part scheduled. Naturally, this imposes some limitation on the possible scheduling decisions, but as soon as the new bandwidth part is activated the device monitors downlink control signaling using the new DCI size and data can be scheduled with full flexibility again.

Although the handling of different bandwidth parts has been described from a downlink perspective above, the same approach of reinterpreting the DCI is applied to the uplink.

14.1.2 DOWNLINK PREEMPTION HANDLING

Dynamic scheduling implies, as discussed above, that a scheduling decision is taken for each time interval. In many cases the time interval is equal to a slot, that is, the scheduling decisions are taken once per slot. The duration of a slot depends on the subcarrier spacing; a higher subcarrier spacing leads to a shorter slot duration. In principle this could be used to support lower-latency transmission, but as the cyclic prefix also shrinks when increasing the subcarrier spacing, it is not a feasible approach in all deployments. Therefore, as discussed in Section 7.2, NR supports a more efficient approach to low latency by allowing for transmission over a fraction of a slot, starting at any OFDM symbol. This allows for very low latency without sacrificing robustness to time dispersion.

---

[2]Strictly speaking, it is sufficient if the size and configuration of the bandwidth part used for PDCCH and PDSCH are the same.

In Fig. 14.3, an example of this is illustrated. Device A has been scheduled with a downlink transmission spanning one slot. During the transmission to device A, latency-critical data for device B arrives to the gNB, which immediately scheduled a transmission to device B. Typically, if there are frequency resources available, the transmission to device B is scheduled using resources not overlapping with the ongoing transmission to device A. However, in the case of a high load in the network, this may not be possible and there is no choice but to use (some of) the resources originally intended for device A for the latency-critical transmission to device B. This is sometimes referred to as the transmission to device B preempting the transmission to device A, which obviously will suffer an impact as a consequence of some of the resources device A assumes contains data for it suddenly containing data for device B.

There are several possibilities to handle this in NR. One approach is to rely on hybrid-ARQ retransmissions. Device A will not be able to decode the data due to the resources being preempted and will consequently report a negative acknowledgment to the gNB, which can retransmit the data at a later time instant. Either the complete transport block is retransmitted, or CBG-based retransmission is used to retransmit only the impacted codeblock groups as discussed in Section 13.1.

There is also a possibility to indicate to device A that some of its resources have been preempted and used for other purposes. This is done by transmitting a *preemption indicator* to device A in a slot after the slot containing the data transmission. The preemption indicator uses DCI format 2-1 (see Chapter 10 for details on different DCI formats) and contains a bitmap of 14 bits. Interpretation of the bitmap is configurable such that each bit represents one OFDM symbol in the time domain and the full bandwidth part, or two OFDM symbols in the time domain and one half of the bandwidth part. Furthermore, the monitoring periodicity of the preemption indicator is configured in the device, for example, every nth slot.

The behavior of the device when receiving the preemption indicator is not specified, but a reasonable behavior could be to flush the part of the soft buffer which corresponds to the preempted time—frequency region to avoid soft-buffer corruption for future retransmissions. From a soft-buffer handling perspective in the device, the more frequent the monitoring of the preemption indicator, the better (ideally, it should come immediately after the preemption occurred).

14.2 DYNAMIC UPLINK SCHEDULING

The basic function of the *uplink scheduler* in the case of dynamic scheduling is similar to its downlink counterpart, namely to dynamically control which devices are to transmit, on which uplink resources, and with what transmission parameters.

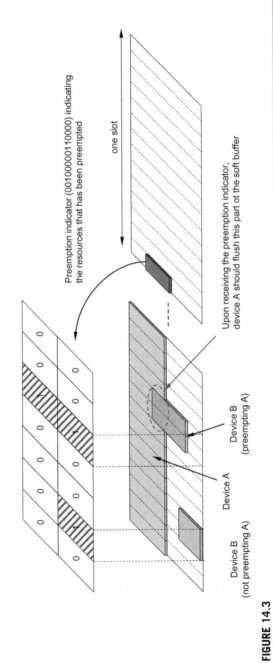

Preemption indicator (0010000011 0000) indicating the resources that has been preempted

one slot

Device B
(not preempting A)

Device A

Device B
(preempting A)

Upon receiving the preemption indicator, device A should flush this part ot the soft buffer

FIGURE 14.3

Downlink preemption indication.

The general downlink scheduling discussion is applicable to the uplink as well. However, there are some fundamental differences between the two. For example, the uplink power resource is *distributed* among the devices, while in the downlink the power resource is *centralized* within the base station. Furthermore, the maximum uplink transmission power of a single device is often significantly lower than the output power of a base station. This has a significant impact on the scheduling strategy. Even in the case of a large amount of uplink data to transmit there might not be sufficient power available—the uplink is basically power limited and not bandwidth limited, while in the downlink the situation can typically be the opposite. Hence, uplink scheduling typically results in a larger degree of frequency multiplexing of different devices than in the downlink.

Each scheduled device is provided with a *scheduling grant* indicating the set of time/frequency/spatial resources to use for the UL-SCH as well as the associated transport format. Uplink data transmissions only take place in the case that the device has a valid grant. Without a grant, no data can be transmitted.

The uplink scheduler is in complete control of the transport format the device shall use, that is, the device has to follow the scheduling grant. The only exception is that the device will not transmit anything, regardless of the grant, if there are no data in the transmission buffer. This reduces the overall interference by avoiding unnecessary transmissions in the case that the network scheduled a device with no data pending transmission.

Logical channel multiplexing is controlled by the device according to a set of rules (see Section 14.2.1). Thus, the scheduling grant does not explicitly schedule a certain logical channel but rather the device as such—uplink scheduling is primarily *per device* and not per radio bearer (although the priority handling mechanism discussed below in principle can be configured to obtain scheduling per radio bearer). Uplink scheduling is illustrated in the right part of Fig. 14.4, where the scheduler controls the transport format and the device controls the logical channel multiplexing. This allows the scheduler to tightly control the uplink activity to maximize the resource usage compared to schemes where the device autonomously selects the data rate, as autonomous schemes typically require some margin in the scheduling decisions. A consequence of the scheduler being responsible for selection of the transport format is that accurate and detailed knowledge about the device situation with respect to buffer status and power availability is accentuated compared to schemes where the device autonomously controls the transmission parameters.

The time during which the device should transmit in the uplink is indicated as part of the DCI as described in Section 10.1.11. Unlike in the downlink case, where the scheduling assignment typically is transmitted close in time to the data, this is not necessarily the case in the uplink. Since the grant is transmitted using downlink control signaling, a half-duplex device needs to change the transmission direction before transmitting in the uplink. Furthermore, depending on the uplink−downlink allocation, multiple uplink slots may need to be scheduled using

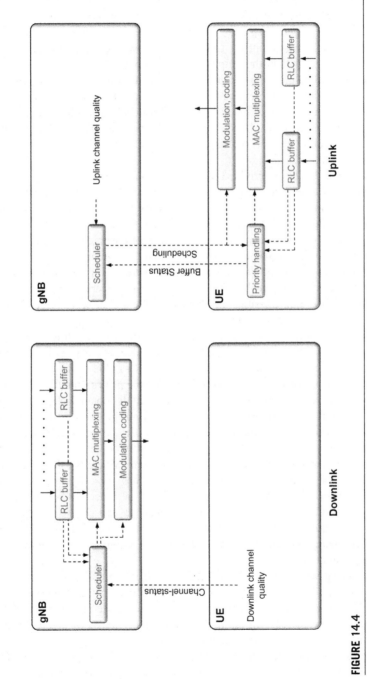

FIGURE 14.4

Downlink and uplink scheduling in NR.

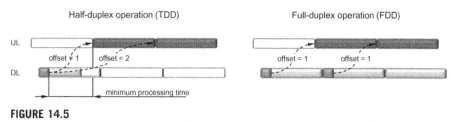

FIGURE 14.5

Example of uplink scheduling into future slots.

Table 14.1 Minimum Processing Time in OFDM Symbols From Grant Reception to Data Transmission

Device Capability	Subcarrier Spacing				LTE Rel 8
	15 kHz	30 kHz	60 kHz	120 kHz	
Baseline	0.71 ms	0.43 ms	0.41 ms	0.32 ms	3 ms
Aggressive	0.18−0.39 ms	0.08−0.2 ms			

multiple grants transmitted at the same downlink occasion. Hence, the timing field in the uplink grant is important.

The device also needs a certain amount of time to prepare for the transmission as outlined in Fig. 14.5. From an overall performance perspective, the shorter the time the better. However, from a device complexity perspective the processing time cannot be made arbitrarily short. In LTE, more than 3 ms was provided for the device to prepare the uplink transmission. For NR, a more latency-focused design, for example, the updated MAC and RLC header structure, as well as technology development in general has considerably reduced this time. The delay from the reception of a grant to the transmission of uplink data is summarized in Fig. 14.1. As seen from these numbers, the processing time depends on the subcarrier spacing, although it is not purely scaled in proportion to the subcarrier spacing. It is also seen that two device capabilities are specified. All devices need to fulfill the baseline requirements, but a device may also declare whether it is capable of a more aggressive processing time line which can be useful in latency-critical applications (Table 14.1).

Similar to the downlink case, the uplink scheduler can benefit from information on channel conditions, buffer status, and power availability. However, the transmission buffers reside in the device, as does the power amplifier. This calls for the reporting mechanisms described below to provide the information to the scheduler, unlike the downlink case where the scheduler, power amplifier, and transmission buffers all are in the same node. Uplink priority handling is, as already touched upon, another area where uplink and downlink scheduling differ.

14.2.1 UPLINK PRIORITY HANDLING

Multiple logical channels of different priorities can be multiplexed into the same transport block using the MAC multiplexing functionality. Except for the case when the uplink scheduling grant provides resources sufficient to transmit all data on all logical channels, the multiplexing needs to prioritize between the logical channels. However, unlike the downlink case, where the prioritization is up to the scheduler implementation, the uplink multiplexing is done according to a set of well-defined rules in the device with parameters set by the network. The reason for this is that a scheduling grant applies to a specific uplink carrier of a device, not explicitly to a specific logical channel within the carrier.

A simple approach would be to serve the logical channels in strict priority order. However, this could result in starvation of lower-priority channels—all resources would go to the high-priority channel until the buffer is empty. Typically, an operator would instead like to provide at least some throughput for low-priority services as well. Furthermore, as NR is designed to handle a mix of a wide range of traffic types, a more elaborate scheme is needed. For example, traffic due to a file upload should not necessarily exploit a grant intended for a latency-critical service.

The starvation problem is present already in LTE where it is handled by assigning a guaranteed data rate to each channel. The logical channels are then served in decreasing priority order up to their guaranteed data rate, which avoids starvation as long as the scheduled data rate is at least as large as the sum of the guaranteed data rates. Beyond the guaranteed data rates, channels are served in strict priority order until the grant is fully exploited, or the buffer is empty.

NR applies a similar approach. However, given the large flexibility of NR in terms of different transmission durations and a wider range of traffic types supported, a more advanced scheme is needed. One possibility would be to define different profiles, each outlining an allowed combination of logical channels, and explicitly signal the profile to use in the grant. However, in NR the profile to use is implicitly derived from other information available in the grant rather than explicitly signaled.

Upon reception of an uplink grant, two steps are performed. First, the device determines which logical channels are eligible for multiplexing using this grant. Second, the device determines the fraction of the resources that should be given to each of the logical channels.

The first step determines the logical channels from which data can be transmitted with the given grant. This can be seen as an implicitly derived profile. For each logical channel, the device can be configured with:

- The set of allowed subcarrier spacings this logical channel is allowed to use;
- The maximum PUSCH duration which is possible to schedule for this logical channel; and
- The set of serving cell, that is, the set of uplink component carriers the logical channel is allowed to be transmitted upon.

Only the logical channels for which the scheduling grant meets the restrictions configured are allowed to be transmitted using this grant, that is, are eligible for multiplexing at this particular time instant. In addition, the logical channel multiplexing can also be restricted for transmission without a dynamic grant.

Coupling the multiplexing rule to the PUSCH duration is in 3GPP motivated by the possibility to control whether latency-critical data should be allowed to exploit a grant intended for less time-critical data.

As an example, assume there are two data flows, each on a different logical channel. One logical channel carries latency-critical data and is given a high priority, while the other logical channel carries non-latency-critical data and is given a low priority. The gNB takes scheduling decisions based on, among other aspects, information about the buffer status in the device provided by the device. Assume that the gNB scheduled a relatively long PUSCH duration based on information that there is only nontime-critical information in the buffers. During the reception of the scheduling grant, time-critical information arrives to the device. Without the restriction on the maximum PUSCH duration, the device would transmit the latency-critical data, possibly multiplexed with other data, over a relatively long transmission duration and potentially not meeting the latency requirements set up for the particular service. Instead, a better approach would be to separately request a transmission during a short PUSCH duration for the latency critical data, something which is possible by configuring the maximum PUSCH duration appropriately. Since the logical channel carrying the latency-critical traffic has been configured with a higher priority than the channel carrying the non-latency-critical service, the noncritical service will not block transmission of the latency-critical data during the short PUSCH duration.

The reason to also include the subcarrier spacing is similar to the duration. In the case of multiple subcarrier spacings configured for a single device, a lower subcarrier spacing implies a longer slot duration and the reasoning above can also be applied in this case.

Restricting the uplink carriers allowed for a certain logical channel is motivated by the possibly different propagation conditions for different carriers and by dual connectivity. Two uplink carriers at vastly different carrier frequencies can have different reliability. Data which are critical to receive might be better to transmit on a lower carrier frequency to ensure good coverage, while less-sensitive data can be transmitted on a carrier with a higher carrier frequency and possibly spottier coverage. Another motivation is duplication, that is, the same data transmitted on multiple logical channels, to obtain diversity as mentioned in Section 6.4.2. If both logical channels would be transmitted on the same uplink carrier, the original motivation for duplication—to obtain a diversity effect—would be gone.

At this point in the process, the set of logical channels from which data are allowed to be transmitted given the current grant is established, based on the mapping-related parameters configured. Multiplexing of the different logical channels also needs to answer the question of how to distribute resources between the logical channels having data to transmit and eligible for transmission. This is done based on a set of priority-related parameters configured for each local channel:

- Priority;
- Prioritized bit rate (PBR); and
- Bucket size duration (BSD).

The prioritized bit rate and the bucket size duration together serve a similar purpose as the guaranteed bit rate in LTE but can account for the different transmission durations possible in NR. The product of the prioritized bit rate and the bucket size duration is in essence a bucket of bits that at a minimum should be transmitted for the given logical channel during a certain time. At each transmission instant, the logical channels are served in decreasing priority order, while trying to fulfill the requirement on the minimum number of bits to transmit. Excess capacity when all the logical channels are served up to the bucket size is distributed in strict priority order.

Priority handling and logical channel multiplexing are illustrated in Fig. 14.6.

14.2.2 SCHEDULING REQUEST

The uplink scheduler needs knowledge of devices with data to transmit and that therefore need to be scheduled. There is no need to provide uplink resources to a device with no data to transmit. Hence, as a minimum, the scheduler needs to know whether the device has data to transmit and should be given a grant. This is known as a *scheduling request*. Scheduling requests are used for devices not having a valid scheduling grant; devices that have a valid grant provide more detailed scheduling information to the gNB as discussed in the next section.

A scheduling request is a flag, raised by the device to request uplink resources from the uplink scheduler. Since the device requesting resources by definition has no PUSCH resource, the scheduling request is transmitted on the PUCCH using preconfigured and periodically reoccurring PUCCH resources dedicated to the device. With a dedicated scheduling-request mechanism, there is no need to provide the identity of the device requesting to be scheduled as the identity is implicitly known from the resources upon which the request is transmitted. When data with higher priority than already existing in the transmit buffers arrive at the device and the device has no grant and hence cannot transmit the data, the device transmits a scheduling request at the next possible instant and the gNB can assign a grant to the device upon reception of the request (see Fig. 14.7).

This is similar to the approach taken by LTE; however, NR supports configuration of *multiple* scheduling requests from a single device. A logical channel can be mapped to zero or more scheduling request configurations. This provides the gNB not only with information that there are data awaiting transmission in the device, but also *what type* of data are awaiting transmission. This is useful information for the gNB given the wider range of traffic types the NR is designed to

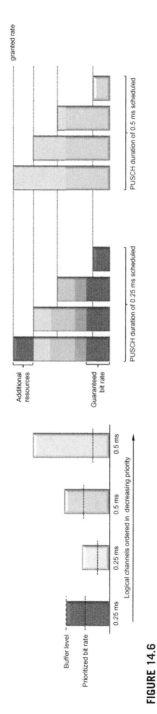

FIGURE 14.6

Example of logical channel prioritization for four different scheduled data rates and two different PUSCH durations.

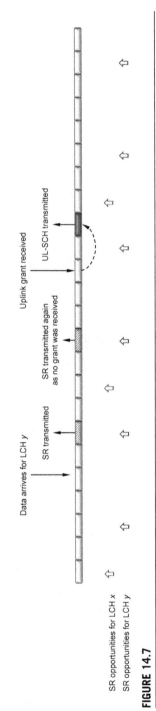

FIGURE 14.7

Example of scheduling request operation.

handle. For example, the gNB may want to schedule a device for transmission of latency-critical information but not for non-latency-critical information.

Each device can be assigned dedicated PUCCH scheduling request resources with a periodicity ranging from every second OFDM symbol to support very latency-critical services up to every 80 ms for low overhead. Only one scheduling request can be transmitted at a given time, that is, in the case of multiple logical channels having data to transmit a reasonable behavior is to trigger the scheduling request corresponding to the highest-priority logical channel. A scheduling request is repeated in subsequent resources, up to a configurable limit, until a grant is received from the gNB. It is also possible to configure a prohibit timer, controlling how often a scheduling request can be transmitted. In the case of multiple scheduling-request resources in a device, both of these configurations are done as per scheduling request resource.

A device which has not been configured with scheduling request resources relies on the random-access mechanism to request resources. This can be used to create a contention-based mechanism for requesting resources. Basically, contention-based designs are suitable for situations where there is a large number of devices in the cell and the traffic intensity, and hence the scheduling intensity, is low. In the case of higher traffic intensities, it is beneficial to set up at least one scheduling request resource for the device.

14.2.3 BUFFER STATUS REPORTS

Devices that already have a valid grant do not need to request uplink resources. However, to allow the scheduler to determine the amount of resources to grant to each device in the future, information about the buffer situation, discussed in this section, and the power availability, discussed in the next section, is useful. This information is provided to the scheduler as part of the uplink transmission through MAC control elements (see Section 6.4.4.1 for a discussion on MAC control elements and the general structure of a MAC header). The LCID field in one of the MAC subheaders is set to a reserved value indicating the presence of a buffer status report, as illustrated in Fig. 14.8.

From a scheduling perspective, buffer information for each logical channel is beneficial, although this could result in a significant overhead. Logical channels are therefore grouped into up to eight logical-channel groups and the reporting is done per group. The buffer-size field in a buffer-status report indicates the amount of data awaiting transmission across all logical channels in a logical-channel group. Four different formats for buffer status reports are defined, differing in how many logical-channel groups are included in one report and the resolution of the buffer status report. A buffer-status report can be triggered for the following reasons:

- Arrival of data with higher priority than currently in the transmission buffer—that is, data in a logical-channel group with higher priority than the one currently being transmitted—as this may impact the scheduling decision.

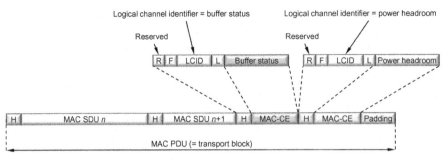

FIGURE 14.8

MAC control elements for buffer status reporting and power headroom reports.

- Periodically as controlled by a timer.
- Instead of padding. If the amount of padding required to match the scheduled transport block size is larger than a buffer-status report, a buffer-status report is inserted as it is better to exploit the available payload for useful scheduling information instead of padding if possible.

14.2.4 POWER HEADROOM REPORTS

In addition to buffer status, the amount of transmission power available in each device is also relevant for the uplink scheduler. There is little reason to schedule a higher data rate than the available transmission power can support. In the downlink, the available power is immediately known to the scheduler as the power amplifier is in the same node as the scheduler. For the uplink, the power availability, or *power headroom*, needs to be provided to the gNB. Power headroom reports are therefore transmitted from the device to the gNB in a similar way as the buffer-status reports—that is, only when the device is scheduled to transmit on the UL-SCH. A power headroom report can be triggered for the following reasons:

- Periodically as controlled by a timer;
- Change in path loss (the difference between the current power headroom and the last report is larger than a configurable threshold);
- Instead of padding (for the same reason as buffer-status reports).

It is also possible to configure a prohibit timer to control the minimum time between two power-headroom reports and thereby the signaling load on the uplink.

There are three different types of power-headroom reports defined in NR, *Type 1*, *Type 2*, and *Type 3*. In the case of carrier aggregation or dual connectivity, multiple power headroom reports can be contained in a single message (MAC control element).

Type 1 power headroom reporting reflects the power headroom assuming PUSCH-only transmission on the carrier. It is valid for a certain component

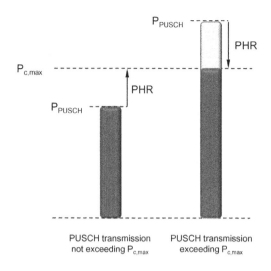

FIGURE 14.9

Illustration of power headroom reports.

carrier, assuming that the device was scheduled for PUSCH transmission during a certain duration, and includes the power headroom and the corresponding value of the *maximum per-carrier transmit power* for component carrier c is denoted, $P_{CMAX,c}$. The value of $P_{CMAX,c}$ is explicitly configured and should hence be known to the gNB, but since it can be separately configured for a normal uplink carrier and a supplementary uplink carrier, both belonging to the same cell (that is, having the same associated downlink component carrier), the gNB needs to know which value the device used and hence which carrier the report belongs to.

It can be noted that the power headroom is not a measure of the difference between the maximum per-carrier transmit power and the actual carrier transmit power. Rather, the power headroom is a measure of the difference between $P_{CMAX,c}$ and the transmit power that would have been used assuming that there would have been no upper limit on the transmit power (see Fig. 14.9). Thus, the power headroom can very well be negative, indicating that the per-carrier transmit power was limited by $P_{CMAX,c}$ at the time of the power headroom reporting—that is, the network has scheduled a higher data rate than the device can support given the available transmission power. As the network knows what modulation-and-coding scheme and resource size the device used for transmission in the time duration to which the power-headroom report corresponds, it can determine the valid combinations of modulation-and-coding scheme and resource size allocation, assuming that the downlink path loss is constant.

Type 1 power headroom can also be reported when there is no actual PUSCH transmission. This can be seen as the power headroom assuming a default transmission configuration corresponding to the minimum possible resource assignment.

Type 2 power headroom reporting is similar to type 1, but assumes simultaneous PUSCH and PUCCH reporting, a feature that is not fully supported in the first release of the NR specifications but planned for finalization in later releases.

Type 3 power headroom reporting is used to handle SRS switching, that is, SRS transmissions on an uplink carrier where the device is not configured to transmit PUSCH. The intention with this report is to be able to evaluate the uplink quality of alternative uplink carries and, if deemed advantageous, (re) configure the device to use this carrier for uplink transmission instead.

Compared to power control, which can operate different power-control processes for different beam-pair links (see Chapter 15), the power-headroom report is per carrier and does not explicitly take beam-based operation into account. One reason is that the network is in control of the beams used for transmission and hence can determine the beam arrangement corresponding to a certain power-headroom report.

14.3 SCHEDULING AND DYNAMIC TDD

One of the key features of NR is the support for dynamic TDD, where the scheduler dynamically determines the transmission direction. Although the description uses the term dynamic TDD, the framework can in principle be applied to half-duplex operation in general, including half-duplex FDD. Since a half-duplex device cannot transmit and receive simultaneously, there is a need to split the resources between the two directions. As mentioned in Chapter 7 three different signaling mechanisms can provide information to the device on whether the resources are used for uplink or downlink transmission:

- Dynamic signaling for the scheduled device;
- Semistatic signaling using RRC; and
- Dynamic slot-format indication shared by a group of devices, primarily intended for nonscheduled devices.

The scheduler is responsible for the dynamic signaling for the scheduled device, that is, the first of the three bullets above.

In the case of a device capable of full-duplex operation, the scheduler can schedule uplink and downlink independently of each other and there is limited, if any, need for the uplink and downlink scheduler to coordinate their decisions.

In the case of a half-duplex device, on the other hand, it is up to the scheduler to ensure that a half-duplex device is not requested to simultaneously receive and transmit. If a semi-static uplink—downlink pattern has been configured, the schedulers obviously need to obey this pattern as well as it cannot, for example, schedule an uplink transmission in a slot configured for downlink usage only.

14.4 **TRANSMISSION WITHOUT A DYNAMIC GRANT**

Dynamic scheduling, as described above, is the main mode of operation in NR. For each transmission interval, for example, a slot, the scheduler uses control signaling to instruct the device to transmit or receive. It is flexible and can adopt to rapid variations in the traffic behavior, but obviously requires associated control signaling; control signaling that in some situations it is desirable to avoid. NR therefore also supports transmission schemes not relying on dynamic grants.

In the downlink, *semi-persistent scheduling* is supported where the device is configured with a periodicity of the data transmissions using RRC signaling. Activation of semi-persistent scheduling is done using the PDCCH as for dynamic scheduling but with the CS-RNTI instead of the normal C-RNTI.[3] The PDCCH also carries the necessary information in terms of time−frequency resources and other parameters needed in a similar way as dynamic scheduling. The hybrid-ARQ process number is derived from the time when the downlink data transmission starts according to a formula. Upon activation of semi-persistent scheduling, the device receives downlink data transmission periodically according to the RRC-configured periodicity using the transmission parameters indicated on the PDCCH activating the transmission. Hence, control signaling is only used once and the overhead is reduced. After enabling semi-persistent scheduling, the device continues to monitor the set of candidate PDCCHs for uplink and downlink scheduling commands. This is useful in the case that there are occasional transmissions of large amounts of data for which the semi-persistent allocation is not sufficient. It is also used to handle hybrid-ARQ retransmissions which are dynamically scheduled.

In the uplink, two schemes for transmission without a dynamic grant are supported, differing in the ways they are activated (see Fig. 14.10):

- *Configured grant type 1*, where an uplink grant is provided by RRC, including activation of the grant; and
- *Configured grant type 2*, where the transmission periodicity is provided by RRC and L1/L2 control signaling is used to activate/deactivate the transmission in a similar way as in the downlink case.

The benefits for the two schemes are similar, namely to reduce control signaling overhead and, to some extent, to reduce the latency before uplink data transmission as no scheduling request−grant cycle is needed prior to data transmission.

Type 1 sets all the transmission parameters, including periodicity, time offset, and frequency resources as well as modulation-and-coding scheme of possible uplink transmissions, using RRC signaling. Upon receiving the RRC

---

[3]Each device has two identities, the "normal" C-RNTI for dynamic scheduling and the CS-RNTI for activation/deactivation of semipersistent scheduling.

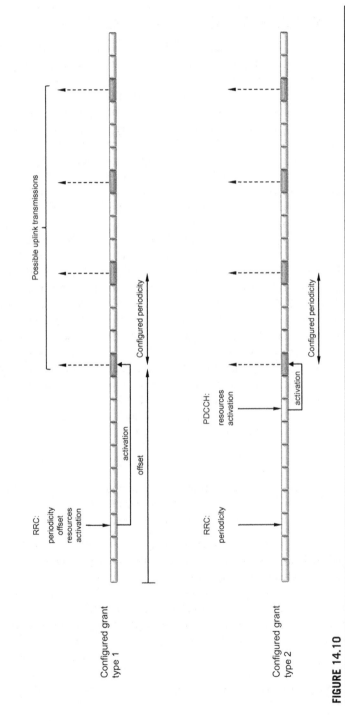

FIGURE 14.10

Uplink transmission without a dynamic grant.

configuration, the device can start to use the configured grant for transmission in the time instant given by the periodicity and offset. The reason for the offset is to control at what time instants the device is allowed to transmit. There is no notion of activation time in the RRC signaling in general; RRC configurations take effect as soon as they are received correctly. This point in time may vary as it depends on whether RLC retransmissions were needed to deliver the RRC command or not. To avoid this ambiguity, a time offset relative to the SFN is included in the configuration.

Type 2 is similar to downlink semi-persistent scheduling. RRC signaling is used to configure the periodicity, while the transmission parameters are provided as part of the activation using the PDCCH. Upon receiving the activation command, the device transmits according to the preconfigured periodicity if there are data in the buffer. If there are no data to transmit, the device will, similarly to type 1, not transmit anything. Note that no time offset is needed in this case as the activation time is well defined by the PDCCH transmission instant.

The device acknowledges the activation/deactivation of the configured grant type 2 by sending a MAC control element in the uplink. If there are no data awaiting transmission when the activation is received, the network would not know if the absence of transmission is due to the activation command not being received by the device or if it is due to an empty transmission buffer. The acknowledgment helps in resolving this ambiguity.

In both these schemes it is possible to configure multiple devices with overlapping time—frequency resources in the uplink. In this case it is up to the network to differentiate between transmissions from the different devices.

14.5 DISCONTINUOUS RECEPTION

Packet-data traffic is often highly bursty, with occasional periods of transmission activity followed by longer periods of silence. From a delay perspective, it is beneficial to monitor the downlink control signaling in each slot (or even more frequently) to receive uplink grants or downlink data transmissions and instantaneously react on changes in the traffic behavior. At the same time this comes at a cost in terms of power consumption at the device; the receiver circuitry in a typical device represents a non-negligible amount of power consumption. To reduce the device power consumption, NR includes mechanisms for *discontinuous reception* (DRX), following the same framework as in LTE with enhancements to handle multiple numerologies. Bandwidth adaptation and carrier activation are two other examples of power-saving mechanisms.

The basic mechanism for DRX is a configurable DRX cycle in the device. With a DRX cycle configured, the device monitors the downlink control signaling only when active, sleeping with the receiver circuitry switched off the remaining time. This allows for a significant reduction in power

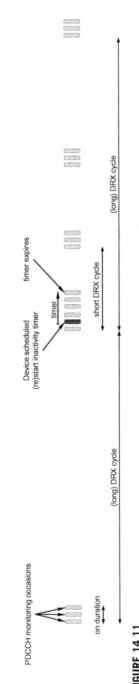

FIGURE 14.11

DRX operation.

consumption: the longer the cycle, the lower the power consumption. Naturally, this implies restrictions to the scheduler as the device can be addressed only when active according to the DRX cycle.

In many situations, if the device has been scheduled and is active with receiving or transmitting data, it is highly likely it will be scheduled again in the near future. One reason could be that it was not possible to transmit all the data in the transmission buffer in using one scheduling occasion and hence additional occasions are needed. Waiting until the next activity period according to the DRX cycle, although possible, would result in additional delays. Hence, to reduce the delays, the device remains in the active state for a certain configurable time after being scheduled. This is implemented by the device (re)starting an inactivity timer every time it is scheduled and remaining awake until the time expires, as illustrated at the top of Fig. 14.11. Due to the fact that NR can handle multiple numerologies, the DRX timers are specified in milliseconds in order not to tie the DRX periodicity to a certain numerology.

Hybrid-ARQ retransmissions are asynchronous in both uplink and downlink. If the device has been scheduled a transmission in the downlink it could not decode, a typical situation is that the gNB retransmits the data at a later time instant, often as soon as possible. Therefore, the DRX functionality has a configurable timer which is started after an erroneously received transport block and used to wake up the device receiver when it is likely for the gNB to schedule a retransmission. The value of the timer is preferably set to match the roundtrip time in the hybrid-ARQ protocol; a roundtrip time that depends on the implementation.

The above mechanism, a (long) DRX cycle in combination with the device remaining awake for some period after being scheduled, is sufficient for most scenarios. However, some services, most notably voice-over-IP, are characterized by periods of regular transmission, followed by periods of no or very little activity. To handle these services, a second short DRX cycle can optionally be used in addition to the long cycle described above. Normally, the device follows the long DRX cycle, but if it has recently been scheduled, it follows a shorter DRX cycle for some time. Handling voice-over-IP in this scenario can be done by setting the short DRX cycle to 20 ms, as the voice codec typically delivers a voice-over-IP packet per 20 ms. The long DRX cycle is then used to handle longer periods of silence between talk spurts.

In addition to the RRC configuration of the DRX parameters, the gNB can terminate an "on duration" and instruct the device to follow the long DRX cycle. This can be used to reduce the device power consumption if the gNB knows that no additional data are awaiting transmission in the downlink and hence there is no need for the device to be active.

Uplink Power and Timing Control

15

Uplink power control and uplink timing control are the topics of this chapter. Power control serves the purpose of controlling the interference, mainly towards other cells as transmissions within the same cell typically are orthogonal. Timing control ensures that different device are received with the same timing, a prerequisite to maintain orthogonality between different transmissions.

15.1 UPLINK POWER CONTROL

NR uplink power control is the set of algorithms and tools by which the transmit power for different uplink physical channels and signals is controlled to ensure that they, to the extent possible, are received by the network at an appropriate power level. In the case of an uplink physical channel, the appropriate power is simply the received power needed for proper decoding of the information carried by the physical channel. At the same time, the transmit power should not be unnecessarily high as that would cause unnecessarily high interference to other uplink transmissions.

The appropriate transmit power will depend on the channel properties, including the channel attenuation and the noise and interference level at the receiver side. It should also be noted that the required received power is directly dependent on the data rate. If the received power is too low one can thus either increase the transmit power or reduce the data rate. In other words, at least in the case of PUSCH transmission, there is an intimate relationship between power control and link adaptation (rate control).

Similar to LTE power control [28], NR uplink power control is based on a combination of:

- *Open-loop* power control, including support for *fractional path-loss compensation*, where the device estimates the uplink path loss based on downlink measurements and sets the transmit power accordingly.
- *Closed-loop* power control based on explicit power-control commands provided by the network. In practice, these power-control commands are determined based on prior network measurements of the received uplink power, thus the term "*closed loop.*"

5G NR: The Next Generation Wireless Access Technology. DOI: https://doi.org/10.1016/B978-0-12-814323-0.00015-6

The main difference, or rather extension, of NR uplink power control is the possibility for beam-based power control (see Section 15.1.2).

15.1.1 BASELINE POWER CONTROL

Power-control for PUSCH transmissions can, somewhat simplified, be described by the following expression:

$$P_{\text{PUSCH}} = min\{P_{CMAX}, P_0(j) + \alpha(j) \cdot PL(q) + 10 \cdot log_{10}(2^\mu \cdot M_{RB}) + \Delta_{TF} + \delta(l)\} \quad (15.1)$$

where

- P_{PUSCH} is the PUSCH transmit power;
- P_{CMAX} is the maximum allowed transmit power per carrier;
- $P_0(.)$ is a network-configurable parameter that can, somewhat simplified, be described as a target received power;
- $PL(.)$ is an estimate of the uplink path loss;
- $\alpha(.)$ is a network-configurable parameter related to fractional path-loss compensation;
- μ relates to the subcarrier spacing Δf used for the PUSCH transmission. More specifically, $\Delta f = 2^\mu \cdot 15$ kHz;
- M_{RB} is the number of resource blocks assigned for the PUSCH transmission;
- Δ_{TF} relates to the modulation scheme and channel-coding rate used for the PUSCH transmission;[1]
- $\delta(.)$ is the power adjustment due to the closed-loop power control.

The above expression describes uplink power control per carrier. If a device is configured with multiple uplink carriers (carrier aggregation and/or supplementary uplink), power control according to expression (15.1) is carried out separately for each carrier. The $min\{P_{CMAX}, \ldots\}$ part of the power-control expression then ensures that the power per carrier does not exceed the maximum allowed transmit power per carrier. However, there will also be a limit on the total device transmit power over all configured uplink carriers. In order to stay below this limit there will, in the end, be a need to coordinate the power setting between the different uplink carriers (see further Section 15.1.4). Such coordination is needed also in the case of LTE/NR dual-connectivity.

We will now consider the different parts of the above power control expression in somewhat more detail. When doing this we will initially ignore the parameters j, q, and l. The impact of these parameters will be discussed in Section 15.1.2.

The expression $P_0 + \alpha \cdot PL$ represents basic open-loop power control supporting fractional path-loss compensation. In the case of full path-loss compensation, corresponding to $\alpha = 1$, and under the assumption that the path-loss estimate PL

---

[1]The abbreviation TF = transport format, a term used in earlier 3GPP technologies but not used explicitly for NR.

is an accurate estimate of the uplink path loss, the open-loop power control adjusts the PUSCH transmit power so that the received power aligns with the "target received power" P_0. The quantity P_0 is provided as part of the power-control configuration and would typically depend on the target data rate but also on the noise and interference level experienced at the receiver.

The device is assumed to estimate the uplink path loss based on measurements on some downlink signal. The accuracy of the path-loss estimate thus partly depends on what extent downlink/uplink reciprocity holds to. Especially, in the case of FDD operation in paired spectra, the path-loss estimate will not be able to capture any frequency-dependent characteristics of the path loss.

In the case of fractional path-loss compensation, corresponding to $\alpha < 1$, the path loss will not be fully compensated for and the received power will even on average vary depending on the location of the device within the cell, with lower received power for devices with higher path loss, in practice for devices at larger distance from the cell site. This must then be compensated for by adjusting the uplink data rate accordingly.

The benefit of fractional path-loss compensation is reduced interference to neighbor cells. This comes at the price of larger variations in the service quality, with reduced data-rate availability for devices closer to the cell border.

The term $10 \cdot log(2^{\mu} \cdot M_{RB})$ reflects the fact that, everything else unchanged, the received power, and thus also the transmit power, should be proportional to the bandwidth assigned for the transmission. Thus, assuming full path-loss compensation ($\alpha = 1$), P_0 can more accurately be described as a *normalized* target received power. Especially, assuming full path-loss compensation, P_0 is the target received power assuming transmission over a single resource block with 15 kHz numerology.

The term Δ_{TF} tries to model how the required received power varies when the number of information bits per resource element varies due to different modulation schemes and channel-coding rates. More precisely

$$\Delta_{TF} = 10 \cdot log\big((2^{1.25 \cdot \gamma} - 1) \cdot \beta\big) \tag{15.2}$$

where γ is the number of information bits in the PUSCH transmission, normalized by the number of resource elements used for the transmission not including resource elements used for demodulation reference symbols.

The factor β equals 1 in the case of data transmission on PUSCH but can be set to a different value in the case that the PUSCH carries layer-1 control signaling (UCI).[2]

It can be noted that, ignoring the factor β, the expression for Δ_{TF} is essentially a rewrite of the Shannon channel capacity $C = W \cdot log_2(1 + SNR)$ with an additional factor 1.25. In other words, Δ_{TF} can be seen as modeling link capacity as 80% of Shannon capacity.

---

[2]Note that one could equally well have described this as a separate term $10 \cdot log(\beta)$ applied when PUSCH carries UCI.

The term Δ_{TF} is not always included when determining the PUSCH transmit power.

- The term Δ_{TF} is only used for single-layer transmission, that is, $\Delta_{TF} = 0$ in the case of uplink multi-layer transmission;
- The term Δ_{TF} can, in general, be disabled. Δ_{TF} should, for example, not be used in combination with fractional power control. Adjusting the transmit power to compensate for different data rates would counteract any adjustment of the data rate to compensate for the variations in received power due to fractional power control as described above.

Finally, the term $\delta(.)$ is the power adjustment related to closed-loop power control. The network can adjust $\delta(.)$ by a certain step given by a *power-control command* provided by the network, thereby adjusting the transmit power based on network measurements of the received power. The power control commands are carried in the TPC field within uplink scheduling grants (DCI formats $0-0$ and $0-1$). Power control commands can also be carried jointly to multiple devices by means of DCI format $2-2$. Each power control command consists of 2 bits corresponding to four different update steps (-1 dB, 0 dB, $+1$ dB, $+3$ dB). The reason for including 0 dB as an update step is that a power-control command is included in every scheduling grant and it is desirable not to have to adjust the PUSCH transmit power for each grant.

15.1.2 BEAM-BASED POWER CONTROL

In the discussion above we ignored the parameter j for the open-loop parameters $P_0(.)$ and $\alpha(.)$, the parameter q in the path loss estimate $PL(.)$, and the parameter l in the closed-loop power adjustment $\delta(.)$. The primary aim of these parameters is to take beam-forming into account for the uplink power control.

15.1.2.1 Multiple Path-Loss-Estimation Processes

In the case of uplink beam-forming, the uplink-path-loss estimate $PL(q)$ used to determine the transmit power according to expression (15.1) should reflect the path loss, including the beam-forming gains, of the uplink beam pair to be used for the PUSCH transmission. Assuming beam correspondence, this can be achieved by estimating the path loss based on measurements on a downlink reference signal transmitted over the corresponding downlink beam pair. As the uplink beam used for the transmission pair may change between PUSCH transmissions, the device may thus have to retain multiple path-loss estimates, corresponding to different candidate beam pairs, in practice, path-loss estimates based on measurements on different downlink reference signals. When actual PUSCH transmission is to take place over a specific beam pair, the path-loss estimate corresponding to that beam pair is then used when determining the PUSCH transmit power according to the power-control expression (15.1).

This is enabled by the parameter q in the path-loss estimate $PL(q)$ of Eq. (15.1). The network configures the device with a set of downlink reference signals (CSI-RS or SS block) on which path loss is to be estimated, with each reference signal being associated with a specific value of q. In order not to put too high requirements on the device, there can be at most four parallel path-loss-estimation processes, each corresponding to a specific value of q. The network also configures a mapping from the possible SRI values provided in the scheduling grant to the up to four different values of q. In the end there is thus a mapping from each of the possible SRI values provided in the scheduling grant to one of up to four configured downlink reference signals and thus, indirectly, a mapping from each of the possible SRI values to one of up to four path-loss estimates reflecting the path loss of a specific beam pair. When a PUSCH transmission is scheduled by a scheduling grant including SRI, the path-loss estimate associated with that SRI is used when determining the transmit power for the scheduled PUSCH transmission.

The procedure is illustrated in Fig. 15.1 for the case of two beam pairs. The device is configured with two downlink reference signals (CSI-RS or SS block) that in practice will be transmitted on the downlink over a first and second beam pair, respectively. The device is running two path-loss-estimation processes in parallel, estimating the path loss $PL(1)$ for the first beam pair based on measurements on reference signal RS-1 and the path loss $PL(2)$ for the second beam pair based on measurements on reference-signal RS-2. The parameter q associates SRI = 1 with RS-1 and thus indirectly with $PL(1)$. Likewise, SRI = 2 is associated with RS-2 and thus indirectly with $PL(2)$. When the device is scheduled for PUSCH transmission with the SRI of the scheduling grant set to 1, the transmit power of the scheduled PUSCH transmission is determined based on the path-loss estimate $PL(1)$ that is, the path-loss estimate based on measurements on RS-1. Thus, assuming beam correspondence the path-loss estimate reflects the path loss of the beam pair over which the PUSCH is transmitted. If the device is instead scheduled for PUSCH transmission with SRI = 2, the path-loss estimate $PL(2)$, reflecting the path loss of the beam pair corresponding to SRI = 2, is used to determine the transmit power for the scheduled PUSCH transmission.

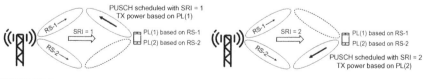

FIGURE 15.1

Use of multiple power-estimation processes to enable uplink power control in the case of dynamic beam management.

15.1.2.2 Multiple Open-Loop-Parameter Sets

In the PUSCH power-control expression (15.1), the open-loop parameters P_0 and α are associated with a parameter j. This simply reflects that there may be multiple open-loop-parameter pairs $\{P_0, \alpha\}$. Partly, different open-loop parameters will be used for different types of PUSCH transmission (random-access "message 3" transmission, see Section 16.2, grant-free PUSCH transmissions, and scheduled PUSCH transmissions). However, there is also a possibility to have multiple pairs of open-loop parameter for scheduled PUSCH transmission, where the pair to use for a certain PUSCH transmission can be selected based on the SRI similar to the selection of path-loss estimates as described above. In practice this implies that the open-loop parameters P_0 and α will depend on the uplink beam.

For the power setting of random-message 3, which in the NR specification corresponds to $j = 0$, α always equals 1. In other words, fractional power control is not used for message-3 transmission. Furthermore, the parameter P_0 can, for message 3, be calculated based on information in the random-access configuration.

For other PUSCH transmissions the device can be configured with different open-loop-parameter pairs $\{P_0(j), \alpha(j)\}$, corresponding to different values for the parameter j. Parameter pair $\{P_0(1), \alpha(1)\}$ should be used in the case of grant-free PUSCH transmission, while the remaining parameter pairs are associated with scheduled PUSCH transmission. Each possible value of the SRI that can be provided as part of the uplink scheduling grant is associated with one of the configured open-loop-parameter pairs. When a PUSCH transmission is scheduled with a certain SRI included in the scheduling grant, the open-loop parameters associated with that SRI are used when determining the transmit power for the scheduled PUSCH transmission.

15.1.2.3 Multiple Closed-Loop Processes

The final parameter is the parameter l for the closed-loop process. PUSCH power control allows for the configuration of two independent closed-loop processes, associated with $l = 1$ and $l = 2$, respectively. Similar to the possibility for multiple path-loss estimates and multiple open-loop-parameter sets, the selection of l, that is, the selection of closed-loop process, can be tied to the SRI included in the scheduling grant by associating each possible value of the SRI to one of the closed-loop processes.

15.1.3 POWER CONTROL FOR PUCCH

Power control for PUCCH follows essentially the same principles as power control for PUSCH, with some minor differences.

First, for PUCCH power control, there is no fractional path-loss compensation, that is, the parameter α always equals one.

Furthermore, for PUCCH power control, the closed-loop power control commands are carried within DCI formats $1-0$ and $1-1$, that is, within downlink scheduling assignments rather than within uplink scheduling grants, which is the case for PUSCH power control. One reason for uplink PUCCH transmissions is the transmission of hybrid-ARQ acknowledgments as a response to downlink transmissions. Such downlink transmissions are typically associated with downlink scheduling assignments on PDCCH and the corresponding power-control commands could thus be used to adjust the PUCCH transmit power prior to the transmission of the hybrid-ARQ acknowledgments. Similar to PUSCH, power-control commands can also be carried jointly to multiple devices by means of DCI format $2-2$.

15.1.4 POWER CONTROL IN THE CASE OF MULTIPLE UPLINK CARRIERS

The above procedures describe how to set the transmit power for a given physical channel in the case of a single uplink carrier. For each such carrier there is a maximum allowed transmit power P_{CMAX} and the $min\{P_{CMAX}, \ldots\}$ part of the power-control expression ensures that the per-carrier transmit power of a carrier does not exceed power P_{CMAX}.[3]

In many cases, a device is configured with multiple uplink carriers:

- Multiple uplink carriers in a carrier aggregation scenario;
- An additional supplementary uplink carrier in the case of SUL.

In addition to the maximum per-carrier transmit power P_{CMAX}, there is a limit P_{TMAX} on the total transmitted power over all carriers. For a device configured for NR transmission on multiple uplink carriers, P_{CMAX} should obviously not exceed P_{TMAX}. However, the sum of P_{CMAX} over all configured uplink carriers may very well, and often will, exceed P_{TMAX}. The reason is that a device will often not transmit simultaneously on all its configured uplink carriers and the device should then preferably still be able to transmit with the maximum allowed power P_{TMAX}. Thus, there may be situations when the sum of the transmit power of each carrier given by the power-control expression (15.1) exceeds P_{TMAX}. In that case, the power of each carrier needs to be scaled down to ensure that the eventual transmit power of the device does not exceed the maximum allowed value.

Another situation that needs to be taken care of is the simultaneous uplink transmission of LTE and NR in the case of a device operating in dual-connectivity between LTE and NR. Note that, at least in an initial phase of NR deployment this will be the normal mode-of-operation as the first release of the NR specifications only support non-standalone NR deployments. In this case, the

---

[3]Note that, in contrast to LTE, at least for NR release 15 there is not simultaneous PUCCH and PUSCH transmission on a carrier, and thus there is at most one physical channel transmitted on an uplink carrier at a given time instant.

transmission on LTE may limit the power available for NR transmission and vice versa. The basic principle is that the LTE transmission has priority, that is the LTE carrier is transmitted with the power given by the LTE uplink power control [28]. The NR transmission can then use whatever power is left up to the power given by the power-control expression (15.1).

The reason for prioritizing LTE over NR is multifold:

- In the specification of NR, including the support for NR/LTE dual connectivity, there has been an aim to as much as possible avoid any impact on the LTE specifications. Imposing restrictions on the LTE power control, due to the simultaneous transmission on NR, would have implied such an impact.
- At least initially, LTE/NR dual-connectivity will have LTE providing the control-plane signaling, that is, LTE is used for the master cell group (MCG). The LTE link is thus more critical in terms of retaining connectivity and it makes sense to priorities that link over the "secondary" NR link.

15.2 UPLINK TIMING CONTROL

The NR uplink allows for uplink intracell orthogonality, implying that uplink transmissions received from different devices within a cell do not cause interference to each other. A requirement for this *uplink orthogonality* to hold is that the uplink slot boundaries for a given numerology are (approximately) time aligned at the base station. More specifically, any timing misalignment between received signals should fall within the cyclic prefix. To ensure such receiver-side time alignment, NR includes a mechanism for *transmit-timing advance*. The mechanism is similar to the corresponding mechanism in LTE, the main difference being the use of different timing advance step sizes for different numerologies.

In essence, timing advance is a negative offset, at the device, between the start of a downlink slot as observed by the device and the start of a slot in the uplink. By controlling the offset appropriately for each device, the network can control the timing of the signals received at the base station from the devices. Devices far from the base station encounter a larger propagation delay and therefore need to start their uplink transmissions somewhat in advance, compared to devices closer to the base station, as illustrated in Fig. 15.2. In this specific example, the first device is located close to the base station and experiences a small propagation delay, $T_{P,1}$. Thus, for this device, a small value of the timing advance offset $T_{A,1}$ is sufficient to compensate for the propagation delay and to ensure the correct timing at the base station. However, a larger value of the timing advance is required for the second device, which is located at a larger distance from the base station and thus experiences a larger propagation delay.

The timing-advance value for each device is determined by the network based on measurements on the respective uplink transmissions. Hence, as long as a

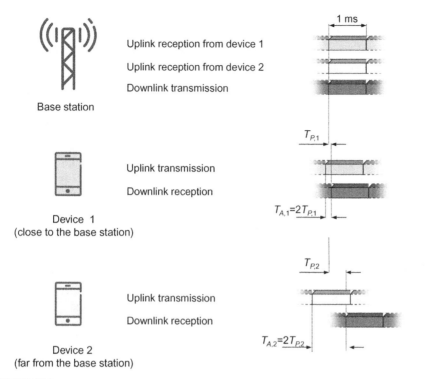

1 ms

Base station

Uplink reception from device 1

Uplink reception from device 2

Downlink transmission

$T_{P,1}$

Device 1
(close to the base station)

Uplink transmission

Downlink reception

$T_{A,1}=2T_{P,1}$

$T_{P,2}$

Device 2
(far from the base station)

Uplink transmission

Downlink reception

$T_{A,2}=2T_{P,2}$

FIGURE 15.2

Uplink timing advance.

device carries out uplink data transmission, this can be used by the receiving base station to estimate the uplink receive timing and thus be a source for the timing-advance commands. Sounding reference signals can be used as a regular signal to measure upon, but in principle the base station can use any signal transmitted from the devices.

Based on the uplink measurements, the network determines the required timing correction for each device. If the timing of a specific device needs correction, the network issues a timing-advance command for this specific device, instructing it to retard or advance its timing relative to the current uplink timing. The user-specific timing-advance command is transmitted as a MAC control element on the DL-SCH. Typically, timing-advance commands to a device are transmitted relatively infrequently—for example, one or a few times per second—but obviously this depends on how fast the device is moving.

The procedure described so far is in essence identical to the one used for LTE. As discussed above, the target of timing advance is to keep the timing mis-alignment within the size of the cyclic prefix and the step size of the timing advance is therefore chosen as a fraction of the cyclic prefix. However, as NR supports multiple numerologies with the cyclic prefix being shorter the higher the subcarrier spacing, the timing advance step size is scaled in proportion to the cyclic prefix length and given by the subcarrier spacing of the active uplink bandwidth part.

If the device has not received a timing-advance command during a (configurable) period, the device assumes it has lost the uplink synchronization. In this case, the device must reestablish uplink timing using the random-access procedure prior to any PUSCH or PUCCH transmission in the uplink.

For carrier aggregation, there may be multiple component carriers transmitted from a single device. A straightforward way of handling this would be to apply the same timing-advance value for all uplink component carriers. However, if different uplink carriers are received at different geographical locations, for example, by using remote radio heads for some carriers but not others, different carriers would need different timing advance values. Dual connectivity with different uplink carriers terminated at different sites is an example when this is relevant. To handle such scenarios, a similar approach as in LTE is taken, namely to group uplink carriers in so-called timing advanced groups (TAGs) and allow for different timing advance commands for different TAGs. All component carriers in the same group are subject to the same timing-advance command. The timing advance step size is determined by the highest subcarriers pacing among the carriers in a timing advance group.

Initial Access

16

Within NR, the initial-access functionality includes:

- The functions and procedures by which a device initially finds a cell when entering the coverage area of a system;
- The functions and procedures by which a device in idle/inactive state accesses the network, typically to request the set-up of a connection and commonly referred to as *random access.*

To quite a large extent, similar functionality is also used in other situations. The basic network signals used to initially find a cell can, for example, also be used to find new cells when a device is moving within the coverage area of the network. Furthermore, when accessing a new cell, the same basic random-access procedure as for initial access may be used. The random-access procedure may also be used by a device in connected state, for example, to request resources for uplink transmission or to re-establish uplink synchronization.

In this chapter, the details of cell search, system-information delivery, and random access are described.

16.1 CELL SEARCH

Cell search covers the functions and procedures by which a device finds new cells. Cell search is carried out when a device is initially entering the coverage area of a system. To enable mobility, cell search is also continuously carried out by devices moving within the system, both when the device is connected to the network and when in idle/inactive state. Here we will describe cell search based on so-called *SS blocks*, which are used for initial cell search as well as idle/inactive-state mobility. Cell search based on SS blocks can also be used for connected-state mobility, although in that case cell search can also be based on CSI-RS explicitly configured for the device.

16.1.1 THE SS BLOCK

To enable devices to find a cell when entering a system, as well as to find new cells when moving within the system, a synchronization signal consisting of two parts, the *Primary Synchronization Signal* (PSS) and the *Secondary*

Synchronization Signal (SSS), is periodically transmitted on the downlink from each NR cell. The PSS/SSS, together with the *Physical Broadcast Channel* (PBCH), is jointly referred to as a *Synchronization Signal Block* or SS block.[1]

The SS block serves a similar purpose and, in many respects, has a similar structure as the PSS/SSS/PBCH of LTE [28].[2] However, there are some important differences between the LTE PSS/SSS/PBCH and the NR SS block. At least partly, the origin of these differences can be traced back to some NR-specific requirements and characteristics including the aim to reduce the amount of "always-on" signals, as discussed in Section 5.2, and the possibility for beam-forming during initial access.

As with all NR downlink transmissions, SS-block transmission is based on OFDM. In other words, the SS block is transmitted on a set of time/frequency resources (resource elements) within the basic OFDM grid discussed in Section 7.3. Fig. 16.1 illustrates the time/frequency structure of a single SS block transmission. As can be seen, the SS block spans four OFDM symbols in the time domain and 240 subcarriers in the frequency domain.

- The PSS is transmitted in the first OFDM symbol of the SS block and occupies 127 subcarriers in the frequency domain. The remaining subcarriers are empty.
- The SSS is transmitted in the third OFDM symbol of the SS block and occupies the same set of subcarriers as the PSS. There are eight and nine empty subcarriers on each side of the SSS.

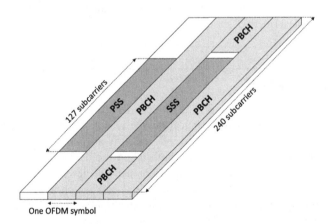

FIGURE 16.1

Time/frequency structure of a single SS block consisting of PSS, SSS, and PBCH.

---

[1]Sometimes only PSS and SSS are included in the term "SS block." Here we will refer to the triplet PSS, SSS, and PBCH as an SS block though.
[2]Even though the terms PSS, SSS, and PBCH are used also in LTE, the term SS block is not used within the context of LTE.

Table 16.1 SS Block Numerologies and Corresponding Frequency Ranges

Numerology (kHz)	SSB Bandwidth[a] (MHz)	SSB Duration (µs)	Frequency Range
15	3.6	≈285	FR1 (<3 GHz)
30	7.2	≈143	FR1
120	28.8	≈36	FR2
240	57.6	≈18	FR2

[a]*The SS-block bandwidth is simply the number of subcarriers used for SS block (240) multiplied by the SS-block subcarrier spacing.*

- The PBCH is transmitted within the second and fourth OFDM symbols of the SS block. In addition, PBCH transmission also uses 48 subcarriers on each side of the SSS.

The total number of resource elements used for PBCH transmission per SS block thus equals 576. Note that this includes resource elements for the PBCH itself but also resource elements for the demodulation reference signals (DMRS) needed for coherent demodulation of the PBCH.

Different numerologies can be used for SS block transmission. However, to limit the need for devices to simultaneously search for SS blocks of different numerologies, there is in many cases only a single SS-block numerology defined for a given frequency band.

Table 16.1 lists the different numerologies applicable for SS-block transmission together with the corresponding SS-block bandwidth and time duration, and the frequency range for which each specific numerology applies.[3] Note that 60 kHz numerology cannot be used for SS-block transmission regardless of frequency range. In contrast, 240 kHz numerology can be used for SS-block transmission although it is currently not supported for other downlink transmissions. The reason to support 240 kHz SS-block numerology is to enable a very short time duration for each SS block. This is relevant in the case of beam-sweeping over many beams with a corresponding large number of time multiplexed SS blocks (see further details in Section 16.1.4).

16.1.2 FREQUENCY-DOMAIN POSITION OF SS BLOCK

In LTE, the PSS and SSS are always located at the center of the carrier. Thus, once an LTE device has found a PSS/SSS, that is, found a carrier, it inherently knows the center frequency of the found carrier. The drawback with this approach, that is, always locating the PSS/SSS at the center of the carrier, is that

---

[3]Note that, although the frequency range for 30 kHz SS-block numerology fully overlaps with the frequency range for 15 kHz numerology, for a given frequency band within the lower frequency range there is in many cases only a single numerology supported.

a device with no a priori knowledge of the frequency-domain carrier position must search for PSS/SSS at all possible carrier positions (the "*carrier raster*").

To allow for faster cell search, a different approach has been adopted for NR. Rather than always being located at the center of the carrier, implying that the possible SS-block locations coincide with the carrier raster, there are, within each frequency band, a more limited set of possible locations of SS block, referred to as the "*synchronization raster*". Instead of searching for an SS block at each position of the carrier raster, a device thus only needs to search for an SS block on the sparser synchronization raster.

As carriers can still be located at an arbitrary position on the more dense carrier raster, the SS block may not end up at the center of a carrier. The SS block may not even end up aligned with the resource-block grid. Hence, once the SS block has been found, the device must be explicitly informed about the exact SS-block frequency-domain position within the carrier. This is done by means of information partly within the SS block itself, more specifically information carried by the PBCH (Section 16.1.5.3), and partly within the remaining broadcast system information (see further Section 6.1.6).

16.1.3 SS BLOCK PERIODICITY

The SS block is transmitted periodically with a period that may vary from 5 ms up to 160 ms. However, devices doing initial cell search, as well as devices in inactive/idle state doing cell search for mobility, can assume that the SS block is repeated at least once every 20 ms. This allows for a device that searches for an SS block in the frequency domain to know how long it must stay on each frequency before concluding that there is no PSS/SSS present and that it should move on to the next frequency within the synchronization raster.

The 20 ms SS-block periodicity is four times longer than the corresponding 5 ms periodicity of LTE PSS/SSS transmission. The longer SS-block period was selected to allow for enhanced NR network energy performance and in general to follow the ultra-lean design paradigm described in Section 5.2. The drawback with a longer SS-block period is that a device must stay on each frequency for a longer time in order to conclude that there is no PSS/SSS on the frequency. However, this is compensated for by the sparse synchronization raster discussed above, which reduces the number of frequency-domain locations on which a device must search for an SS block.

Even though devices doing initial cell search can assume that the SS block is repeated at least once every 20 ms, there are situations when there may be reasons to use either a shorter or longer SS-block periodicity:

- A shorter SS-block periodicity may be used to enable faster cell search for devices in connected mode.
- A longer SS-block periodicity may be used to further enhance network energy performance. A carrier with an SS-block periodicity larger than 20 ms may

not be found by devices doing initial access. However, such a carrier could still be used by devices in connected mode, for example, as a secondary carrier in a carrier-aggregation scenario.

It should be noted that there is even the possibility to deploy secondary carriers without any SS block.

16.1.4 SS BURST SET: MULTIPLE SS BLOCKS IN THE TIME DOMAIN

One key difference between the SS block and the corresponding signals for LTE is the possibility to apply beam-sweeping for SS-block transmission, that is, the possibility to transmit SS blocks in different beams in a time-multiplexed fashion (see Fig. 16.2). The set of SS blocks within a beam-sweep is referred to as an *SS burst set*.[4] Note that the SS-block period discussed in the previous section is the time between SS-block transmissions *within a specific beam*, that is, it is actually the periodicity of the SS burst set. This makes sense as a device located in a certain downlink beam may only "see" a single SS block and be unaware of any other SS blocks transmitted from the cell.

By applying beam-forming for the SS block, the coverage of a single SS-block transmission is increased. Beam-sweeping for SS-block transmission also enables receiver-side beam-sweeping for the reception of uplink random-access transmissions as well as downlink beam-forming for the *random-access response* (see further details in Section 16.2.1.5).

Although the periodicity of the SS burst set is flexible with a minimum period of 5 ms and a maximum period of 160 ms, each SS burst set is always confined to a 5 ms time interval, either in the first or second half of a 10 ms frame.

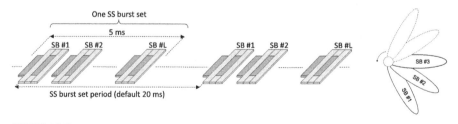

FIGURE 16.2

Multiple time-multiplexed SS blocks within an SS-burst-set period.

---

[4]The term SS *burst set* originates from early 3GPP discussions when SS blocks were assumed to be grouped into *SS bursts* and the SS bursts then grouped into *SS burst sets*. The intermediate SS-burst grouping was eventually not used but the term *SS burst set* for the full set of SS blocks was retained.

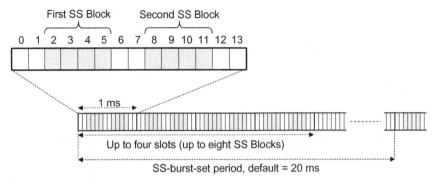

FIGURE 16.3

Possible time-domain locations of an SS block within an SS burst set for 15 kHz numerology.

The maximum number of SS blocks within an SS burst set is different for different frequency bands.

- For frequency bands below 3 GHz, there can be up to four SS blocks within an SS burst set, enabling SS-block beam-sweeping over up to four beams;
- For frequency bands between 3 GHz and 6 GHz, there can be up to eight SS blocks within an SS burst set, enabling beam-sweeping over up to eight beams;
- For higher-frequency bands (FR2) there can be up to 64 SS blocks within an SS burst set, enabling beam-sweeping over up to 64 beams.

There are two reasons why the maximum number of SS blocks within an SS burst set, and thus also the maximum number of beams over which the SS block can be swept, is larger for higher-frequency bands.

- The use of a large number of beams with more narrow beam-width is typically more relevant for higher frequencies;
- As the duration of the SS block depends on the SS-block numerology (see Table 16.1), a large number of SS blocks within an SS burst set would imply a very large SS-block overhead for lower frequencies for which lower SS-block numerology (15 or 30 kHz) must be used.

The set of possible SS-block locations in the time domain differ somewhat between different SS-block numerologies. As an example, Fig. 16.3 illustrates the possible SS-block locations within an SS-burst-set period for the case of 15 kHz numerology. As can be seen, there may be SS-block transmission in any of the first four slots.[5] Furthermore, there can be up to two SS-block transmissions in each of these slot, with the first possible SS-block location corresponding to

---

[5]For operation below 3 GHz, the SS block can only be located within the first two slots.

symbol two to symbol five and the second possible SS-block location corresponding to symbol eight to symbol eleven. Finally, note that the first and last two OFDM symbols of a slot are unoccupied by SS-block transmission. This allows for these OFDM symbols to be used for downlink and uplink control signaling, respectively, for devices already connected to the network. The same is true for all SS-block numerologies.

It should be noted that the SS-block locations outlined in Fig. 16.3 are possible SS-block locations, that is, an SS block is not necessarily transmitted in all the locations outlined in Fig. 16.3. There may be anything from one single SS-block transmission up to the maximum number of SS blocks within an SS burst set depending on the number of beams over which the SS block is to be beam-swept.

Furthermore, if less than the maximum number of SS blocks is transmitted, the transmitted SS blocks do not have to be transmitted in consecutive SS-block locations. Rather, any subset of the possible set of SS-block locations outlined in Fig. 16.3 can be used for actual SS-block transmission. In the case of four SS blocks within an SS burst set these may, for example, be located as two SS blocks within each of the two first slots or as one SS block in each of the four slots of Fig. 16.3.

The PSS and SSS of an SS block only depend on the physical cell identity (see below). Thus, the PSS and SS of all SS blocks within a cell are identical and cannot be used by the device to determine the relative location of an acquired SS block within the set of possible SS-block locations. For this reason, each SS block, more specifically, the PBCH, includes a "time index" that explicitly provides the relative location of the SS block within the sequence of possible SS-block locations (see further details in Section 16.1.5.3). Knowing the relative location of the SS block is important for several reasons:

- It makes it possible for the device to determine frame timing (see Section 16.1.5.3).
- It makes it possible to associate different SS blocks, in practice different beams, with different RACH occasions. This, in turn, is a prerequisite for the use of network-side beam forming during random-access reception (see further details in Section 16.2).

16.1.5 DETAILS OF PSS, SSS, AND PBCH

Above we have described the overall structure of an SS block and how it consists of three parts: PSS, SSS, and PBCH. We have also described how multiple SS blocks in the time domain constitute an SS burst set and how an SS block is mapped to certain OFDM symbols. In this section we will describe the detailed structure of the different SS-block components.

16.1.5.1 The Primary Synchronization Sequence (PSS)

The PSS is the first signal that a device entering the system will search for. At that stage, the device has no knowledge of the system timing. Furthermore, even though the device searches for a cell at a given carrier frequency, there may, due to inaccuracy of the device internal frequency reference, be a relatively large deviation between the device and network carrier frequency. The PSS has been designed to be detectable despite these uncertainties.

Once the device has found the PSS, it has found synchronization up to the periodicity of the PSS. It can then also use transmissions from the network as a reference for its internal frequency generation, thereby to a large extent eliminating any frequency deviation between the device and the network.

As described above, the PSS extends over 127 resource elements onto which a *PSS sequence* $\{x_n\} = x_n(0), x_n(1), \ldots, x_n(126)$ is mapped (see Fig. 16.4).

There are three different PSS sequences $\{x_0\}$, $\{x_1\}$, and $\{x_2\}$, derived as different cyclic shifts of a basic length-127 *M*-sequence [70] $\{x\} = x(0), x(1), \ldots, x(126)$ generated according to the recursive formula (see also Fig. 16.5):

$$x(n) = x(n-7) \oplus x(n-3)$$

By applying different cyclic shifts to the basic *M*-sequence $x(n)$, three different PSS sequences $x_0(n)$, $x_1(n)$, and $x_2(n)$ can be generated according to:

$$x_0(n) = x(n);$$
$$x_1(n) = x(n + 43 \bmod 127);$$
$$x_2(n) = x(n + 86 \bmod 127)$$

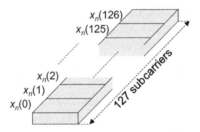

FIGURE 16.4

PSS structure.

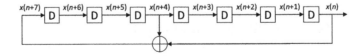

Initial value: [x(6) x(5) x(4) x(3) x(2) x(1) x(0)] = [1 1 1 0 1 1 0]

FIGURE 16.5

Generation of basic *M*-sequence from which three different PSS sequences are derived.

Which of the three PSS sequences to use in a certain cell is determined by the *physical cell identity* (PCI) of the cell. When searching for new cells, a device thus must search for all three PSSs.

16.1.5.2 The Secondary Synchronization Sequence (SSS)

Once a device has detected a PSS it knows the transmission timing of the SSS. By detecting the SSS, the device can determine the PCI of the detected cell. There are 1008 different PCIs. However, already from the PSS detection the device has reduced the set of candidate PCIs by a factor 3. There are thus 336 different SSSs, that together with the already-detected PSS provides the full PCI. Note that, since the timing of the SSS is known to the device, the per-sequence search complexity is reduced compared to the PSS, enabling the larger number of SSS sequences.

The basic structure of the SSS is the same as that of the PSS (Fig. 16.4), that is, the SSS consists of 127 subcarriers to which an SSS sequence is applied.

On an even more detailed level, each SSS is derived from two basic *M*-sequences generated according to the recursive formulas

$$x(n) = x(n - 7) \oplus x(n - 3)$$

$$y(n) = y(n - 7) \oplus y(n - 6)$$

The actual SSS sequence is then derived by adding the two M sequences together, with different shifts being applied to the two sequences.

$$x_{m_1,m_2}(n) = x(n + m_1) + y(n + m_2)$$

16.1.5.3 PBCH

While the PSS and SSS are physical signals with specific structures, the PBCH is a more conventional physical channel on which explicit channel-coded information is transmitted. The PBCH carries the *master information block* (MIB), which contains a small amount of information that the device needs in order to be able to acquire the remaining system information broadcast by the network.[6]

Table 16.2 lists the information carried within the PBCH. Note that the information differs slightly depending on if the carrier is operating in lower-frequency bands (FR1) or higher-frequency bands (FR2).

As already mentioned, the SS-block time index identifies the SS-block location within an SS burst set. As described in Section 16.1.4, each SS block has a well-defined position within an SS burst set which, in turn, is contained within the first or second half of a 5 ms frame. From the SS-block time index, in combination with the *half-frame bit* (see below), the device can thus determine the frame boundary.

---

[6]Some of the information on the PBCH is strictly speaking not part of the MIB (see also below).

Table 16.2 Information Carried Within the PBCH

Information	Number of Bits
SS-block time index	0 (FR1)/3 (FR2)
CellBarred flag	2
1st PDSCH DMRS position	1
SIB1 numerology	1
SIB1 configuration	8
CRB grid offset	5 (FR1)/4 (FR2)
Half-frame bit	1
System frame number (SFN)	10
Cyclic redundancy check (CRC)	24

The SS-block time index is provided to the device as two parts:

- An implicit part encoded in the scrambling applied to the PBCH;
- An explicit part included in the PBCH payload.

Eight different scrambling patterns can be used for the PBCH, allowing for the implicit indication of up to eight different SS-block time indices. This is sufficient for operation below 6 GHz (FR1) where there can be at most eight SS blocks within an SS burst set.[7]

For operation in the higher NR frequency range (FR2) there can be up to 64 SS blocks within an SS burst set, implying the need for three additional bits to indicate the SS-block time index. These three bits, which are thus only needed for operation above 10 GHz, are included as explicit information within the PBCH payload.

The *CellBarred flag* consist of two bits:

- The first bit, which can be seen as the actual CellBarred flag, indicates whether or not devices are allowed to access the cell;
- Assuming devices are not allowed to access the cell, the second bit, also referred to as the *Intra-frequency-reselection* flag, indicates whether or not access is permitted to other cells on the same frequency.

If detecting that a cell is barred and that access to other cells on the same frequency is not permitted, a device can and should immediately re-initiate cell search on a different carrier frequency.

It may seem strange to deploy a cell and then prevent devices from accessing it. Historically this kind of functionality has been used to temporarily prevent access to a certain cell during maintenance. However, the functionality has additional usage within NR due to the possibility for non-standalone NR deployments

---

[7]Only up to four SS blocks for operation below 3 GHz.

for which devices should access the network via the corresponding LTE carrier. By setting the CellBarred flag for the NR carrier in an NSA deployment, the network prevents NR devices from trying to access the system via the NR carrier.

The *1st PDSCH DMRS position* indicates the time-domain position of the first DMRS symbol assuming DMRS Mapping Type A (see Section 9.11).

The *SIB1 numerology* provides information about the subcarrier spacing used for the transmission of the so-called SIB1, which is part of the system information (see Section 16.1.6). The same numerology is also used for the downlink Message 2 and Message 4 that are part of the random-access procedure (see Section 16.2). Although NR supports four different numerologies (15 kHz, 30 kHz, 60 kHz, and 120 kHz) for data transmission, for a given frequency band there are only two possible numerologies. Thus, one bit is sufficient to signal the SIB1 numerology.

The *SIB1 configuration* provides information about the search space, corresponding CORESET, and other PDCCH-related parameters that a device needs in order to monitor for scheduling of SIB1.

The *CRB grid offset* provides information about the frequency offset between the SS block and the common resource block grid. As discussed in Section 16.1.2, the frequency-domain position of the SS block relative to the carrier is flexible and does not even have to be aligned with the carrier CRB grid. However, for SIB1 reception, the device needs to know the CRB grid. Thus, information about the frequency offset between the SS block and the CRB grid must be provided within the PBCH in order to be available to devices prior to SIB1 reception.

Note that the CRB grid offset only provides the offset between the SS block and the CRB grid. Information about the absolute position of the SS block within the overall carrier is then provided within SIB1.

The *half-frame bit* indicates if the SS block is located in the first or second 5 ms part of a 10 ms frame. As mentioned above, the half-frame bit, together with the SS-block time index, allows for a device to determine the cell frame boundary.

All information above, including the CRC, is jointly channel coded and rate-matched to fit the PBCH payload of an SS block.

Although all the information above is carried within the PBCH and is jointly channel coded and CRC-protected, some of the information is strictly speaking not part of the MIB. The MIB is assumed to be the same over an 80 ms time interval (eight subframes) as well as for all SS blocks within an SS burst set. Thus, the SS-block time index, which is inherently different for different SS blocks within an SS burst set, the half-frame bit and the four least significant bits of the SFN are PBCH information carried outside of the MIB.[8]

---

[8]As the SFN is updated every 10 ms it would have been sufficient to place the three least significant bits of the SFN outside of the MIB.

16.1.6 PROVIDING REMAINING SYSTEM INFORMATION

System information is a joint name for all the common (non-device-specific) information that a device needs in order to properly operate within the network. In general, the system information is carried within different *System Information Blocks* (SIBs), each consisting of different types of system information.

In LTE, all system information is periodically broadcast over the entire cell area making it always available but also implying that it is transmitted even if there is no device within the cell.

For NR, a different approach has been adopted where the system information, beyond the very limited information carried within the MIB, has been divided into two parts.

SIB1, sometimes also referred to as the *remaining minimum system information* (RMSI) consists of the system information that a device needs to know before it can access the system. SIB1 is always periodically broadcast over the entire cell area. One important task of SIB1 is to provide the information the device needs in order to carry out an initial random access (see Section 16.2).

SIB1 is provided by means of ordinary scheduled PDSCH transmissions with a periodicity of 160 ms. As described above, the PBCH/MIB provides information about the numerology used for SIB1 transmission as well as the search space and corresponding CORESET used for scheduling of SIB1. Within that CORESET, the device then monitors for scheduling of SIB1 which is indicated by a special *System Information RNTI* (SI-RNTI).

The remaining SIBs, not including SIB1, consist of the system information that a device does not need to know before accessing the system. These SIBs can also be periodically broadcast similar to SIB1. Alternatively, these SIBs can be transmitted *on demand*, that is, only transmitted when explicitly requested by a connected device. This implies that the network can avoid periodic broadcast of these SIBs in cells where no device is currently camping, thereby allowing for enhanced network energy performance.

16.2 RANDOM ACCESS

Once a device has found a cell it may access the cell. This is done as part of the *random-access procedure*.

Similar to LTE, NR uses a four-step random-access procedure consisting of the following steps (see also Fig. 16.6):

- *Step 1:* Device transmission of a *preamble* also referred to as the *Physical Random-Access Channel* (PRACH);
- *Step 2:* Network transmission of a *Random-Access Response* (RAR) indicating reception of the preamble and providing a time-alignment command adjusting

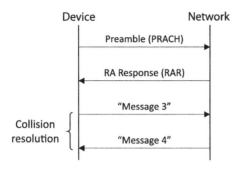

FIGURE 16.6

Four-step random-access procedure.

the transmission timing of the device based on the timing of the received preamble;

- *Steps 3/4:* Device and network exchange of messages (uplink "Message 3" and subsequent downlink "Message 4") with the aim of resolving potential collisions due to simultaneous transmissions of the same preamble from multiple devices within the cell. If successful, Message 4 also transfers the device to connected state.

Once the random-access procedure is completed, the device is in connected state and network-device communication can continue using normal dedicated transmission.

The basic random-access procedure is also used in other contexts within NR, for example:

- For handover, when synchronization needs to be established to a new cell;
- To reestablish uplink synchronization to the current cell if synchronization has been lost due to a too long period without any uplink transmission from the device;
- To request uplink scheduling if no dedicated scheduling-request resource has been configured for the device.

Parts of the basic random-access procedure are also used within the *beam-recovery* procedure (see Section 12.3).

16.2.1 **PREAMBLE TRANSMISSION**

As mentioned above, the random-access preamble is also referred to as the *Physical Random Access Channel* (PRACH) indicating that, in contrast to the other random-access-related transmissions, the preamble corresponds to a special physical channel.

16.2.1.1 Characteristics of Preamble Transmission

Several factors impact the structure of the preamble transmission.

As described in Section 15.2, the transmission timing of NR uplink transmissions is typically controlled by the network by means of regularly provided time-adjustment commands ("closed-loop timing control").

Prior to preamble transmission, there is no such closed-loop timing control in operation. Rather, the device must base the preamble transmission timing on the received timing of some downlink signal, in practice the received timing of the acquired SS block. Consequently, there will be an uncertainty in the preamble reception timing of at least two times the maximum propagation delay within the cell. For cell sizes in the order of a few hundred meters, this uncertainty will be in the order of few microseconds. However, for large cells the uncertainty could be in the order of 100 μs or even more.

In general, it is up to the base-station scheduler to ensure that there are no other transmissions in the uplink resources in which preamble transmissions may take place. When doing this, the network needs to take the uncertainty in the pre-amble reception timing into account. In practice the scheduler needs to provide an extra *guard time* that captures this uncertainty (see Fig. 16.7).

Note that the presence of the guard time is not part of the NR specifications but just a result of scheduling restrictions. Consequently, different guard times can easily be provided to match different uncertainty in the preamble reception timing, for example, due to different cell sizes.

In addition to the lack of closed-loop timing control, there is also no closed-loop power control in operation prior to preamble transmission. Rather, similar to the transmission timing, the device must determine its transmit power based on the received power of some downlink signal, in practice the received power of the acquired SS block. The lack of closed-loop power control may lead to a relatively large uncertainty in the received preamble power for several reasons:

- Estimates of the absolute received power are inherently uncertain;
- Especially in the case of FDD operation with downlink and uplink in different frequency bands, there can be significant differences in the instantaneous uplink and downlink path loss.

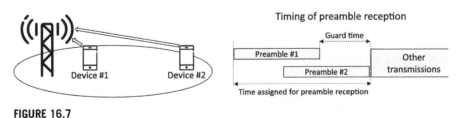

FIGURE 16.7

Guard-time for preamble transmission.

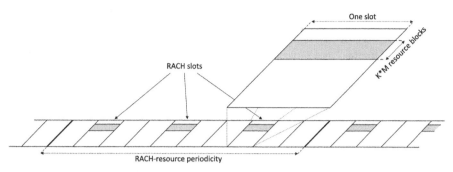

FIGURE 16.8

Overall RACH resource consisting of a set of consecutive resource blocks within a set of RACH slots and where the slot pattern repeats every *RACH-resource period*.

Finally, while normal uplink transmissions are typically based on explicit scheduling grants, thereby enabling contention-free access, initial random access is inherently contention-based, implying that multiple devices may initiate preamble transmission simultaneously. The preamble should preferably be able to handle such a situation and as much as possible allow for correct preamble reception when such "collisions" occur.

16.2.1.2 RACH Resources

Within a cell, preamble transmission can take place within a configurable subset of slots (the *RACH slots*) that repeats itself every *RACH configuration period* (see Fig. 16.8).[9]

Furthermore, within these "*RACH slots*", there may be multiple frequency-domain *RACH occasions* jointly covering $K \cdot M$ consecutive resource blocks where M is the preamble bandwidth measured in number of resource blocks and K is the number of frequency-domain RACH occasions.

For a given preamble type, corresponding to a certain preamble bandwidth, the overall available time/frequency RACH resource within a cell can thus be described by:

- A configurable *RACH periodicity* that can range from 10 ms up to 160 ms;
- A configurable set of RACH slots within the RACH period;
- A configurable frequency-domain RACH resource given by the index of the first resource block in the resource and the number of frequency-domain RACH occasions.

---

[9]As will be seen in Section 16.1.5, a preamble transmission may actually extend outside RACH slots, that is, strictly speaking the RACH slots define possible starting points of preamble transmissions.

16.2.1.3 Basic Preamble Structure

Fig. 16.9 illustrates the basic structure for generating NR random-access preambles. A preamble is generated based on a length-L *preamble sequence* $p_0, p_1, \ldots, p_{L-1}$ which is DFT precoded before being applied to a conventional OFDM modulator. The preamble can thus be seen as a DFTS-OFDM signal. It should be noted though that one could equally well see the preamble as a conventional OFDM signal based on a frequency-domain sequence $P_0, P_1, \ldots, P_{L-1}$ being the discrete Fourier transform of the sequence $p_0, p_1, \ldots, p_{L-1}$.

The output of the OFDM modulator is then repeated N times, after which a cyclic prefix is inserted. For the preamble, the cyclic prefix is thus not inserted per OFDM symbol but only once for the block of N repeated symbols.

Different preamble sequences can be used for the NR preambles. Similar to, for example, uplink SRS, the preamble sequences are based on Zadoff–Chu sequences [25]. As described in Section 8.3.1, for prime-length ZC sequences, which is the case for the sequences used as a basis for the NR preamble sequences, there are $L-1$ different sequences, with each sequence corresponding to a unique *root index*.

Different preamble sequences can be generated from different Zadoff–Chu sequences corresponding to different root indices. However, different preamble sequences can also be generated from different cyclic shifts of the same root sequence. As described in Section 8.3.1, such sequences are inherently orthogonal to each other. However, this orthogonality is retained at the receiver side only if the relative cyclic shift between two sequences is larger than any difference in their respective receive timing. Thus, in practice only a subset of the cyclic shifts can be used to generate different preambles, where the number of available shifts depends on the maximum timing uncertainty which, in turn, depends on, for example, the cell size. For small cell sizes a relatively large number of cyclic shifts can often be used. For larger cells, a smaller number of cyclic shifts will typically be available.

The set of cyclic shifts that can be used within a cell is given by the so-called *zero-correlation zone* parameter which is part of the cell random-access configuration provided within SIB1. In practice, the zero-correlation zone parameter points to a table that indicates the set of cyclic shifts available in the cell. The name "zero-correlation zone" comes from the fact that the different tables indicated by the zero-correlation-zone parameter have different distances between the cyclic shifts, thus providing larger or smaller "zones" in terms of timing error for which orthogonality (=zero correlation) is retained.

16.2.1.4 Long vs Short Preambles

NR defines two types of preambles, referred to as *long preambles* and *short preambles*, respectively. As the name suggests, the two preamble types differ in terms of the length of the preamble sequence. They also differ in the numerology (subcarrier spacing) used for the preamble transmission. The type of preamble is

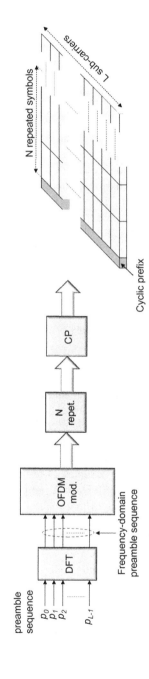

FIGURE 16.9

Basic structure for generation of NR random-access preamble.

Table 16.3 Preamble Formats for Long Preambles

Format	Numerology (kHz)	Number of Repetitions	CP Length (μs)	Preamble Length (Not Including CP) (μs)
0	1.25	1	≈ 100	800
1	1.25	2	≈ 680	1600
2	1.25	4	≈ 15	3200
3	5	1	≈ 100	800

part of the cell random-access configuration, that is, within a cell only one type of preamble can be used for initial access.

Long preambles are based on a sequence length $L = 839$ and a subcarrier spacing of either 1.25 kHz or 5 kHz. The long preambles thus use a numerology different from any other NR transmissions. The long preambles partly originate from the preambles used for LTE random-access [28]. Long preambles can only be used for frequency bands below 6 GHz (FR1).

As illustrated in Table 16.3 there are four different formats for the long preamble where each format corresponds to a specific numerology (1.25 kHz or 5 kHz), a specific number of repetitions (the parameter N in Fig. 16.9), and a specific length of the cyclic prefix. The preamble format is also part of the cell random-access configuration, that is, each cell is limited to a single preamble format. It could be noted that the two first formats of Table 16.3 are identical to the LTE preamble formats 0 and 2 [14].

In the previous section it was described how the overall RACH resource consists of a set of slots and resource blocks in the time-domain and frequency-domain, respectively. For long preambles, which use a numerology that is different from other NR transmissions, the slot and resource block should be seen from a 15 kHz numerology point of view. In the context of long preambles, a slot thus has a length of 1 ms, while a resource-block has a bandwidth of 180 kHz. A long preamble with 1.25 kHz numerology thus occupies six resource blocks in the frequency domain, while a preamble with 5 kHz numerology occupies 24 resource blocks.

It can be observed that preamble format 1 and preamble format 2 in Table 16.3 correspond to a preamble length that exceeds a slot. This may appear to contradict the assumption of preamble transmissions taking place in RACH slots of length 1 ms as discussed in Section 16.2.1.2. However, the RACH slots only indicate the possible *starting* positions for preamble transmission. If a preamble transmission extends into a subsequent slot, this only implies that the scheduler needs to ensure that no other transmissions take place within the corresponding frequency-domain resources within that slot.

Short preambles are based on a sequence length $L = 139$ and use a subcarrier spacing aligned with the normal NR subcarrier spacing. More specifically, short preambles use a subcarrier spacing of:

- 15 kHz or 30 kHz in the case of operation below 6 GHz (FR1);

- 60 kHz or 120 kHz in the case of operation in the higher NR frequency bands (FR2).

In the case of short preambles, the RACH resource described in Section 16.2.1.2 is based on the same numerology as the preamble. A short preamble thus always occupies 12 resource blocks in the frequency domain regardless of the preamble numerology.

Table 16.4 lists the preamble formats available for short preambles. The labels for the different preamble formats originate from the 3GPP standardization discussions during which an even larger set of preamble formats were discussed. The table assumes a preamble subcarrier spacing of 15 kHz. For other numerologies, the length of the preamble as well as the length of the cyclic prefix scale correspondingly, that is, with the inverse of the subcarrier spacing.

The short preambles are, in general, shorter than the long preambles and often span only a few OFDM symbols. In most cases it is therefore possible to have multiple preamble transmissions multiplexed in time within a single RACH slot. In other words, for short preambles there may not only be multiple RACH occasions in the frequency domain but also in the time domain within a single RACH slot (see Table 16.5).

It can be noted that Table 16.5 includes additional formats A1/B1, A2/B2, and A3/B3. These formats correspond to the use of a mix of the "A" and "B" formats of Table 16.4, where the A format is used for all except the last RACH occasion within a RACH slot. Note that the A and B preamble formats are identical except for a somewhat shorter cyclic prefix for the B formats.

Table 16.4 Preamble Formats for Short Preambles

Format	Number of Repetitions	CP Length (μs)	Preamble Length (Not Including CP) (μs)
A1	2	9.4	133
A2	4	18.7	267
A3	6	28.1	400
B1	2	7.0	133
B2	4	11.7	267
B3	6	16.4	400
B4	12	30.5	800
C0	1	40.4	66.7
C2	4	66.7	267

Table 16.5 Number of RACH Time-Domain Occasions Within a RACH Slot for Short Preambles

	A1	A2	A3	B1	B4	C0	C2	A1/B1	A2/B2	A3/B3
Number of RACH occasions	6	3	2	7	1	7	2	7	3	2

For the same reason there are no explicit formats B2 and B3 in Table 16.5 as these formats are always used in combination with the corresponding A formats (A2 and A3) according to the above.

16.2.1.5 Beam Establishment During Initial Access

A key feature of the NR initial access is the possibility to establish a suitable beam pair already during the initial-access phase and to apply receiver-side analog beam-sweeping for the preamble reception.

This is enabled by the possibility of associating different SS-block time indices with different RACH time/frequency occasions and/or different preamble sequences. As different SS-block time indices in practice correspond to SS-block transmissions in different downlink beams, this means that the network, based on the received preamble, will be able to determine the downlink beam in which the corresponding device is located. This beam can then be used as an initial beam for subsequent downlink transmissions to the device.

Furthermore, if the association between SS-block time index and RACH occasion is such that a given time-domain RACH occasion corresponds to one specific SS-block time index, the network will know when, in time, preamble transmission from devices within a specific downlink beam will take place. Assuming beam correspondence, the network can then focus the uplink receiver beam in the corresponding direction for beam-formed preamble reception. In practice this implies that the receiver beam will be swept over the coverage area synchronized with the corresponding downlink beam sweep for the SS-block transmission.

Note that beam-sweeping for preamble transmission is only relevant when analog beam-forming is applied at the receiver side. If digital beam-forming is applied, beam-formed preamble reception can be done from multiple directions simultaneously.

To associate a certain SS-block time index with a specific random-access occasion and a specific set of preambles, the random-access configuration of the cell specifies the number of SS-block time indices per RACH time/frequency occasion. This number can be larger than one, indicating that multiple SS-block time indices correspond to a single RACH time/frequency occasion. However, it can also be smaller than one, indicating that one single SS-block time index corresponds to multiple RACH time/frequency occasions.

SS-block time indices are then associated with RACH occasions in the following order:

- First in the frequency domain;
- Then in the time domain within a slot, assuming the preamble format configured for the cell allows for multiple time-domain RACH occasions within a slot (only relevant for short preambles);
- Finally in the time domain between RACH slots.

Fig. 16.10 exemplifies the association between SS-block time indices and RACH occasions under the following assumptions:

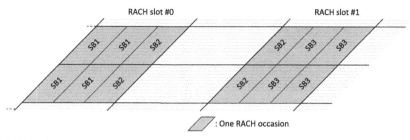

FIGURE 16.10

Association between SS-block time indices and RACH occasions (example).

- Two RACH frequency occasions;
- Three RACH time occasions per RACH slot;
- Each SS-block time index associated with four RACH occasions.

16.2.1.6 Preamble Power Control and Power Ramping

As discussed above, preamble transmission will take place with a relatively large uncertainty in the required preamble transmit power. Preamble transmission therefore includes a *power-ramping* mechanism where the preamble may be repeatedly transmitted with a transmit power that is increased between each transmission.

The device selects the initial preamble transmit power based on estimates of the downlink path loss in combination with a target received preamble power configured by the network. The path loss should be estimated based on the received power of the SS block that the device has acquired and from which it has determined the RACH resource to use for the preamble transmission. This is aligned with an assumption that if the preamble transmission is received by means of beam-forming the corresponding SS block is transmitted with a corresponding beam-shaper. If no random-access response (see below) is received within a predetermined window, the device can assume that the preamble was not correctly received by the network, most likely due to the fact that the preamble was transmitted with too low power. If this happens, the device repeats the preamble transmission with the preamble transmit power increased by a certain configurable offset. This power ramping continues until a random-access response has been received or until a configurable maximum number of retransmissions has been carried out, alternatively a configurable maximum preamble transmit power has been reached. In the two latter cases, the random-access attempt is declared as a failure.

16.2.2 RANDOM-ACCESS RESPONSE

Once a device has transmitted a random-access preamble, it waits for a random-access response, that is, a response from the network that it has properly received the preamble. The random-access response is transmitted as a conventional downlink PDCCH/PDSCH transmission with the corresponding PDCCH transmitted within the common search space.

The random-access response includes the following:

- Information about the random-access preamble sequence the network detected and for which the response is valid;
- A timing correction calculated by the network based on the preamble receive timing;
- A scheduling grant, indicating resources the device will use for the transmission of the subsequent Message 3 (see below);
- A temporary identity, the TC-RNTI, used for further communication between the device and the network.

If the network detects multiple random-access attempts (from different devices), the individual response messages can be combined in a single transmission. Therefore, the response message is scheduled on the DL-SCH and indicated on a PDCCH using an identity reserved for random-access response, the RA-RNTI. The use of the RA-RNTI is also necessary as a device may not have a unique identity in the form of a C-RNTI allocated. All devices that have transmitted a preamble monitor the L1/L2 control channels for random-access response within a configurable time window. The timing of the response message is not fixed in the specification in order to be able to respond to many simultaneous accesses. It also provides some flexibility in the base-station implementation. If the device does not detect a random-access response within the time window, the preamble will be retransmitted with higher power according to the preamble power ramping described above.

As long as the devices that performed random access in the same resource used different preambles, no collision will occur and from the downlink signaling it is clear to which device(s) the information is related. However, there is a certain probability of contention—that is, multiple devices using the same random-access preamble at the same time. In this case, multiple devices will react upon the same downlink response message and a collision occurs. Resolving these collisions is part of the subsequent steps, as discussed below.

Upon reception of the random-access response, the device will adjust its uplink transmission timing and continue to the third step. If contention-free random access using a dedicated preamble is used, then this is the last step of the random-access procedure as there is no need to handle contention in this case. Furthermore, the device already has a unique identity allocated in the form of a C-RNTI.

In the case of downlink beam-forming, the random-access response should follow the beam-forming used for the SS block which was acquired during the initial cell search. This is important as the device may use receive-side beam-forming and it needs to know how to direct the receiver beam. By transmitting the random-access response using the same beam as the SS block, the device knows that it can use the same receiver beam as identified during the cell search.

16.2.3 **MESSAGE 3: CONTENTION RESOLUTION**

After the second step, the uplink of the device is time synchronized. However, before user data can be transmitted to/from the device, a unique identity within the cell, the C-RNTI, must be assigned to the device (unless the device already has a C-RNTI assigned). Depending on the device state, there may also be a need for additional message exchange for setting up the connection.

In the third step, the device transmits the necessary messages to the gNB using the UL-SCH resources assigned in the random-access response in the second step.

An important part of the uplink message is the inclusion of a device identity, as this identity is used as part of the contention-resolution mechanism in the fourth step. If the device is already known by the radio-access network, that is, in RRC_CONNECTED or RRC_INACTIVE state, the already-assigned C-RNTI is used as the device identity.[10] Otherwise, a core-network device identifier is used and the gNB needs to involve the core network prior to responding to the uplink message in step 4 (see below).

16.2.4 **MESSAGE 4: CONTENTION RESOLUTION AND CONNECTION SET UP**

The last step in the random-access procedure consists of a downlink message for contention resolution. Note that, from the second step, multiple devices performing simultaneous random-access attempts using the same preamble sequence in the first step listen to the same response message in the second step and therefore have the same temporary identifier. Hence, the fourth step in the random-access procedure is a contention-resolution step to ensure that a device does not incorrectly use another device's identity. The contention resolution mechanism differs somewhat depending on whether the device already has a valid identity in the form of a C-RNTI or not. Note that the network knows from the uplink message received in step 3 whether the device has a valid C-RNTI or not.

If the device already had a C-RNTI assigned, contention resolution is handled by addressing the device on the PDCCH using the C-RNTI. Upon detection of its C-RNTI on the PDCCH the device will declare the random-access attempt successful and there is no need for contention-resolution-related information on the DL-SCH. Since the C-RNTI is unique to one device, unintended devices will ignore this PDCCH transmission.

If the device does not have a valid C-RNTI, the contention resolution message is addressed using the TC-RNTI and the associated DL-SCH contains the contention-resolution message. The device will compare the identity in the message with the identity transmitted in the third step. Only a device which observes a match between the identity received in the fourth step and the identity transmitted as part of the third step will declare the random-access procedure successful

---

[10]The device identity is included as a MAC control element on the UL-SCH.

and promote the TC-RNTI from the second step to the C-RNTI. Since uplink synchronization has already been established, hybrid ARQ is applied to the downlink signaling in this step and devices with a match between the identity they transmitted in the third step and the message received in the fourth step will transmit a hybrid-ARQ acknowledgment in the uplink.

Devices that do not detect PDCCH transmission with their C-RNTI or do not find a match between the identity received in the fourth step and the respective identity transmitted as part of the third step are considered to have failed the random-access procedure and need to restart the procedure from the first step. No hybrid-ARQ feedback is transmitted from these devices. Furthermore, a device that has not received the downlink message in step 4 within a certain time from the transmission of the uplink message in step 3 will declare the random-access procedure as failed and need to restart from the first step.

16.2.5 RANDOM ACCESS FOR SUPPLEMENTARY UPLINK

Section 7.7 discussed the concept of supplementary uplink (SUL), that is, that a downlink carrier may be associated with two uplink carriers (the non-SUL carrier and the SUL carrier), where the SUL carrier is typically located in lower-frequency bands thereby providing enhanced uplink coverage.

That a cell is an SUL cell, that is, includes a complementary SUL carrier, is indicated as part of SIB1. Before initially accessing a cell, a device will thus know if the cell to be accessed is an SUL cell or not. If the cell is an SUL cell and the device supports SUL operation for the given band combination, initial random access may be carried out using either the SUL carrier or the non-SUL uplink carrier. The cell system information provides separate RACH configurations for the SUL carrier and the non-SUL carrier and a device capable of SUL determines what carrier to use for the random access by comparing the measured RSRP of the selected SS block with a *carrier-selection threshold* also provided as part of the cell system information.

- If the RSRP is above the threshold, random access is carried out on the non-SUL carrier.
- If the RSRP is below the threshold, random access is carried out on the SUL carrier.

In practice the SUL carrier is thus selected by devices with a (downlink) path-loss to the cell that is larger than a certain value.

The device carrying out a random-access transmission will transmit the random-access message 3 on the same carrier as used for the preamble transmission.

For other scenarios when a device may do a random access, that is, for devices in connected mode, the device can be explicitly configured to use either the SUL carrier or the non-SUL carrier for the uplink random-access transmissions.

LTE/NR Interworking and Coexistence

17

The initial deployment of a new generation of mobile-communication technology typically takes place in areas with high traffic density and with high demands for new service capabilities. This is then followed by a gradual further build-out that can be more or less rapid depending on the operator strategy. During this subsequent gradual deployment, ubiquitous coverage to the operator network will be provided by a mix of new and legacy technology, with devices continuously moving in and out of areas covered by the new technology. Seamless handover between new and legacy technology has therefore been a key requirement at least since the introduction of the first 3G networks.

Furthermore, even in areas where a new technology has been deployed, earlier generations must typically be retained and operated in parallel for a relatively long time in order to ensure continued service for legacy devices not supporting the new technology. The majority of users will migrate to new devices supporting the latest technology within a few years. However, a limited amount of legacy devices may remain for a long time. This becomes even more the case with an increasing number of mobile devices not being directly used by persons but rather being an integrated part of other equipment, such as parking meters, card readers, surveillance cameras, etc. Such equipment may have a life time of more than 10 years and will be expected to remain connectable during this life time. This is actually one important reason why many second-generation GSM networks are still in operation even though both 3G and 4G networks have subsequently been deployed.

However, the interworking between NR and LTE goes further than just enabling smooth handover between the two technologies and allowing for their parallel deployment.

- NR allows for *dual-connectivity* with LTE, implying that devices may have simultaneous connectivity to both LTE and NR. As already mentioned in Chapter 5, the first release of NR actually relies on such dual-connectivity, with LTE providing the control plane and NR only providing additional user-plane capacity;
- NR can be deployed in the same spectrum as LTE in such a way that the overall spectrum capacity can be dynamically shared between the two

technologies. Such *spectrum coexistence* allows for a more smooth introduction of NR in spectra already occupied by LTE.

17.1 LTE/NR DUAL-CONNECTIVITY

The basic principle of LTE/NR dual-connectivity is the same as LTE dual-connectivity [28], see also Fig. 17.1):

- A device has simultaneous connectivity to multiple nodes within the radio-access network (eNB in the case of LTE, gNB in the case of NR);
- There is one *master node* (in the general case either an eNB or a gNB) responsible for the radio-access control plane. In other words, on the network side the signaling radio bearer terminates at the master node which then also handles all RRC-based configuration of the device;
- There is one, or in the general case multiple, *secondary node(s)* (eNB or gNB) that provides additional user-plane links for the device.

17.1.1 DEPLOYMENT SCENARIOS

In the case of LTE dual-connectivity, the multiple nodes to which a device has simultaneous connectivity are typically geographically separated. The device may, for example, have simultaneous connectivity to a small-cell layer and an overlaid macro layer.

The same scenario, that is, simultaneous connectivity to a small-cell layer and an overlaid macrolayer, is a highly relevant scenario also for LTE/NR dual-connectivity. Especially, NR in higher-frequency bands may be deployed as a small-cell layer under an existing macro-layer based on LTE (see Fig. 17.2). The LTE macro layer would then provide the master nodes, ensuring that the control plane is retained even if the connectivity to the high-frequency small-cell layer is temporarily lost. In this case, the NR layer provides very high capacity and very high data rates, while dual-connectivity to the lower-frequency LTE-based macro

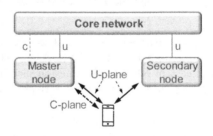

FIGURE 17.1

Basic principle of dual-connectivity.

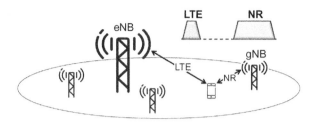

FIGURE 17.2

LTE/NR dual-connectivity in a multi-layer scenario.

layer provides additional robustness to the inherently less robust high-frequency small-cell layer. Note that this is essentially the same scenario as the LTE dual-connectivity scenario described above, except for the use of NR instead of LTE in the small-cell layer.

However, LTE/NR dual-connectivity is also relevant in the case of co-sited LTE and NR network nodes (Fig. 17.3).[1] As an example, for initial NR deployment an operator may want to reuse an already deployed LTE site grid also for NR to avoid the cost of deploying additional sites. In this scenario, dual-connectivity enables higher end-user data rates by allowing for aggregation of the throughput of the NR and LTE carriers. In the case of a single radio-access technology, such aggregation between carriers transmitted from the same node would be more efficiently realized by means of *carrier aggregation* (see Section 7.6). However, NR does not support carrier aggregation with LTE and thus dual-connectivity is needed to support aggregation of the LTE and NR throughput.

Co-sited deployments are especially relevant when NR is operating in lower-frequency spectrum, that is, in the same or similar spectrum as LTE. However, co-sited deployments can also be used when the two technologies are operating in very different spectra, including the case when NR is operating in mm-wave bands (Fig. 17.4). In this case, NR may not be able to provide coverage over the entire cell area. However, the NR part of the network could still capture a large part of the overall traffic, thereby allowing for the LTE part to focus on providing service to devices in poor-coverage locations.

In the scenario in Fig. 17.4, the NR carrier would typically have much wider bandwidth compared to LTE. As long as there is coverage, the NR carrier would therefore, in most cases, provide significantly higher data rates compared to LTE, making throughput aggregation less important. Rather, the main benefit of dual-connectivity in this scenario would, once again, be enhanced robustness for the higher-frequency deployment.

---

[1] Note that three would in this case still be two different logical nodes (an eNB and a gNB) although these could very well be implemented in the same physical hardware.

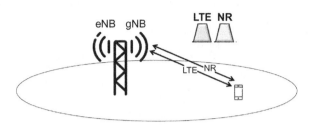

FIGURE 17.3

LTE/NR dual-connectivity, co-sited deployment.

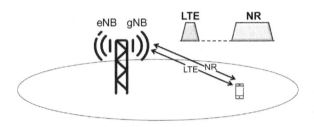

FIGURE 17.4

LTE/NR dual-connectivity, co-sited deployment in different spectrum.

17.1.2 ARCHITECTURE OPTIONS

Due to the presence of two different radio-access technologies (LTE and NR) as well as the future availability of a new 5G core network as an alternative to the legacy 4G core network (EPC), there are several different alternatives, or *options*, for the architecture of LTE/NR dual-connectivity (see Fig. 17.5). The labeling of the different options in Fig. 17.5 originates from early 3GPP discussions on possible NR architecture options where a number of different alternatives were on the table, a subset of which was eventually agreed to be supported (see Chapter 6 for some additional, non-dual-connectivity, options).

It can be noted that LTE/NR dual-connectivity using EPC with NR providing the master node is not included among the options outlined in Fig. 17.5. At the time of the writing of this book, the possible support for this alternative is still under discussion.

17.1.3 SINGLE-TX OPERATION

In the case of dual-connectivity between LTE and NR there will be multiple uplink carriers (at least one LTE uplink carrier and one NR uplink carrier) transmitted from the same device. Due to non-linearities in the RF circuitry, simultaneous transmission on two carriers will create intermodulation products at the transmitter output. Depending on the specific carrier frequencies of the

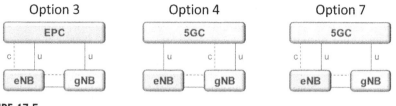

FIGURE 17.5

Different architecture options for LTE/NR dual-connectivity.

transmitted signals, some of these intermodulation products may end up within the device receiver band causing "self-interference," also referred to as *intermodulation distortion* (IMD). The IMD will add to the receiver noise and lead to a degradation of the receiver sensitivity. The impact from IMD can be reduced by imposing tighter linearity requirements on the device. However, this will have a corresponding negative impact on device cost and energy consumption.

To reduce the impact of IMD without imposing very tight RF requirements on all devices, NR allows for *single-TX* dual-connectivity for "difficult band combinations." In this context, difficult band combinations correspond to specifically identified combinations of LTE and NR frequency bands for which lower-order intermodulation products between simultaneously transmitted LTE and NR uplink carriers may fall into a corresponding downlink band. Single-TX operation implies that there will not be simultaneous transmission on the LTE and NR uplink carriers within a device even though the device is operating in LTE/NR dual-connectivity.

It is the task of the LTE and NR schedulers to jointly prevent simultaneous transmission on the LTE and NR uplink carriers in the case of single-TX operation. This requires coordination between the schedulers, that is between an eNB and a gNB. The 3GPP specifications include explicit support for the interchange of standardized inter-node messages for this purpose.

Single TX operation inherently leads to time multiplexing between the LTE and NR uplink transmissions within a device, with none of the uplinks being continuously available. However, it is still desirable to be able to retain full utilization of the corresponding downlink carriers.

For NR, with its high degree of scheduling and hybrid-ARQ flexibility, this can easily be achieved with no additional impact on the NR specifications. For the LTE part of the connection the situation is somewhat different though. LTE FDD is based on synchronous HARQ, where uplink HARQ feedback is to be transmitted a specified number of subframes after the reception of the corresponding downlink transmission. With a single-TX constraint, not all uplink subframes will be available for transmission of HARQ feedback, potentially restricting the subframes in which downlink transmission can take place.

However, the same situation may already occur within LTE itself, more specifically in the case of FDD/TDD carrier aggregation with the TDD carrier being the primary cell [28]. In this case, the TDD carrier, which is inherently not continuously available for uplink transmission, carries uplink HARQ feedback corresponding to downlink transmissions on the FDD carrier. To handle this situation, LTE release 13 introduced so-called *DL/UL reference configurations* [28] allowing for a TDD-like timing relation, for example for uplink feedback, for an FDD carrier. The same functionality can be used to support continuous LTE downlink transmission in the case of LTE/NR dual-connectivity constrained by single-TX operation.

In the LTE FDD/TDD carrier-aggregations scenario, the uplink constraints are due to cell-level downlink/uplink configurations. On the other hand, in the case of single-TX dual-connectivity the constraints are due to the need to avoid simultaneous transmission on the LTE and NR uplink carriers, but without any tight interdependency between different devices. The set of unavailable uplink subframes may thus not need to be the same for different devices. To enable a more even load on the LTE uplink, the DL/UL reference configurations in the case of single-TX operation can therefore be shifted in time on a per-device basis.

17.2 **LTE/NR COEXISTENCE**

The introduction of earlier generations of mobile communication has always been associated with the introduction of a new spectrum in which the new technology can be deployed. This is the case also for NR, for which the support for operation in mm-wave bands opens up for the use of a spectrum range never before applied to mobile communication.

Even taking into account the use of antenna configurations with a large number of antenna elements enabling extensive beam forming, operation in such high-frequency spectrum is inherently disadvantageous in terms of coverage. Rather, to provide truly wide-area NR coverage, lower-frequency spectrum must be used.

However, most lower-frequency spectrum is already occupied by current technologies, primarily LTE. Furthermore, an additional low-frequency spectrum is planned to be deployed with LTE in the relatively near future. In many cases NR deployments in lower-frequency spectrum will therefore need to take place in spectrum already used by LTE.

The most straightforward way to deploy NR in a spectrum already used by LTE is static frequency-domain sharing, where part of the LTE spectrum is migrated to NR (see Fig. 17.6).

There are two drawbacks with this approach though.

At least at an initial stage, the main part of the traffic will still be via LTE. At the same time, the static frequency-domain sharing reduces the spectrum available for LTE, making it more difficult to satisfy the traffic demands.

Furthermore, static frequency-domain sharing will lead to less bandwidth being available for each technology, leading to a reduced peak data rate per

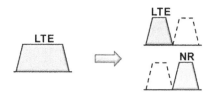

FIGURE 17.6

Migration of LTE spectrum to NR.

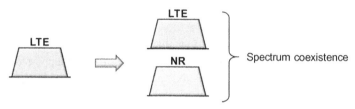

FIGURE 17.7

LTE/NR spectrum coexistence.

carrier. The possible use of LTE/NR dual-connectivity may compensate for this for new devices capable of such operation. However, at least for legacy LTE devices there will be a direct impact on the achievable data rates.

A more attractive solution is to have NR and LTE dynamically share the same spectrum as illustrated in Fig. 17.7. Such spectrum coexistence will retain the full bandwidth and corresponding peak data rates for each technology. Furthermore, the overall spectrum capacity could be dynamically assigned to match the traffic conditions on each technology.

The fundamental tool to enable such LTE/NR spectrum coexistence is the dynamic scheduling of both LTE and NR. However, there are several other NR features that play a role in the overall support for LTE/NR spectrum coexistence:

- The availability of the LTE-compatible 15 kHz NR numerology that allows for LTE and NR to operate on a common time/frequency grid;
- The general NR forward-compatibility design principles listed in Section 5.1.3. This also includes the possibility to define reserved resources based on bitmaps as described in Section 9.10;
- A possibility for NR PDSCH mapping to avoid resource elements corresponding to LTE cell-specific reference signals (see further details below).

As already mentioned in Section 5.1.11 there are two main scenarios for LTE/NR coexistence (see also Fig. 17.8):

- Coexistence in both downlink and uplink;
- Uplink-only coexistence.

FIGURE 17.8

Downlink/uplink coexistence vs uplink-only coexistence.

A typical use case for uplink-only coexistence is the deployment of a supplementary uplink carrier (see Section 7.7).

In general, coexistence in the uplink direction is more straightforward compared to the downlink direction and can, to a large extent, be supported by means of scheduling coordination/constraints. NR and LTE uplink scheduling should be coordinated to avoid collision between LTE and NR PUSCH transmissions. Furthermore, the NR scheduler should be constrained to avoid resources used for LTE uplink layer 1 control signaling (PUCCH) and vice versa. Depending on the level of interaction between the eNB and gNB, such coordination and constraints can be more or less dynamic.

Also for the downlink, scheduling coordination should be used to avoid collision between scheduled LTE and NR transmissions. However, the LTE downlink also includes several non-scheduled "always-on" signals that cannot be readily scheduled around. This includes (see [28] for details):

- The LTE PSS and SSS, which are transmitted over two OFDM symbols and six resource blocks in the frequency domain once every fifth subframe;
- The LTE PBCH, which is transmitted over four OFDM symbols and six resource blocks in the frequency domain once every frame (10 subframes);
- The LTE CRS, which is transmitted regularly in the frequency domain and in four or six symbols in every subframe depending on the number of CRS antenna ports.[2]

Rather than being avoided by means of scheduling, the concept or reserved resources (see Section 9.10) can be used to rate match the NR PDSCH around these signals.

Rate matching around the LTE PSS/SSS can be done by defining reserved resources according to bitmaps as described in Section 9.10. More specifically a single reserved resource given by a {bitmap-1, bitmap-2, bitmap-3} triplet could be defined as follows (see also Fig. 17.9):

- A bitmap-1 of a length equal to the number of NR resource blocks in the frequency domain, indicating the six resource blocks within which LTE PSS and SSS are transmitted;

---

[2]Only one or two symbols in case of so-called MBSFN subframes.

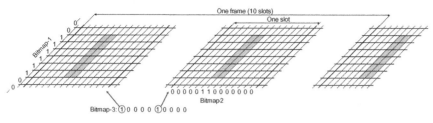

FIGURE 17.9

Configuration of reserved resource to enable PDSCH rate matching around LTE PSS/SS.
Note that the figure assumes 15 kHz NR numerology.

- A bitmap-2 of length 14 (one slot), indicating the two OFDM symbols within
  which the PSS and SSS are transmitted within an LTE subframe;
- A bitmap-3 of length 10 indicating the two subframes within which the PSS
  and SSS are transmitted within a 10 ms frame.

This assumes a 15 kHz NR numerology. Note though that the use of reserved
resources based on bitmaps is not limited to 15 kHz numerology and, in principle,
a similar approach to rate match around LTE PSS and SSS could be used also
with, for example, a 30 kHz NR numerology.

The same approach can be used to rate match around the LTE PBCH with the
only difference that bitmap-2 would, in this case, indicate the four symbols within
which PBCH is transmitted, while bitmap-3 would indicate a single subframe.

Regarding the LTE CRS, the NR specification includes explicit support for
PDSCH rate matching around resource elements corresponding to CRS of an
overlaid LTE carrier. In order to be able to properly receive such a rate-matched
PDSCH, the device is configured with the following information:

- The LTE carrier bandwidth and frequency domain location, to allow for LTE/
  NR coexistence even though the LTE carrier may have a different bandwidth
  and a different center-carrier location, compared to the NR carrier;
- The LTE MBSFN subframe configuration, as this will influence the set of
  OFDM symbols in which CRS transmission takes place within a given LTE
  subframe;
- The number of LTE CRS antenna ports as this will impact the set of OFDM
  symbols on which CRS transmission takes place as well as the number of
  CRS resource elements per resource block in the frequency domain;
- The LTE CRS shift, that is, the exact frequency-domain position of the LTE
  CRS.

Rate matching around LTE CRS is only possible for the 15 kHz NR
numerology.

RF Characteristics

The RF characteristics of NR are strongly tied to the spectrum available for 5G as described in Chapter 3 and the spectrum flexibility required to operate in those spectrum allocations. While spectrum flexibility has been a cornerstone for previous generations of mobile systems, this becomes even more accentuated for NR. It consists of several components, including deployment in different-sized spectrum allocations and diverse frequency ranges, both in paired and unpaired frequency bands and with aggregation of different frequency allocations within and between different bands. NR will also have the capability to operate with mixed numerology on the same RF carrier and will have an even higher flexibility than LTE in terms of frequency domain scheduling and multiplexing of devices within a base station RF carrier. It is the use of OFDM in NR that gives flexibility both in terms of the size of the spectrum allocation needed and in the instantaneous transmission bandwidth used, and that enables frequency-domain scheduling.

Implementation of Active Antenna Systems (AAS) and multiple antennas in devices has been in use for LTE, but is taken one step further in NR, which will support massive MIMO and beam-forming applications both in existing bands and in the new mm-wave bands. Beyond the physical layer implications, this impacts the RF implementation in terms of filters, amplifiers, and all other RF components that are used to transmit and receive the signal and must be defined taking also the spectrum flexibility into account. These are further discussed in Chapter 19.

Note that for the purpose of defining RF characteristics, the physical representation of the gNB is called a base station (BS). A base station is defined with interfaces where the RF requirements are defined, either as conducted requirements at an antenna port or as radiated requirements over-the-air (OTA).

18.1 SPECTRUM FLEXIBILITY IMPLICATIONS

Spectrum flexibility was a fundamental requirement for LTE and it had major implications for how LTE was specified. The need for spectrum flexibility is even higher for NR, because of the diverse spectrum where NR needs to operate and the way the physical layer is designed to meet the key characteristics required for 5G. The following are some important aspects impacting how the RF characteristics are defined:

5G NR: The Next Generation Wireless Access Technology. DOI: https://doi.org/10.1016/B978-0-12-814323-0.00018-1

- *Diverse spectrum allocations:* The spectrum used for 3G and 4G is already very diverse in terms of the sizes of the frequency of operation, bandwidth allocations, how they are arranged (paired and unpaired), and what the related regulation is. For NR it will be even more diverse, with the fundamental frequency varying from below 1 GHz up to 40–50 GHz and above; the maximum frequency presently under study in ITU-R is 86 GHz. The size of allocated bands where NR is to be deployed varies from 5 MHz to 3 GHz, with both paired and unpaired allocations, where the intention is to use some allocations as supplementary downlinks or uplinks together with other paired bands. The spectrum that is planned and under investigation to be used for 5G and the related operating bands defined for NR are described in Chapter 3.
- *Various spectrum block definitions:* Within the diverse spectrum allocations, spectrum blocks will be assigned for NR deployment, usually through operator licenses. The exact frequency boundaries of the blocks can vary between countries and regions and it must be possible to place the RF carriers in positions where the blocks are used efficiently without wasting spectrum. This puts specific requirements on the channel raster to use for placing carriers.
- *LTE-NR coexistence:* The LTE/NR coexistence in the same spectrum makes it possible to deploy NR with in-carrier coexistence in both uplink and downlink of existing LTE deployments. This is further described in Chapter 17. Since the coexisting NR and LTE carriers need to be subcarrier-aligned, this poses restrictions on the NR channel raster in order to align the placing of NR and LTE carriers.
- *Multiple and mixed numerologies:* As described in Section 7.1, the transmission scheme for NR has a high flexibility and supports multiple numerologies with subcarrier spacings ranging from 15 to 240 kHz, with direct implications for the time and frequency domain structure. The subcarrier spacing has implications for the RF in terms of the roll-off of the transmitted spectrum, which impact the resulting guard bands that are needed between the transmitted resource blocks and the RF carrier edge defined for the purpose of defining RF requirements (see Section 18.3). NR also supports mixed numerologies on the same carrier, which has further RF impacts since the guard bands may need to be different at the two edges of the RF carrier.
- *Independent channel bandwidth definitions:* NR devices in general do not receive or transmit using the full channel bandwidth of the BS but can be assigned what is called a bandwidth part (see Section 7.4). While the concept does not have any direct RF implications, it is important to note that BS and device channel bandwidth are defined independently and that the device bandwidth capability does not have to match the BS channel bandwidth.
- *Variation of duplex schemes:* As shown in Section 7.2, a single frame structure is defined in NR that supports TDD, FDD, and half-duplex FDD. The duplex method is specifically defined for each operating band defined for NR as shown in Chapter 3. Some bands are also defined as supplementary

downlink (SDL) or supplementary uplink (SUL) to be used in FDD operation. This is further described in Section 7.7.

Many of the frequency bands identified for deployment of NR are existing bands identified for IMT (see Chapter 3) and they may already have 2G, 3G, and/or 4G systems deployed. Many bands are also in some regions defined and regulated in a "technology-neutral" manner, which means that coexistence between different technologies is a requirement. The capability to operate in this wide range of bands for any mobile system, including NR, has direct implications for the RF requirements and how those are defined, in order to support the following:

- *Coexistence between operators in the same geographical area in the band:* Operators in the same band may deploy NR or other IMT technologies, such as LTE, UTRA, or GSM/EDGE. There may in some cases also be non-IMT technologies. Such coexistence requirements are to a large extent developed within 3GPP, but there may also be regional requirements defined by regulatory bodies in certain cases.
- *Co-location of base-station equipment between operators:* There are in many cases limitations to where base-station equipment can be deployed. Often, sites must be shared between operators or an operator will deploy multiple technologies in one site. This puts additional requirements on both base-station receivers and transmitters to operate in close proximity to other base stations.
- *Coexistence with services in adjacent frequency bands and across country borders:* The use of the RF spectrum is regulated through complex international agreements, involving many interests. There will therefore be requirements for coordination between operators in different countries and for coexistence with services in adjacent frequency bands. Most of these are defined in different regulatory bodies. In some cases, the regulators request that 3GPP includes such coexistence limits in the 3GPP specifications.
- *Coexistence between operators of TDD systems* in the same band is in general provided by inter-operator synchronization, in order to avoid interference between downlink and uplink transmissions of different operators. This means that all operators need to have the same downlink/uplink configurations and frame synchronization, which is not in itself an RF requirement, but it is implicitly assumed in the 3GPP specifications. RF requirements for unsynchronized systems become much stricter.
- *Release-independent frequency-band principles:* Frequency bands are defined regionally, and new bands are added continuously for each generation of mobile systems. This means that every new release of 3GPP specifications will have new bands added. Through the "release independence" principle, it is possible to design devices based on an early release of 3GPP specifications that support a frequency band added in a later release. The first set of NR bands (see Chapter 3) is defined in release 15 and additional bands will be added in a release-independent way.

- *Aggregation of spectrum allocations:* Operators of mobile systems have quite diverse spectrum allocations, which in many cases do not consist of a block that easily fits exactly within one carrier. The allocation may even be non-contiguous, consisting of multiple blocks spread out in a band or in multiple bands. For these scenarios, the NR specifications supports *carrier aggregation*, where multiple carriers within a band, or in multiple bands, can be combined to create larger transmission bandwidths.

18.2 RF REQUIREMENTS IN DIFFERENT FREQUENCY RANGES

As discussed above and in Chapter 3, there will be a very wide range of diverse spectrum allocations where NR can operate. The allocations vary in block size, channel bandwidth and duplex spacing supported, but what really differentiates NR from previous generations is the wide frequency range over which requirements need to be defined, where not only the requirement limits but also the definitions and conformance testing aspects may be quite different at different frequencies. Measurement equipment, such as spectrum analyzers, becomes more complex and expensive at higher frequencies and for the highest frequencies considered, including the harmonics of the highest possible carrier frequencies, requirements may not even be possible to test in a reasonable way.

For this reason, the RF requirements for both devices and base stations are divided into *frequency ranges* (FRs), where presently two are defined (FR1 and FR2) in 3GPP release 15 as shown in Table 18.1. The frequency range concept is not intended to be static. If new NR band(s) are added that are outside the existing frequency ranges, one of them could be extended to cover the new band(s) if the requirements will align well with that range. If there are large differences compared to existing FR, a new frequency range could be defined for the new band.

The frequency ranges are also illustrated in Fig. 18.1 on a logarithmic scale, where the related bands identified for IMT (in at least one region) are shown. FR1 starts at 450 MHz at the first IMT allocation and ends at 6 GHz. FR2 covers a subset of the bands that are presently under study for IMT identification in the ITU-R (see Section 3.1). The subset ends at 52.6 GHz, which is the highest frequency within the scope of the specification work in 3GPP release 15.

Table 18.1 Frequency Ranges Defined in 3GPP Release 15

Frequency Range Designation	Corresponding Frequency Range
Frequency range 1 (FR1)	450–6,000 MHz
Frequency range 2 (FR2)	24,250–52,600 MHz

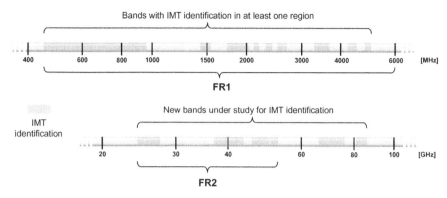

FIGURE 18.1

Frequency ranges FR1 and FR2 and corresponding IMT identifications. Note that the frequency scales are logarithmic.

All existing LTE bands are within FR1 and NR is expected to coexist with LTE and previous generations of systems in many of the FR1 bands. It is only in what is often referred to as the "mid bands" around 3.5 GHz (in fact spanning 3.3−5 GHz) that NR will to a larger extent be deployed in a "new" spectrum, that is a spectrum previously not exploited for mobile services. FR2 covers a part of what is often referred to as the mm-wave band (strictly, mm-wave starts at 30 GHz with 10 mm wavelength). At such high frequencies compared to FR1, propagation properties are different, with less diffraction, higher penetration losses, and in general higher path losses. This can be compensated for by having more antenna elements both at the transmitter and receiver, to be used for narrower antenna beams with higher gain and for massive MIMO. This gives overall different coexistence properties and therefore leads to different RF requirements for coexistence. mm-wave RF implementation for FR2 bands will also have different complexity and performance compared to FR1 bands, impacting all components including A/D and D/A converters, LO generation, PA efficiently, filtering, etc. This is further discussed in Chapter 19.

18.3 CHANNEL BANDWIDTH AND SPECTRUM UTILIZATION

The operating bands defined for NR have a very large variation in bandwidth, as shown in Chapter 3. The spectrum available for uplink or downlink can be as small as 5 MHz in some LTE re-farming bands, while it is up to 900 MHz in "new" bands for NR in frequency range 1, and up to several GHz in frequency range 2. The spectrum blocks available for a single operator will often be smaller than this. Furthermore, the migration to NR in operating bands currently used for other radio-access technologies such as LTE, must often take place gradually to ensure that a sufficient amount of spectrum remains to support the existing users. Thus, the amount of spectrum that can initially be migrated to NR can be relatively small but may then gradually increase. The variation of the size of spectrum blocks and possible spectrum scenarios implies a requirement for very high spectrum flexibility for NR in terms of the transmission bandwidths supported.

The fundamental bandwidth of an NR carrier is called the channel bandwidth ($BW_{Channel}$) and is a fundamental parameter for defining most of the NR RF requirements. The spectrum flexibility requirement points out the need for NR to be scalable in the frequency domain over a large range. In order to limit implementation complexity, only a limited set of bandwidths is defined in the RF specifications. A range of channel bandwidths from 5 to 400 MHz is supported.

The bandwidth of a carrier is related to the spectrum utilization, which is the fraction of a channel bandwidth occupied by the physical resource blocks. In LTE, the maximum spectrum utilization was 90%, but a higher number has been targeted for NR to achieve a higher spectrum efficiency. Considerations however must be taken for the numerology (subcarrier spacing), which impacts the OFDM waveform roll-off, and for the implementation of filtering and windowing solutions. In addition, spectrum utilization is related to the achievable error vector magnitude (EVM) and transmitter unwanted emissions, and also to receiver performance including adjacent channel selectivity (ACS). The spectrum utilization is specified as a maximum number of physical resource blocks, N_{RB}, which will be the maximum possible *transmission bandwidth configuration*, defined separately for each possible channel bandwidth.

What the spectrum utilization ultimately defines is a guard band at each edge of the RF carrier, as shown in Fig. 18.2. Outside of the guard band and thereby outside the RF channel bandwidth, the "external" RF requirements such as unwanted emissions are defined, while only requirements on the actual RF carrier such as EVM are defined inside. For a channel bandwidth $BW_{Channel}$, the guard band will be

$$W_{Guard} = \frac{BW_{Channel} - N_{RB} \cdot 12 \cdot \Delta f - \Delta f}{2} \tag{18.1}$$

where N_{RB} is the maximum number of resource blocks possible and Δf is the subcarrier spacing. The extra $\Delta f/2$ guard applied on each side of the carrier is due to the relation to the RF channel raster, which has a subcarrier-based granularity and is defined independently of the actual spectrum blocks. It may therefore not be possible to place a carrier exactly in the center of a spectrum block and an extra guard will be required to make sure RF requirements can be met.

As shown in Eq. (18.1), the guard band and thereby the spectrum utilization will depend on the numerology applied. As described in Section 7.3, different bandwidths will be possible depending on the subcarrier spacing of the numerology, since the maximum value for N_{RB} is 275. In order to have reasonable spectrum utilization, values of N_{RB} below 11 are not used either. The result is a range of possible channel bandwidths and corresponding spectrum utilization numbers

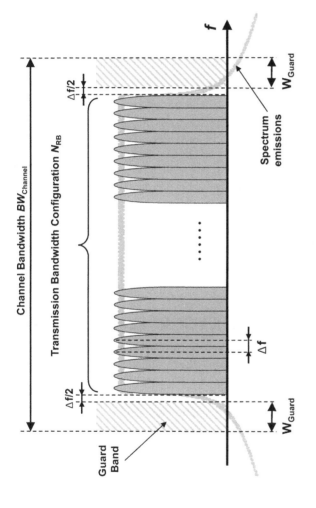

FIGURE 18.2

The channel bandwidth for one RF carrier and the corresponding transmission bandwidth configuration.

Table 18.2 Range of Channel Bandwidths and Spectrum Utilization Numbers Defined for the Different Numerologies and Frequency Ranges

Frequency Range	Set of $BW_{Channel}$ Used in Frequency Range (MHz)	SCS (kHz)	Range of Possible $BW_{Channel}$ Per SCS (MHz)	Corresponding Range for Spectrum Utilization (N_{RB})
FR1	5, 10, 15, 20, 25, 30, 40, 50, 60, 70, 80, 90, 100	15	5–50	25–270
		30	5–100	11–273
		60	10–100	11–135
FR2	50, 100, 200, 400	60	50–200	66–264
		120	50–400	32–264

defined for NR, as shown in Table 18.2. Note that the subcarrier spacing used differs between frequency ranges 1 and 2. The spectrum utilization expressed as a fraction is up to 98% for the widest channel bandwidths and it is above 90% for all cases, except for the smaller bandwidths, where $N_{RB} \le 25$.

Since the channel bandwidth is defined independently for base stations and devices (see above and in Section 7.4), the actual channel bandwidths that are supported by the base station and device specifications will also be different. For a specific bandwidth, the supported spectrum utilization is however the same for base station and device, if the combination of bandwidth and subcarrier spacing is supported by both.

18.4 OVERALL STRUCTURE OF DEVICE RF REQUIREMENTS

The differences in coexistence properties and implementation between FR1 and FR2 means that device RF requirements for NR are defined separately for FR1 and FR2. For a more detailed discussion of the implementation aspects in FR2 using mm-wave technology for devices and base stations, see Chapter 19.

For LTE and previous generations, RF requirements have in general been specified as conducted requirements that are defined and measured at an antenna connector. Since antennas are normally not detachable on a device, this is done at an antenna test port. Device requirements in FR1 are defined in this way.

With the higher number of antenna elements for operation in FR2 and the high level of integration expected when using mm-wave technology, conducted requirements are no longer seen as feasible. FR2 will therefore be specified with radiated requirements and testing will have to be done OTA. While this is an extra challenge when defining requirements, in particular for testing, it is seen as a necessity for FR2.

There will also be a set of device requirements for interworking with other radios within the same device. This concerns primarily interworking with

E-UTRA for non-standalone (NSA) operation and interworking between FR1 and FR2 radios for carrier aggregation.

Finally, there is a set of device performance requirements, which set the baseband demodulation performance of physical channels of the device receiver across a range of conditions, including propagation in different environments.

Because of the differences between the different types of requirements, the specification for device RF characteristics is separated into four different parts, where the device is called *user equipment* (UE) in 3GPP specifications:

- TS 38.101-1 [5]: UE radio transmission and reception, FR1;
- TS 38.101-2 [6]: UE radio transmission and reception, FR2;
- TS 38.101-3 [7]: UE radio transmission and reception, interworking with other radios;
- TS 38.101-4 [8]: UE radio transmission and reception, performance requirements.

The conducted RF requirements for FR1 are described in Sections 18.6–18.11.

18.5 OVERALL STRUCTURE OF BASE-STATION RF REQUIREMENTS

18.5.1 CONDUCTED AND RADIATED RF REQUIREMENTS FOR NR BS

For the continuing evolution of mobile systems, AAS have an increasing importance. While there were several attempts to develop and deploy base stations with passive antenna arrays of different kinds for many years, there have been no specific RF requirements associated with such antenna systems. With RF requirements in general defined at the base station RF antenna connector, the antennas have also not been seen as part of the base station, at least not from a standardization point of view.

Requirements specified at an antenna connector are referred to as *conducted requirements*, usually defined as a power level (absolute or relative) measured at the antenna connector. Most emission limits in regulation are defined as conducted requirements. An alternative way is to define a *radiated requirement*, which is assessed including the antenna, often accounting for the antenna gain in a specific direction. Radiated requirements demand more complex OTA test procedures, using for example an anechoic chamber. With OTA testing, the spatial characteristics of the whole BS, including the antenna system, can be assessed.

For base stations with *AAS*, where the active parts of the transmitter and receiver may be an integral part of the antenna system, it is not always suitable to maintain the traditional definition of requirements at the antenna connector. For this purpose, 3GPP developed RF requirements in release 13 for AAS base stations in a set of separate RF specifications that are applicable to both LTE and UTRA equipment.

For NR, radiated RF requirements and OTA testing will be a part of the specifications from the start, both in FR1 and FR2. Much of the work from AAS has therefore been taken directly into the NR specifications. The term AAS as such is not used within the NR base-station RF specification [4], however requirements are instead defined for different *BS types*.

The AAS BS requirements are based on a generalized AAS BS radio architecture, as shown in Fig. 18.3. The architecture consists of a *transceiver unit array* that is connected to a *composite antenna* that contains a *radio distribution network* and an *antenna array*. The transceiver unit array contains multiple transmitter and receiver units. These are connected to the composite antenna through a number of connectors on the *transceiver array boundary* (TAB). These TAB connectors correspond to the antenna connectors on a non-AAS base station and serve as a reference point for conducted requirements. The radio distribution network is passive and distributes the transmitter outputs to the corresponding antenna elements and vice versa for the receiver inputs. Note that the actual implementation of an AAS BS may look different in terms of physical location of the different parts, array geometry, type of antenna elements used, etc.

Based on the architecture in Fig. 18.3, there are two types of requirements:

- *Conducted requirements* are defined for each RF characteristic at an individual or a group of TAB connectors. The conducted requirements are defined in such a way that they are in a sense "equivalent" to the corresponding conducted requirement for a non-AAS base station, that is, the

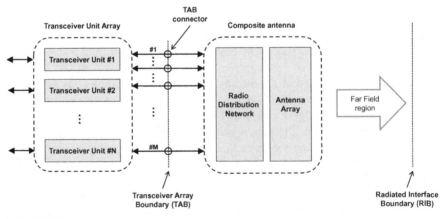

FIGURE 18.3

Generalized radio architecture of an active antenna system (AAS), used also for the NR-radiated requirements.

performance of the system or the impact on other systems is expected to be the same.

- *Radiated requirements* are defined OTA in the far field of the antenna system. Since the spatial direction becomes relevant in this case, it is detailed for each requirement how it applies. Radiated requirements are defined with reference to a *radiated interface boundary* (RIB), somewhere in the far-field region.

18.5.2 BS TYPES IN DIFFERENT FREQUENCY RANGES FOR NR

A number of different base-station design possibilities have to be considered for the RF requirements. First in FR1, there are base stations built in a way similar to "classical" 3G and 4G base stations with antenna connectors through which external antennas are connected. Then we have base stations with AAS, but where antenna connectors can still be accessed for definition and testing of some RF requirements. Finally, we have base stations with highly integrated antenna systems where all requirements must be assessed OTA, since there are no antenna connectors. It is assumed that in FR2 where mm-wave technology is used for implementation of the antenna systems, only the latter type of base station needs to be specified.

3GPP has defined four base-station types based on the above assumptions, with reference to the architecture defined above in Fig. 18.3:

- *BS type 1-C:* NR base station operating in FR1, specified only with conducted requirements defined at individual antenna connectors.
- *BS type 1-O:* NR base station operating in FR1, specified only with conducted (OTA) requirements defined at the RIB.
- *BS type 1-H:* NR base station operating at FR1, specified with a "hybrid" set of requirements consisting of both conducted requirements defined at individual TAB connectors and some OTA requirements defined at the RIB.
- *BS type 2-O:* NR base station operating in FR2, specified only with conducted (OTA) requirements defined at the RIB.

BS type 1-C has requirements defined in the same way as for UTRA or LTE conducted requirements. These are described in Sections 18.6–18.11.

BS type 1-H corresponds to the first type of AAS base stations specified for LTE/UTRA in 3GPP Release 13, where two radiated requirements are defined (radiated transmit power and OTA sensitivity), while all others are defined as conducted requirements, as described in Sections 18.6–18.11. Many conducted requirements, such as unwanted emission limits, are for BS type 1-H defined in two steps. First a *basic limit* is defined, which is identical to the conducted limit at an individual antenna connector for BS type 1-C and thereby equivalent to the limit at a TAB connector for BS type 1-H. In a second step, the basic limit is converted to a radiated limit at the RIB through a scaling factor based on the number of active transmitter units. The scaling is capped at a maximum of 8 (9 dB), which is the maximum number of antenna elements used in defining certain regulatory limits. Note that the maximum scaling may vary depending on regional regulation.

BS type 1-O and BS type 2-O have all requirements defined as radiated. BS type 1-O has many requirements defined with reference to the corresponding FR1 conducted requirements, where unwanted emission limits also have a scaling applied as for BS type 1-H. The overall differences in coexistence properties and implementation between FR1 and FR2 mean that BS type 2-O has separate FR2 requirements defined that in many cases are different from the FR1 requirements for BS type 1-O.

An overview of the radiated requirements used for BS types 1-O and 2-O, and to some extent for BS type 1-H, is given in Section 18.12.

18.6 OVERVIEW OF CONDUCTED RF REQUIREMENTS FOR NR

The RF requirements define the receiver and transmitter RF characteristics of a base station or device. The base station is the physical node that transmits and receives RF signals on one or more antenna connectors. Note that an NR base station is not the same thing as a gNB, which is the corresponding logical node in the radio-access network (see Chapter 6). The device is denoted UE in all RF specifications. Conducted RF requirements are defined for operating bands in FR1, while only radiated (OTA) requirements are defined for FR2 (see Section 18.12).

The set of conducted RF requirements defined for NR is fundamentally the same as those defined for LTE or any other radio system. Some requirements are also based on regulatory requirements and are more concerned with the frequency band of operation and/or the place where the system is deployed, than with the type of system.

What is particular to NR is the flexible channel bandwidths and multiple numerologies of the system, which makes some requirements more complex to define. These properties have special implications for the transmitter requirements on unwanted emissions, where the definition of the limits in international regulation depend on the channel bandwidth. Such limits are harder to define for a system where the base station may operate with multiple channel bandwidths and where the device may vary its channel bandwidth of operation. The properties of the flexible OFDM-based physical layer also have implications for specifying the transmitter modulation quality and how to define the receiver selectivity and blocking requirements. Note that the channel bandwidth in general is different for the BS and the device as discussed in Section 18.3.

The type of transmitter requirements defined for the device is very similar to what is defined for the base station, and the definitions of the requirements are often similar. The output power levels are, however, considerably lower for a device, while the restrictions on the device implementation are much higher. There is tight pressure on cost and complexity for all telecommunications equipment, but this is much more pronounced for devices, due to the scale of the total market, being close to *two billion* devices per year. In cases where there are differences in how requirements are defined between device and base station, they are treated separately in this chapter.

The detailed background of the conducted RF requirements for NR is described in Refs. [74] and [75]. The conducted RF requirements for the base station are specified in Ref. [4] and for the device in Ref. [5]. The RF requirements are divided into transmitter and receiver characteristics. There are also *performance characteristics* for base stations and devices that define the receiver baseband performance for all physical channels under different propagation conditions. These are not strictly RF requirements, though the performance will also depend on the RF to some extent.

Each RF requirement has a corresponding test defined in the NR test specifications for the base station and the device. These specifications define the test setup, test procedure, test signals, test tolerances, etc. needed to show compliance with the RF and performance requirements.

18.6.1 CONDUCTED TRANSMITTER CHARACTERISTICS

The transmitter characteristics define RF requirements for the wanted signal transmitted from the device and the base station, but also for the unavoidable unwanted emissions outside the transmitted carrier(s). The requirements are fundamentally specified in three parts:

- *Output power level* requirements set limits for the maximum allowed transmitted power, for the dynamic variation of the power level, and in some cases for the transmitter OFF state;
- *Transmitted signal quality* requirements define the "purity" of the transmitted signal and also the relation between multiple transmitter branches;
- *Unwanted emissions* requirements set limits to all emissions outside the transmitted carrier(s) and are tightly coupled to regulatory requirements and coexistence with other systems.

A list of the device and base-station transmitter characteristics arranged according to the three parts defined above is shown in Table 18.3. A more detailed description of the specific requirements can be found later in this chapter.

18.6.2 CONDUCTED RECEIVER CHARACTERISTICS

The set of receiver requirements for NR is quite similar to what is defined for other systems such as LTE and UTRA. The receiver characteristics are fundamentally specified in three parts:

- *Sensitivity and dynamic range* requirements for receiving the wanted signal;

Table 18.3 Overview of Conducted NR Transmitter Characteristics

	Base-Station Requirement	Device Requirement
Output power level	Maximum output power Output power dynamics ON/OFF power (TDD only)	Transmit power Output power dynamics Power control
Transmitted signal quality	Frequency error Error vector magnitude (EVM) Time alignment between transmitter branches	Frequency error Transmit modulation quality In-band emissions
Unwanted emissions	Operating band unwanted emissions Adjacent channel leakage ratio (ACLR and CACLR) Spurious emissions Occupied bandwidth Transmitter intermodulation	Spectrum emission mask Adjacent channel leakage ratio (ACLR and CACLR) Spurious emissions Occupied bandwidth Transmit intermodulation

- *Receiver susceptibility to interfering signals* defines receivers' susceptibility to different types of interfering signals at different frequency offsets;
- *Unwanted emissions* limits are also defined for the receiver.

A list of the device and base-station receiver characteristics arranged according to the three parts defined above is shown in Table 18.4. A more detailed description of each requirement can be found later in this chapter.

18.6.3 REGIONAL REQUIREMENTS

There are a number of regional variations to the RF requirements and their application. The variations originate in different regional and local regulations of the spectrum and its use. The most obvious regional variation is the different frequency bands and their use, as discussed above. Many of the regional RF requirements are also tied to specific frequency bands.

When there is a regional requirement on, for example, spurious emissions, this requirement should be reflected in the 3GPP specifications. For the base station it is entered as an optional requirement and is marked as "regional." For the device, the same procedure is not possible, since a device may roam between different regions and will therefore have to fulfill all regional requirements that are tied to an operating band in the regions where the band is used. For NR (and also for LTE), this becomes more complex than for UTRA, since there is an additional variation in the transmitter (and receiver) bandwidth used, making some regional requirements difficult to meet as a mandatory requirement. The concept of *network signaling* of RF requirements is therefore introduced for NR, where a device can be informed at call setup of whether some specific RF requirements apply when the device is connected to a network.

Table 18.4 Overview of Conducted NR Receiver Characteristics

	Base-Station Requirement	Device Requirement
Sensitivity and dynamic range	Reference sensitivity	Reference sensitivity power level
	Dynamic range	Maximum input level
	In-channel selectivity	
Receiver susceptibility to interfering signals	Out-of-band blocking	Out-of-band blocking
		Spurious response
	In-band blocking	In-band blocking
	Narrowband blocking	Narrowband blocking
	Adjacent channel selectivity	Adjacent channel selectivity
	Receiver intermodulation	Intermodulation characteristics
Unwanted emissions from the receiver	Receiver spurious emissions	Receiver spurious emissions

18.6.4 BAND-SPECIFIC DEVICE REQUIREMENTS THROUGH NETWORK SIGNALING

For the device, the channel bandwidths supported are a function of the NR operating band, and also have a relation to the transmitter and receiver RF requirements. The reason is that some RF requirements may be difficult to meet under conditions with a combination of maximum power and high number of transmitted and/or received resource blocks.

In both NR and LTE, some additional RF requirements apply for the device when a specific network signaling value (NS_x) is signaled to the device as part of the cell handover or broadcast message. For implementation reasons, these requirements are associated with restrictions and variations to RF parameters such as device output power, maximum channel bandwidth, and number of transmitted resource blocks. The variations of the requirements are defined together with the NS_x in the device RF specification, where each value corresponds to a specific condition. The default value for all bands is NS_01. NS_x values are connected to an allowed power reduction called *additional maximum power reduction* (A-MPR) and may apply for transmission using a certain minimum number of resource blocks, depending also on the channel bandwidth.

18.6.5 BASE-STATION CLASSES

In order to accommodate different deployment scenarios for base stations, there are multiple sets of RF requirements for NR base stations, each applicable to a *base station class.* When the RF requirements were derived for NR, base-station classes were introduced that were intended for macrocell, microcell, and picocell scenarios. The terms macro, micro, and pico relate to the deployment scenario and are not used in 3GPP to identify the base-station classes, instead the following terminology is used:

- *Wide area base stations:* This type of base station is intended for macrocell scenarios, with a BS-to-device minimum distance along the ground equal to 35 m. This is the typical large cell deployment with high-tower or above-rooftop installations, giving wide area outdoor coverage, but also indoor coverage.
- *Medium range base stations:* This type of base station is intended for microcell scenarios, with a BS-to-device minimum distance along the ground equal to 5 m. Typical deployments are outdoor below-rooftop installations, giving both outdoor hotspot coverage and outdoor-to-indoor coverage through walls.
- *Local area base stations:* This type of base station is intended for picocell scenarios, defined with a BS-to-device minimum distance along the ground equal to 2 m. Typical deployments are indoor offices and indoor/outdoor hotspots, with the BS mounted on walls or ceilings.

The local area and medium range base station classes have modifications to a number of requirements compared to wide area base stations, mainly due to the assumption of a lower minimum base station to device distance, giving a lower minimum coupling loss:

- Maximum base station power is limited to 38 dBm output power for medium range base stations and 24 dBm output power for local area base stations. This power is defined per antenna and carrier. There is no maximum base station power defined for wide area base stations.
- The spectrum mask (operating band unwanted emissions) has lower limits for medium range and local area, in line with the lower maximum power levels.
- Receiver reference sensitivity limits are higher (more relaxed) for medium range and local area. Receiver dynamic range and in-channel selectivity (ICS) are also adjusted accordingly.
- Limits for co-location for medium range and local area are relaxed compared to wide area BS, corresponding to the relaxed reference sensitivity for the base station.
- All medium range and local area limits for receiver susceptibility to interfering signals are adjusted to take the higher receiver sensitivity limit and the lower assumed minimum coupling loss (base station-to-device) into account.

18.7 CONDUCTED OUTPUT POWER LEVEL REQUIREMENTS

18.7.1 BASE-STATION OUTPUT POWER AND DYNAMIC RANGE

There is no general maximum output power requirement for base stations. As mentioned in the discussion of base-station classes above, there is, however, a maximum output power limit of 38 dBm for medium range base stations and 24 dBm for local area base stations. In addition to this, there is a tolerance specified, defining how much the actual maximum power may deviate from the power level declared by the manufacturer.

The base station also has a specification of the total power control dynamic range for a resource element, defining the power range over which it should be possible to configure. There is also a dynamic range requirement for the total base-station power.

For TDD operation, a power mask is defined for the base-station output power, defining the off power level during the uplink subframes and the maximum time for the *transmitter transient period* between the transmitter on and off states.

18.7.2 DEVICE OUTPUT POWER AND DYNAMIC RANGE

The device output power level is defined in three steps:

- *UE power class* defines a *nominal* maximum output power for QPSK modulation. It may be different in different operating bands, but the main device power class is today set at 23 dBm for all bands.
- *Maximum power reduction* (MPR) defines an allowed reduction of maximum power level for certain combinations of modulation used and resource block allocation.
- *Additional maximum power reduction* (A-MPR) may be applied in some regions and is usually connected to specific transmitter requirements such as regional emission limits and to certain carrier configurations. For each such set of requirements, there is an associated network signaling value NS_*x* that identifies the allowed A-MPR and the associated conditions, as explained in Section 18.6.4.

A minimum output power level setting defines the device dynamic range. There is a definition of the transmitter off power level, applicable to conditions when the device is not allowed to transmit. There is also a general on/off time mask specified, plus specific time masks for PRACH, PUCCH, SRS, and for PUCCH/PUSCH/SRS transitions.

The device transmit power control is specified through requirements for the *absolute power tolerance* for the initial power setting, the *relative power tolerance* between two subframes, and the *aggregated power tolerance* for a sequence of power-control commands.

18.8 TRANSMITTED SIGNAL QUALITY

The requirements for transmitted signal quality specify how much the transmitted base station or device signal deviates from an "ideal" modulated signal in the signal and frequency domains. Impairments on the transmitted signal are introduced by the transmitter RF parts, with the non-linear properties of the PA being a major contributor. The signal quality is assessed for the base station and device through requirements on *EVM* and *frequency error*. An additional device requirement is device in-band emissions.

18.8.1 EVM AND FREQUENCY ERROR

While the theoretical definitions of the signal quality measures are quite straightforward, the actual assessment is a very elaborate procedure, described in great detail in the 3GPP specification. The reason is that it becomes a multidimensional optimization problem, where the best match for the timing, the frequency, and the signal constellation are found.

The EVM is a measure of the error in the modulated signal constellation, taken as the root mean square of the error vectors over the active subcarriers, considering all symbols of the modulation scheme. It is expressed as a percentage value in relation to the power of the ideal signal. The EVM fundamentally defines the maximum SINR that can be achieved at the receiver, if there are no additional impairments to the signal between transmitter and receiver.

Since a receiver can remove some impairments of the transmitted signal such as time dispersion, the EVM is assessed after cyclic prefix removal and equalization. In this way, the EVM evaluation includes a standardized model of the receiver. The frequency offset resulting from the EVM evaluation is averaged and used as a measure of the *frequency error* of the transmitted signal.

18.8.2 DEVICE IN-BAND EMISSIONS

In-band emissions are emissions within the channel bandwidth. The requirement limits how much a device can transmit into non-allocated resource blocks within the channel bandwidth. Unlike the out-of-band (OOB) emissions, the in-band emissions are measured after cyclic prefix removal and FFT, since this is how a device transmitter affects a real base-station receiver.

18.8.3 BASE-STATION TIME ALIGNMENT

Several NR features require the base station to transmit from two or more antennas, such as transmitter diversity and MIMO. For carrier aggregation, the carriers may also be transmitted from different antennas. In order for the device to properly receive the signals from multiple antennas, the timing relation between any two transmitter branches is specified in terms of a maximum time alignment error between transmitter branches. The maximum allowed error depends on the feature or combination of features in the transmitter branches.

18.9 CONDUCTED UNWANTED EMISSIONS REQUIREMENTS

Unwanted emissions from the transmitter are divided into *OOB emissions* and *spurious emissions* in ITU-R recommendations [42]. OOB emissions are defined as emissions on a frequency close to the RF carrier, which results from the modulation process. Spurious emissions are emissions outside the RF carrier that may be reduced without affecting the corresponding transmission of information. Examples of spurious emissions are harmonic emissions, intermodulation products, and frequency conversion products. The frequency range where OOB emissions are normally defined is called the *OOB domain*, whereas spurious emission limits are normally defined in the *spurious domain*.

ITU-R also defines the boundary between the OOB and spurious domains at a frequency separation from the carrier center of 2.5 times the necessary bandwidth, which corresponds to 2.5 times the channel bandwidth for NR. This division of the requirements is easily applied for systems that have a fixed channel bandwidth. It does, however, become more difficult for NR, which is a flexible bandwidth system, implying that the frequency range where requirements apply would then vary with the channel bandwidth. The approach taken for defining the boundary in 3GPP is slightly different for base-station and device requirements.

With the recommended boundary between OOB emissions and spurious emissions set at 2.5 times the channel bandwidth, third- and fifth-order intermodulation products from the carrier will fall inside the OOB domain, which will cover a frequency range of twice the channel bandwidth on each side of the carrier. For the OOB domain, two overlapping requirements are defined for both base station and device: *spectrum emissions mask* (SEM) and *adjacent channel leakage ratio* (ACLR). The details of these are further explained below.

18.9.1 IMPLEMENTATION ASPECTS

The spectrum of an OFDM signal decays rather slowly outside of the transmission bandwidth configuration. Since the transmitted signal for NR occupies up to 98% of the channel bandwidth, it is not possible to meet the unwanted emission limits directly outside the channel bandwidth with a "pure" OFDM signal. The techniques used for achieving the transmitter requirements are, however, not specified or mandated in NR specifications. Time-domain windowing is one method commonly used in OFDM-based transmission systems to control spectrum emissions. Filtering is always used, both time-domain digital filtering of the baseband signal and analog filtering of the RF signal.

The non-linear characteristics of the *power amplifier* (PA) used to amplify the RF signal must also be taken into account, since it is the source of intermodulation products outside the channel bandwidth. Power back-off to give a more linear operation of the PA can be used, but at the cost of a lower power efficiency. The power back-off should therefore be kept to a minimum. For this reason, additional linearization schemes can be employed. These are especially important for the base station, where there are fewer restrictions on implementation complexity and use of advanced linearization schemes is an essential part of controlling spectrum emissions. Examples of such techniques are feed-forward, feedback, predistortion, and postdistortion.

18.9.2 EMISSION MASK IN THE OOB DOMAIN

The emission mask defines the permissible OOB spectrum emissions outside the necessary bandwidth. As explained above, how to take the flexible channel bandwidth into account when defining the frequency boundary between OOB emissions and spurious emissions is done differently for the NR base station and device. Consequently, the emission masks are also based on different principles.

18.9.2.1 Base-Station Operating Band Unwanted Emission Limits

For the NR base station, the problem of the implicit variation of the boundary between OOB and spurious domain with the varying channel bandwidth is handled by not defining an explicit boundary. The solution is a unified concept of *operating band unwanted emissions* (OBUEs) for the NR base station instead of the spectrum mask usually defined for OOB emissions. The operating band unwanted emissions requirement applies over the whole base-station transmitter operating band, plus an additional 10—40 MHz on each side, as shown in Fig. 18.4. All requirements outside of that range are set by the regulatory spurious emission limits, based on the ITU-R recommendations [42]. As seen in Fig. 18.4, a large part of the operating band unwanted emissions is defined over a frequency range that for smaller channel bandwidths can be both in spurious and OOB domains. This means that the limits for the frequency ranges that may be in the spurious domain also have to align with the regulatory limits from the ITU-R. The shape of the mask is generic for all channel bandwidths, with a mask that consequently has to align with the ITU-R limits starting 10—40 MHz from the channel edges. The operating band unwanted emissions are defined with a 100 kHz measurement bandwidth and align to a large extent with the corresponding masks for LTE.

In the case of carrier aggregation for a base station, the OBUE requirement (as other RF requirements) applies as for any multicarrier transmission, where the OBUE will be defined relative to the carriers on the edges of the RF bandwidth. In the case of non-contiguous carrier aggregation, the OBUE within a sub-block gap is partly calculated as the cumulative sum of contributions from each sub-block.

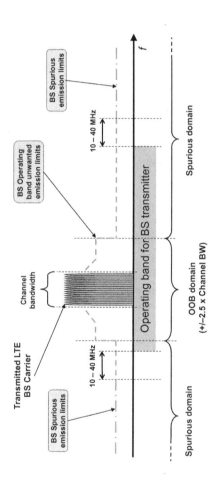

FIGURE 18.4

Frequency ranges for operating band unwanted emissions and spurious emissions applicable to NR base station (FR1).

There are also special limits defined to meet a specific regulation set by the FCC (Federal Communications Commission, Title 47) for the operating bands used in the USA and by the ECC for some European bands. These are specified as separate limits in addition to the operating band unwanted emission limits.

18.9.2.2 Device Spectrum Emission Mask

For implementation reasons, it is not possible to define a generic device spectrum mask that does not vary with the channel bandwidth, so the frequency ranges for OOB limits and spurious emissions limits do not follow the same principle as for the base station. The SEM extends out to a separation Δf_{OOB} from the channel edges, as illustrated in Fig. 18.5. For 5 MHz channel bandwidth, this point corresponds to 250% of the necessary bandwidth as recommended by the ITU-R, but for higher channel bandwidths it is set closer than 250%.

The SEM is defined as a general mask and a set of additional masks that can be applied to reflect different regional requirements. Each additional regional mask is associated with a specific network signaling value NS_x.

18.9.3 ADJACENT CHANNEL LEAKAGE RATIO

In addition to a spectrum emissions mask, the OOB emissions are defined by an ACLR requirement. The ACLR concept is very useful for analysis of coexistence between two systems that operate on adjacent frequencies. The ACLR defines the ratio of the power transmitted within the assigned channel bandwidth to the power of the unwanted emissions transmitted on an adjacent channel. There is a

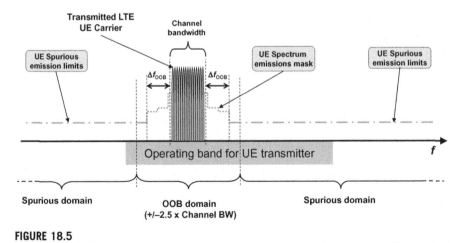

FIGURE 18.5

Frequency ranges for spectrum emission mask and spurious emissions applicable to an NR device.

corresponding receiver requirement called ACS, which defines a receiver's ability to suppress a signal on an adjacent channel.

The definitions of ACLR and ACS are illustrated in Fig. 18.6 for a wanted and an interfering signal received in adjacent channels. The interfering signal's leakage of unwanted emissions at the wanted signal receiver is given by the ACLR and the ability of the receiver of the wanted signal to suppress the interfering signal in the adjacent channel is defined by the ACS. The two parameters when combined define the total leakage between two transmissions on adjacent channels. That ratio is called the *adjacent channel interference ratio* (ACIR) and is defined as the ratio of the power transmitted on one channel to the total interference received by a receiver on the adjacent channel, due to both transmitter (ACLR) and receiver (ACS) imperfections.

This relation between the adjacent channel parameters is [11]:

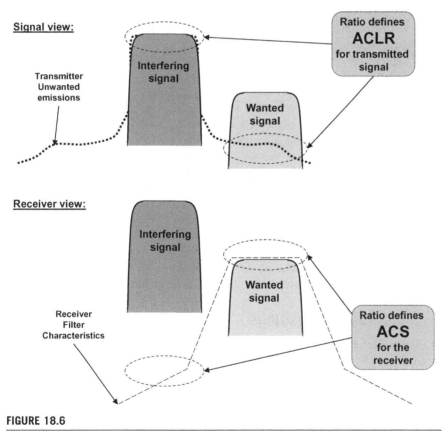

FIGURE 18.6

Illustration of ACLR and ACS, with example characteristics for an "aggressor" interferer and a receiver for a "victim" wanted signal.

$$ACIR = \frac{1}{(1/ACLR) + (1/ACS)} \tag{18.2}$$

ACLR and ACS can be defined with different channel bandwidths for the two adjacent channels, which is the case for some requirements set for NR due to the bandwidth flexibility. Eq. (18.2) will also apply for different channel bandwidths, but only if the same two channel bandwidths are used for defining all three parameters, ACIR, ACLR, and ACS, used in the equation.

The ACLR limits for NR device and base station are derived based on extensive analysis [11] of NR coexistence with NR or other systems on adjacent carriers.

For an NR base station, there are ACLR requirements both for an adjacent channel with an NR receiver of the same channel bandwidth and for an adjacent LTE receiver. The ACLR requirement for NR BS is set to 45 dB. This is considerably more strict than the ACS requirement for the device, which according to Eq. (18.2) implies that in the downlink, the device receiver performance will be the limiting factor for ACIR and consequently for coexistence between base stations and devices. From a system point of view, this choice is cost-efficient since it moves implementation complexity to the BS, instead of requiring all devices to have high-performance RF.

In the case of carrier aggregation for a base station, the ACLR (as other RF requirements) apply as for any multicarrier transmission, where the ACLR requirement will be defined for the carriers on the edges of the RF bandwidth. In the case of non-contiguous carrier aggregation where the sub-block gap is so small that the ACLR requirements at the edges of the gap will "overlap," a special *cumulative ACLR* requirement (CACLR) is defined for the gap. For CACLR, contributions from carriers on both sides of the sub-block gap are accounted for in the CACLR limit. The CACLR limit is the same as the ACLR for the base station at 45 dB.

ACLR limits for the device are set both with assumed NR and an assumed UTRA receiver on the adjacent channel. In the case of carrier aggregation, the device ACLR requirement applies to the aggregated channel bandwidth instead of per carrier. The ACLR limit for NR devices is set to 30 dB. This is considerably relaxed compared to the ACS requirement for the BS, which according to Eq. (18.2) implies that in the uplink, the device transmitter performance will be the limiting factor for ACIR and consequently for coexistence between base stations and devices.

18.9.4 SPURIOUS EMISSIONS

The limits for base station spurious emissions are taken from international recommendations [42], but are only defined in the region outside the frequency range of operating band unwanted emission limits as illustrated in Fig. 18.4—that is, at frequencies that are separated from the base-station transmitter operating band by at least 10−40 MHz. There are also additional regional or optional limits for protection of other systems that NR may coexist with or even be co-located with. Examples of other systems considered in those additional spurious emissions requirements are GSM, UTRA FDD/TDD, CDMA2000, and PHS.

Device spurious emission limits are defined for all frequency ranges outside the frequency range covered by the SEM. The limits are generally based on international regulations [42] but there are also additional requirements for coexistence with other bands when the device is roaming. The additional spurious emission limits can have an associated network signaling value.

In addition, there are base-station and device emission limits defined for the receiver. Since receiver emissions are dominated by the transmitted signal, the receiver spurious emission limits are only applicable when the transmitter is not active, and also when the transmitter is active for an NR FDD base station that has a separate receiver antenna connector.

18.9.5 OCCUPIED BANDWIDTH

Occupied bandwidth is a regulatory requirement that is specified for equipment in some regions, such as Japan and the USA. It was originally defined by the ITU-R as a maximum bandwidth, outside of which emissions do not exceed a certain percentage of the total emissions. The occupied bandwidth is for NR equal to the channel bandwidth, outside of which a maximum of 1% of the emissions are allowed (0.5% on each side).

18.9.6 TRANSMITTER INTERMODULATION

An additional implementation aspect of an RF transmitter is the possibility of intermodulation between the transmitted signal and another strong signal transmitted in the proximity of the base station or device. For this reason, there is a requirement for *transmitter intermodulation.*

For the base station, the requirement is based on a stationary scenario with a co-located other base-station transmitter, with its transmitted signal appearing at the antenna connector of the base station being specified but attenuated by 30 dB. Since it is a stationary scenario, there are no additional unwanted emissions allowed, implying that all unwanted emission limits also have to be met with the interferer present.

For the device, there is a similar requirement based on a scenario with another device-transmitted signal appearing at the antenna connector of the device being specified but attenuated by 40 dB. The requirement specifies the minimum attenuation of the resulting intermodulation product below the transmitted signal.

18.10 CONDUCTED SENSITIVITY AND DYNAMIC RANGE

The primary purpose of the *reference sensitivity requirement* is to verify the receiver *noise figure*, which is a measure of how much the receiver's RF signal chain degrades the SNR of the received signal. For this reason, a low-SNR transmission scheme using QPSK is chosen as a reference channel for the reference sensitivity test. The reference sensitivity is defined at a receiver input level where the throughput is 95% of the maximum throughput for the reference channel.

For the device, reference sensitivity is defined for the full channel bandwidth signals and with all resource blocks allocated for the wanted signal.

The intention of the *dynamic range requirement* is to ensure that the receiver can also operate at received signal levels considerably higher than the reference sensitivity. The scenario assumed for base-station dynamic range is the presence of increased interference and corresponding higher wanted signal levels, thereby testing the effects of different receiver impairments. In order to stress the receiver, a higher SNR transmission scheme using 16QAM is applied for the test. In order to further stress the receiver to higher signal levels, an interfering AWGN signal at a level 20 dB above the assumed noise floor is added to the received signal. The dynamic range requirement for the device is specified as a *maximum signal level* at which the throughput requirement is met.

18.11 RECEIVER SUSCEPTIBILITY TO INTERFERING SIGNALS

There is a set of requirements for base station and device, defining the receiver's ability to receive a wanted signal in the presence of a stronger interfering signal. The reason for the multiple requirements is that, depending on the frequency offset of the interferer from the wanted signal, the interference scenario may look very different and different types of receiver impairments will affect the performance. The intention of the different combinations of interfering signals is to model as far as possible the range of possible scenarios with interfering signals of different bandwidths that may be encountered inside and outside the base-station and device receiver operating band.

While the types of requirements are very similar between base station and device, the signal levels are different, since the interference scenarios for the base station and device are very different. There is also no device requirement corresponding to the base-station ICS requirement.

The following requirements are defined for NR base station and device, starting from interferers with large frequency separation and going close in (see also Fig. 18.7). In all cases where the interfering signal is an NR signal, it has the same or smaller bandwidth than the wanted signal, but at most 20 MHz.

- *Blocking:* This corresponds to the scenario with strong interfering signals received outside the operating band (out-of-band blocking) or inside the operating band (in-band blocking), but not adjacent to the wanted signal. In-band blocking includes interferers in the first 20−60 MHz outside the operating band for the base station and the first 15 MHz for the device. The scenarios are modeled with a *continuous wave* (CW) signal for the out-of-band case and an NR signal for the in-band case. There are additional (optional) base-station blocking requirements for the scenario when the base station is co-located with another base station in a different operating band. For the device, a fixed number of *exceptions* are allowed from the out-of-band blocking requirement, for each assigned frequency channel and at the respective *spurious response frequencies*. At those frequencies, the device must comply with the more relaxed spurious response requirement.

- *Adjacent channel selectivity:* The ACS scenario is a strong signal in the channel adjacent to the wanted signal and is closely related to the corresponding ACLR requirement (see also the discussion in Section 18.9.3). The adjacent interferer is an NR signal. For the device, the ACS is specified for two cases with a lower and a higher signal level.
- *Narrowband blocking:* The scenario is an adjacent strong narrowband interferer, which in the requirement is modeled as a single resource block NR signal for the base station and a CW signal for the device.
- *In-channel selectivity (ICS):* The scenario is multiple received signals of different received power levels inside the channel bandwidth, where the performance of the weaker "wanted" signal is verified in the presence of the stronger "interfering" signal. ICS is only specified for the base station.
- *Receiver intermodulation:* The scenario is *two* interfering signals near to the wanted signal, where the interferers are one CW and one NR signal (not shown in Fig. 18.7). The purpose of the requirement is to test receiver linearity. The interferers are placed in frequency in such a way that the main intermodulation product falls inside the wanted signal's channel bandwidth. There is also a *narrowband intermodulation* requirement for the base station where the CW signal is very close to the wanted signal and the NR interferer is a single RB signal.

For all requirements except ICS, the wanted signal uses the same reference channel as in the corresponding reference sensitivity requirement. With the interference added, the same 95% relative throughput is met as for the reference sensitivity, but at a "desensitized" higher wanted signal level.

18.12 RADIATED RF REQUIREMENTS FOR NR

Many of the radiated RF requirements defined for devices and base stations are derived directly from the corresponding conducted RF requirements. Unlike conducted requirements, the radiated requirements will account also for the antenna. When defining emission levels such as base station output power and unwanted emissions, this can be done either by incorporating the antenna gain as a directional requirement using an *effective isotropic radiated power* (EIRP) or by definition of limits using *total radiated power* (TRP). Two new radiated requirements

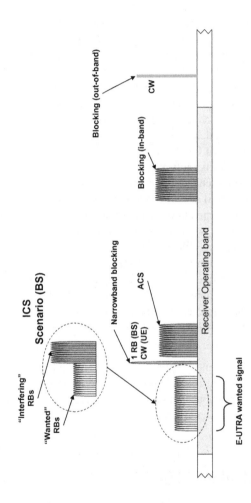

FIGURE 18.7

Base-station and device requirements for receiver susceptibility to interfering signals in terms of blocking, ACS, narrowband blocking, and ICS (BS only).

are defined as directional for the base station (see Section 18.12.2), but most radiated device and base station requirements for NR are defined with limits expressed as TRP. There are several reasons for this choice [19].

TRP and EIRP are directly related through the number of radiating antennas and depend also on specific base station implementation, considering the geometry of the antenna array and the correlation between unwanted emission signals from different antenna ports. The implication is that an EIRP limit could result in different levels of total radiated unwanted emission power depending on the implementation. An EIRP limit will thus not give control of the total amount of interference in the network, while a TRP requirement limits the total amount of interference injected in the network regardless of the specific BS implementation.

Another relevant element behind the 3GPP choice of defining unwanted emission as TRP is the different behavior between passive and AAS. In the case of passive systems, the antenna gain does not vary much between the wanted signal and unwanted emissions. Thus, EIRP is directly proportional to TRP and can be used as a substitute. For an active system such as NR, the EIRP could vary greatly between the wanted signal and unwanted emissions and also between implementations, so EIRP is not proportional to TRP and using EIRP to substitute TRP would be incorrect.

The radiated RF requirements for device and base station are described below.

18.12.1 RADIATED DEVICE REQUIREMENTS IN FR2

As described in Section 18.4, the RF requirements in FR2 operating bands are described in a separate specification [6] for devices, because of the higher number of antenna elements for operation in FR2 and the high level of integration expected when using mm-wave technology. The set of requirements is basically the same as the conducted RF requirements defined for FR1 operating bands. The limits for many requirements are however different. The difference in coexistence at mm-wave frequencies leads to lower requirements on, for example, ACLR and spectrum mask. This is demonstrated through co-existence studies performed in 3GPP and documented in [11]. The possibility for different limits has also been demonstrated in academia [73].

The implementation using mm-wave technologies is more challenging than using the more mature technologies in the frequency bands below 6 GHz (FR1). The mm-wave RF implementation aspects are further discussed in Chapter 19.

It should also be noted that the channel bandwidths and numerologies defined for FR2 are in general different from FR1, making it not possible to compare requirement levels, especially for receiver requirements.

The following is a brief overview of the radiated RF requirements in FR2:

- *Output power level requirements:* Maximum output power is of the same order as in FR1 but is expressed both as TRP and EIRP. The minimum output power and transmitter OFF power levels are higher than in FR1. Radiated transmit power is an additional radiated requirement, which unlike the maximum output power is directional.

- *Transmitted signal quality:* Frequency error and EVM requirements are defined similar to what is done in FR1 and mostly with the same limits.
- *Radiated unwanted emissions requirements:* Occupied bandwidth, ACLR, spectrum mask, and spurious emissions are defined in the same way as for FR1. The latter two are based on TRP. Many limits are less strict than in FR1. ACLR is on the order of 10 dB relaxed compared to FR1, due to more favorable coexistence.
- *Reference sensitivity and dynamic range:* Defined in the same way as in FR1, but levels are not comparable.
- *Receiver susceptibility to interfering signals:* ACS, in-band and out-of-band blocking are defined as for FR1, but there is no narrow-band blocking scenario defined since there are only wideband systems in FR2. ACS is on the order of 10 dB relaxed compared to FR1, due to more favorable coexistence.

18.12.2 RADIATED BASE-STATION REQUIREMENTS IN FR1

As described in Section 18.5, the RF requirements for BS type 1-O consisted of only radiated (OTA) requirements. These are in general based on the corresponding conducted requirements, either directly or through scaling. Two additional radiated requirements defined are *radiated transmit power* and *OTA sensitivity*, described further below.

BS type 1-H is defined with a "hybrid" set of requirements consisting mostly of conducted requirements and in addition two radiated requirements, which are the same as for BS type 1-O:

- *Radiated transmit power* is defined accounting for the antenna array beam-forming pattern in a specific direction as EIRP for each beam that the base station is declared to transmit. In a way similar to BS output power, the actual requirement is on the accuracy of the declared EIRP level.
- *OTA sensitivity* is a directional requirement based on a quite elaborate declaration by the manufacturer of one or more *OTA sensitivity direction declarations* (OSDDs). The sensitivity is in this way defined accounting for the antenna array beam-forming pattern in a specific direction as declared *equivalent isotropic sensitivity* (EIS) level towards a receiver target. The EIS limit is to be met not only in a single direction but within a *range of angle of arrival* (RoAoA) in the direction of the receiver target. Depending on the level of adaptivity for the AAS BS, two alternative declarations are made:
  - If the receiver is adaptive to direction, so that the receiver target can be redirected, the declaration contains a *receiver target redirection range* in a specified *receiver target direction*. The EIS limit should be met within the redirection range, which is tested at five declared sensitivity RoAoA within that range.

• If the receiver is not adaptive to direction and thus cannot redirect the receiver target, the declaration consists of a single sensitivity RoAoA in a specified receiver target direction, in which the EIS limit should be met.

Note that the OTA sensitivity is defined in addition to the reference sensitivity requirement, which exists both as conducted (for BS type 1-H) and radiated (for BS type 1-O).

18.12.3 RADIATED BASE-STATION REQUIREMENTS IN FR2

As described in Section 18.5, the RF requirements for BS type 2-O are radiated requirements for base stations in FR2 operating bands. These are described separately, together with the radiated requirements for BS type 1-O, but in the same specification [4] as the conducted base-station RF requirements.

The set of requirements is identical to the radiated RF requirements defined for FR1 operating bands described above, but the limits for many requirements are different. As for the device, the difference in coexistence at mm-wave frequencies leads to lower requirements on, for example, ACLR, ACS as demonstrated through 3GPP coexistence studies [11]. The implementation using mm-wave technologies is also more challenging than using the more mature technologies in the frequency bands below 6 GHz (FR1) as further discussed in Chapter 19.

The following is a brief overview of the radiated RF requirements in FR2:

• *Output power level requirements:* Maximum output power is the same for FR1 and FR2, but is scaled from the conducted requirement and expressed as TRP. There is in addition a directional radiated transmit power requirement. The dynamic range requirement is defined similarly to FR1.
• *Transmitted signal quality:* Frequency error, EVM, and time-alignment requirements are defined similar to what is done in FR1 and mostly with the same limits.
• *Radiated unwanted emissions requirements:* Occupied bandwidth, spectrum mask, ACLR, and spurious emissions are defined in the same way as for FR1. The three latter are based on TRP and also have less strict limits than in FR1. ACLR is on the order of 15 dB, relaxed compared to FR1, due to more favorable coexistence.
• *Reference sensitivity and dynamic range:* Defined in the same way as in FR1, but levels are not comparable. There is in addition a directional OTA sensitivity requirement.
• *Receiver susceptibility to interfering signals:* ACS, in-band, and out-of-band blocking are defined as for FR1, but there is no narrow-band blocking scenario defined since there are only wideband systems in FR2. ACS is relaxed compared to FR1, due to more favorable coexistence.

18.13 ONGOING DEVELOPMENTS OF RF REQUIREMENTS FOR NR

The first set of NR specifications in 3GPP release 15 does not have full support for some RF deployment options that exist for LTE. Multistandard radio (MSR) base stations, multiband base stations, and non-contiguous operation are features that are under development in 3GPP and will have full support in the final release 15 specifications, or in some cases in release 16. A short description of those features is given below, where a more detailed description (applicable to LTE) can be found in Ref. [28].

18.13.1 MULTISTANDARD RADIO BASE STATIONS

Traditionally the RF specifications have been developed separately for the different 3GPP radio-access technologies GSM/EDGE, UTRA, LTE, and NR. The rapid evolution of mobile radio and the need to deploy new technologies alongside the legacy deployments has, however, led to implementation of different radio-access technologies (RAT) at the same sites, sharing antennas and other parts of the installation. In addition, operation of multiple RATs is often done within the same base-station equipment. The evolution to multi-RAT base stations is fostered by the evolution of technology. While multiple RATs have traditionally shared parts of the site installation, such as antennas, feeders, backhaul, or power, the advance of both digital baseband and RF technologies enables a much tighter integration.

3GPP defines an MSR base station, as a base station where the receiver and the transmitter are capable of simultaneously processing multiple carriers of different RATs in common *active* RF components. The reason for this stricter definition is that the true potential of multi-RAT base stations, and the challenge in terms of implementation complexity, comes from having a common RF. This principle is illustrated in Fig. 18.8 with an example base station capable of both NR and LTE. Some of the NR and LTE baseband functionality may be separate in the base station but is possibly implemented in the same hardware. The RF must, however, be implemented in the same active components as shown in the figure.

While development of MSR BS specifications including NR is part of the work in 3GPP release 15, the set of specifications first issued for NSA operation will in the first step not cover the MSR BS specifications. It is expected that NR will be added as a new RAT for MSR BS during 2018. The main advantages of an MSR base station implementation for NR are twofold:

- Migration between RATs in a deployment, for example, from previous mobile generations to NR, is possible using the same base-station hardware. The operation of NR can then be introduced gradually over time when parts of the spectrum used for previous generations is freed up for NR.

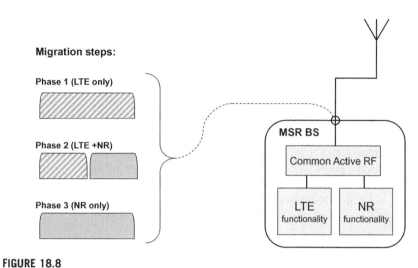

FIGURE 18.8

Example of migration from LTE to NR using an MSR base station for all migration phases.

- A single base station designed as an MSR base station can be deployed in various environments for single-RAT operation for each RAT supported, as well as for multi-RAT operation, where that is required by the deployment scenario. This is also in line with the recent technology trends seen in the market, with fewer and more generic base-station designs. Having fewer varieties of base station is an advantage both for the base-station vendor and for the operator, since a single solution can be developed and implemented for a variety of scenarios.

The MSR concept has a substantial impact for many requirements, while others remain completely unchanged. A fundamental concept introduced for MSR base stations is *RF bandwidth*, which is defined as the total bandwidth over the set of carriers transmitted and received. Many receiver and transmitter requirements are usually specified relative to the carrier center or the channel edges. For an MSR base station, they are instead specified relative to the *RF bandwidth edges*, in a way similar to carrier aggregation. By introducing the RF bandwidth concept and introducing generic limits, the requirements for MSR shift from being carrier centric towards being frequency block centric, thereby embracing technology neutrality by being independent of the access technology or operational mode.

For the specification of MSR base stations, the operating bands are divided into *band categories* (BC) depending on what RATs are supported in the band. There are presently three band categories, BC1−BC3, covering paired bands without GSM operation, paired bands with GSM operation and unpaired bands, respectively. It is not yet determined whether new band categories could be needed when NR is added as additional RAT.

Another important concept for MSR base stations is the supported *capability set* (CS), which is part of the declaration of base-station capabilities by the vendor. The capability set defines all supported single RATs and multi-RAT combinations. There are currently 15 capability sets, CS1−CS15, defined in the MSR BS test specification [2]. When NR is added as a new RAT, it is expected that the new CS will be added to cover RAT combinations that include NR operation.

Carrier aggregation is also applicable to MSR base stations. Since the MSR specification has most of the concepts and definitions in place for defining multi-carrier RF requirements, whether aggregated or not, the differences for the MSR requirements compared to non-aggregated carriers are very minor.

More details on the RF requirements for MSR base stations supporting LTE, UTRA, and GSM/EDGE operation are given in Section 22.5 of Ref. [28].

18.13.2 MULTIBAND-CAPABLE BASE STATIONS

The 3GPP specifications have been continuously developed to support larger RF bandwidths for transmission and reception through multicarrier and multi-RAT operation and carrier aggregation within a band and with requirements defined for one band at a time. This has been made possible with the evolution of RF technology supporting larger bandwidths for both transmitters and receivers. From 3GPP release 11, there is support in the LTE and MSR base-station specifications for simultaneous transmission and/or reception in two bands through a common radio. Such a multiband base station covers multiple bands over a frequency range of a few 100 MHz. Support for more than two bands is given from 3GPP release 14.

While development of NR specifications for multiband base stations is not excluded from the work in 3GPP release 15, the set of specifications first issued for NSA operation does not have full descriptions of multiband operation of NR for bands in frequency range 2.

One obvious application for multiband base stations is for interband carrier aggregation. It should however be noted that base stations supporting multiple bands were in existence long before carrier aggregation was introduced in LTE and UTRA. Already for GSM, dual-band base stations were designed to enable more compact deployments of equipment at base-station sites, but they were really two separate sets of transmitters and receivers for the bands that were integrated in the same equipment cabinet. The difference for "true" multiband-capable base stations is that the signals for the bands are transmitted and received in common *active* RF in the base station.

An example base station is illustrated in Fig. 18.9, which shows a base station with a common RF implementation of both transmitter and receiver for two operating bands X and Y. Through a duplex filter, the transmitter and receiver are connected to a common antenna connector and a common antenna. The example is also a multi-RAT-capable MB-MSR base station, with LTE + GSM configured in band X and LTE configured in band Y. Note that the figure has only one

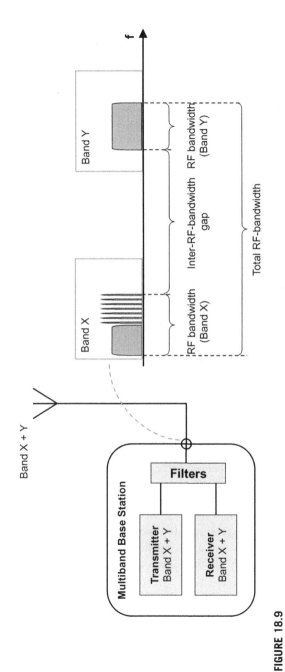

FIGURE 18.9

Example of multiband base station with multiband transmitter and receiver for two bands with one common antenna connector.

diagram showing the frequency range for the two bands, which could either be the receiver or transmitter frequencies.

While having only a single antenna connector and a common feeder that connects to a common antenna is desirable to reduce the amount of equipment needed in a site, it is not always possible. It may also be desirable to have separate antenna connectors, feeders, and antennas for each band. An example of a multiband base station with separate connectors for two operating bands X and Y is shown in Fig. 18.10. Note that while the antenna connectors are separate for the two bands, the RF implementation for transmitter and receiver is in this case common for the bands. The RF for the two bands is separated into individual paths for band X and band Y before the antenna connectors through a filter. As for multiband base stations with a common antenna connector for the bands, it is also here possible to have either the transmitter or receiver be a single-band implementation, while the other is multiband.

Further possibilities are base station implementations with separate antenna connectors for receiver and transmitter, in order to give better isolation between the receiver and transmitter paths. This may be desirable for a multiband base station, considering the large total RF bandwidths, which will in fact also overlap between receiver and transmitter.

For a multiband base station, with a possible capability to operate with multiple RATs and several alternative implementations with common or separate antenna connectors for the bands and/or for the transmitter and receiver, the declaration of the base station capability becomes quite complex. What requirements will apply to such a base station and how they are tested will also depend on these declared capabilities.

More details on the RF requirements for multiband base stations supporting LTE operation is given in Section 22.12 of [28].

18.13.3 OPERATION IN NON-CONTIGUOUS SPECTRUM

Some spectrum allocations consist of fragmented parts of spectrum for different reasons. The spectrum may be a recycled 2G spectrum, where the original licensed spectrum was "interleaved" between operators. This was quite common for original GSM deployments, for implementation reasons (the original combiner filters used were not easily tuned when spectrum allocations were expanded). In some regions, operators have also purchased spectrum licenses on auctions and have for different reasons ended up with multiple allocations in the same band that are not adjacent.

For deployment of non-contiguous spectrum allocations there are a few implications:

- If the full spectrum allocation in a band is to be operated with a single base station, the base station has to be capable of operation in non-contiguous spectrum.

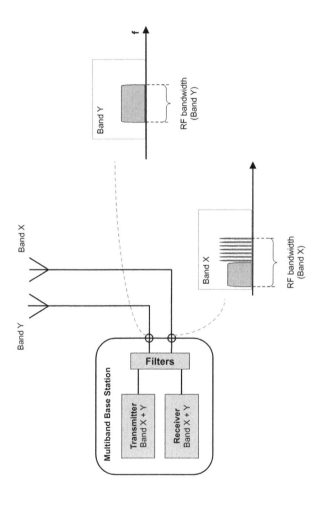

FIGURE 18.10

Multiband base station with multiband transmitter and receiver for two bands with separate antenna connectors for each band.

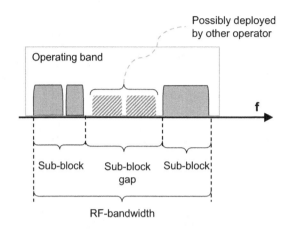

FIGURE 18.11

Example of non-contiguous spectrum operation, illustrating the definitions of RF bandwidth, sub-block, and sub-block gap.

- If a larger transmission bandwidth is to be used than what is available in each of the spectrum fragments, both the device and the base station have to be capable of *intraband non-contiguous carrier aggregation* in that band.

Note that the capability for the base station to operate in non-contiguous spectrum is not directly coupled to carrier aggregation as such. From an RF point of view, what will be required by the base stations is to receive and transmit carriers over an RF bandwidth that is split in two (or more) separate sub-blocks, with a sub-block gap in-between, as shown in Fig. 18.11. The spectrum in the sub-block gap can be deployed by any other operator, which means that the RF requirements for the base station in the sub-block gap will be based on coexistence for uncoordinated operation. This has a few implications for some of the base station RF requirements within an operating band.

While development of NR BS specifications for non-contiguous spectrum is not excluded from the work in 3GPP Release 15, the set of specifications first issued for NSA operation does not have a full description of non-contiguous operation. More details on the RF requirements for non-contiguous operation for LTE are given in Section 22.4 of Ref. [28].

RF Technologies at mm-Wave Frequencies

19

The existing 3GPP specifications for 2G, 3G, and 4G mobile communications are applicable to frequency ranges below 6 GHz and the corresponding RF requirements consider the technology aspects related to below 6 GHz operation. NR also operates in those frequency ranges (identified as frequency range 1) but will in addition be defined for operation above 24.25 GHz (frequency range 2 or FR2), also referred to as mm-wave frequencies. A fundamental aspect for defining the RF performance and setting RF requirements for NR base stations and devices is the change in technologies used for RF implementation in order to support operation in those higher frequencies. In this chapter, some important and fundamental aspects related to mm-wave technologies are presented in order to better understand the performance that mm-wave technology can offer, but also what the limitations are.

In this chapter, Analog-to-Digital/Digital-to-Analog converters and power amplifiers are discussed, including aspects such as the achievable output power versus efficiency and linearity. In addition, some detailed insights are provided into receiver essential metrics such as noise figure, bandwidth, dynamic range, power dissipation, and the dependencies between metrics. The mechanism for frequency generation and the related phase noise aspects are also covered. Filters for mm-waves are another important part, indicating the achievable performance for various technologies and the feasibility of integrating filters into NR implementations.

The data sets used in this chapter indicate the current state-of-the-art capability and performance and are either published elsewhere or have been presented as part of the 3GPP study for developing NR [11]. Note that neither the 3GPP specifications nor the discussion here mandate any restrictions, specific models, or implementations for NR in frequency range 2. The discussion highlights and analyzes different possibilities for RF implementation of mm-wave receivers and transmitters.

An additional aspect is that essentially all operation in Frequency Range 2 will be with Active Antenna System base stations using large antenna array sizes and devices with multi-antenna implementations. While this is enabled by the smaller scale of antennas at mm-wave frequencies, it also drives complexity. The compact building practice needed for mm-wave systems with many transceivers and antennas requires careful and often complex consideration regarding the power efficiency and heat dissipation within a small area or volume. These considerations directly affect the achievable performance and possible RF

5G NR: The Next Generation Wireless Access Technology. DOI: https://doi.org/10.1016/B978-0-12-814323-0.00019-3

requirements. The discussion here in many aspects applies for both NR base stations and NR devices, noting also that the mm-wave transceiver implementation between device and base station will have less differences compared to frequency bands below 6 GHz.

19.1 ADC AND DAC CONSIDERATIONS

The larger bandwidths available at mm-wave communication will challenge the data conversion interfaces between analog and digital domains in both receivers and transmitters. The signal-to-noise-and-distortion ratio (SNDR)-based Schreier Figure-of-Merit (FoM) is a widely accepted metric for Analog-to-Digital Converters (ADCs) defined by [61]

$$\text{FoM} = \text{SNDR} + 10 \ \log_{10}\left(\frac{f_s/2}{P}\right)$$

with SNDR in dB, power consumption P in W, and Nyquist sampling frequency f_s in Hz. Fig. 19.1 shows the Schreier FoM for a large number of ADCs vs the Nyquist sampling frequency f_s ($=2 \times \text{BW}$ for most converters), published at the two most acknowledged conferences [62] in this field of research. The dashed

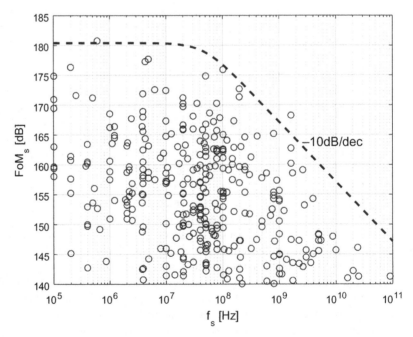

FIGURE 19.1

Schreier figure-of-merit for published ADCs [62].

line indicates the FoM envelope which is constant at roughly 180 dB for sampling frequencies below some 100 MHz. With constant FoM, the power consumption doubles for every doubling of bandwidth or 3 dB increase in SNDR. Above 100 MHz there is an additional 10 dB/decade penalty, and this means that a doubling of bandwidth will increase power consumption by a factor of 4.

Although the FoM envelope is expected to be slowly pushed towards higher frequencies by continued development of integrated circuit technology, RF bandwidths in the GHz range inevitably give poor power efficiency in the analog-to-digital conversion. The large bandwidths and array sizes assumed for NR at mm-wave will thus lead to a large ADC power footprint and it is important that specifications driving SNDR requirements are not unnecessarily high. This applies to devices as well as base stations.

Digital-to-Analog Converters (DACs) are typically less complex than their ADC counterparts for the same resolution and speed. Furthermore, while ADC operation commonly involves iterative processes, the DACs do not. DACs also attract substantially less interest in the research community. While structurally quite different from their ADC counterparts they can still be benchmarked using the same FoM and render similar numbers as for ADCs. In the same way as for ADC, a larger bandwidth and unnecessarily high SNDR requirement on the transmitter will result in higher DAC power footprint.

19.2 LO GENERATION AND PHASE NOISE ASPECTS

Local Oscillator (LO) is an essential component in all modern communication systems for shifting carrier frequency up- or downwards in transceivers. A parameter featuring the LO quality is the so-called phase noise (PN) of the signal generated by the LO. In plain words, phase noise is a measure of how stable the signal is in frequency domain. Its value is given in dBc/Hz for an offset frequency Δf and it describes the likelihood that the oscillation frequency deviates by Δf from the desired frequency.

LO phase noise may significantly impact system performance; this is illustrated in Fig. 19.2, though somewhat exaggerated for a single-carrier example, where the constellation diagram for a 16-QAM signal is compared for cases with and without phase noise, including in both cases an Additive White Gaussian Noise (AWGN) signal modeling thermal noise. For a given symbol error rate, phase noise limits the highest modulation scheme that may be utilized, as demonstrated in Fig. 19.2. In other words, different modulation schemes pose different requirements on the LO phase noise level.

19.2.1 PHASE NOISE CHARACTERISTICS OF FREE-RUNNING OSCILLATORS AND PLLS

The most common circuit solution for frequency generation is to use a Voltage-Controlled Oscillator (VCO). Fig. 19.3 shows a model and the characteristic PN

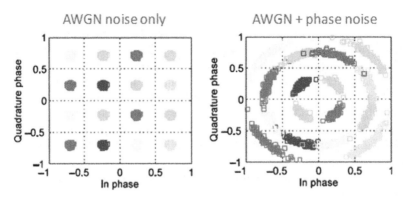

FIGURE 19.2

Constellation diagram of a single-carrier 16-QAM signal without (left) and with (right) LO phase noise.

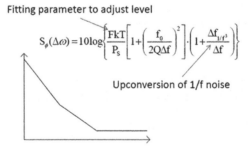

FIGURE 19.3

Phase noise characteristic for a typical free-running VCO [57]: phase noise in dBc/Hz (y-axis) versus offset frequency in Hz (x-axis, logarithmic scale).

behavior of a free-running VCO in different offset frequency regions, where f_0 is the oscillation frequency, Δf is the offset frequency from f_0, P_s is the signal strength, Q is the loaded quality factor of the resonator, F is an empirical fitting parameter but has physical meaning of noise figure, and $\Delta f_{1/f^3}$ is the 1/f-noise corner frequency of the active device in use [57].

The following can be concluded from the Leeson formula in Fig. 19.3:

1. PN increases by 6 dB per every doubling of the oscillation frequency f_0;
2. PN is inversely proportional to signal strength, P_s;
3. PN is inversely proportional to the square of the loaded quality factor of the resonator, Q;
4. 1/f noise up-conversion gives rise to close-to-carrier PN increase (at small offset).

Thus, there are several parameters that may be used for design trade-offs in VCO development. To make performance comparison of the VCOs made in different semiconductor technologies and circuitry topologies, a Figure-of-Merit (FoM) is often used which takes into account power consumption and thus allows for a fair comparison:

$$\text{FoM} = \text{PN}_{\text{VCO}}(\Delta f) - 20\log\left(\frac{f_0}{\Delta f}\right) + 10\log\left(P_{\text{DC}}/1\text{mW}\right)$$

Here $\text{PN}_{\text{VCO}}(\Delta f)$ is the phase noise of the VCO in dBc/Hz and P_{DC} is the power consumption in watt. One noticeable result of this expression is that both phase noise and power consumption in linear power are proportional to $f_0{}^2$. Thus, to maintain a phase noise level at a certain offset while increasing f_0 by a factor N would require the power to be increased by N^2 (assuming a fixed FoM value).

A common way to suppress the phase noise is to apply a Phase Locked Loop (PLL) [18]. Basic PLL building blocks contain a VCO, frequency divider, phase detector, loop filter, and a low-frequency reference source of high stability, such as a crystal oscillator. The total phase noise of the PLL output is composed of contributions from the VCO outside the loop bandwidth and the reference oscillator inside the loop. A significant noise contribution is also added by the phase detector and the divider.

As an example for the typical behavior of an mm-wave LO, Fig. 19.4 shows the measured phase noise from a 28 GHz LO produced by applying a PLL at a

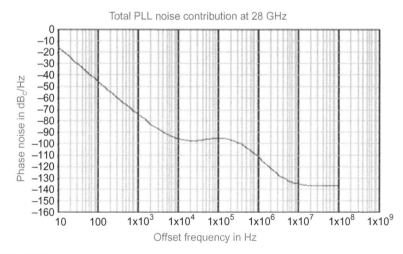

FIGURE 19.4

Example of measured phase noise behavior for a phase locked VCO multiplied to 28 GHz.

Ericsson AB, used with permission.

lower frequency and then multiplying up to 28 GHz. There are four different off-set ranges that show distinctive characteristics:

1. f_1, for small offset, <10 kHz: ~30 dB/decade roll-off, due to 1/f noise up-conversion;
2. f_2, for offset within the PLL bandwidth: relatively flat and composed of several contributions;
3. f_3, for offset larger than PLL bandwidth: ~20 dB/decade roll-off, dominant by VCO phase noise;
4. f_4, for even larger offset, >10 MHz: flat, due to finite noise floor.

19.2.2 CHALLENGES WITH MM-WAVE SIGNAL GENERATION

As phase noise increases with frequency, increasing the oscillation frequency from 3 GHz to 30 GHz, for instance, will result in fundamental PN degradation of 20 dB at a given offset frequency. This will certainly limit the highest order of PN-sensitive modulation schemes usable at mm-wave and thus poses a limitation on achievable spectrum efficiency for mm-wave communications.

Millimeter-wave LOs also suffer from the degradation in quality factor Q and the signal power P_s. Leeson's equation tells us that in order to achieve low phase noise, Q and P_s need to be maximized, while minimizing the noise figure of the active device. Unfortunately, these three factors contribute in an unfavorable manner when oscillation frequency increases. In monolithic VCO implementation, the Q-value of the on-chip resonator decreases rapidly with frequency increases due mainly to (1) the increase of parasitic losses such as metal loss and/or substrate loss and (2) the decrease of varactor Q. Meanwhile, the signal strength of the oscillator becomes increasingly limited when going to higher frequencies. This is because higher-frequency operation requires more advanced semiconductor devices whose breakdown voltage decreases as their feature size shrinks. This is manifested by the observed reduction in power capability versus frequency for power amplifiers (−20 dB per decade) as detailed in Section 19.3. For this reason, a method widely applied in mm-wave LO implementation is to generate a lower-frequency PLL and then multiply the signal up to the target frequency.

Except for the challenges discussed above, up-conversion of the 1/f noise creates an added slope close to the carrier. The 1/f noise is strongly technology-dependent, where planar devices such as CMOS and HEMT (High Electron Mobility Transistor) generally show higher 1/f noise than vertical bipolar devices such as bipolar and HBTs. Technologies used in fully integrated MMIC/RFIC VCO and PLL solution range from CMOS and BiCMOS to III−V materials where InGaP HBT is popular due to its relatively low 1/f noise and high break-down. Occasionally also pHEMT devices are used, even if suffering from severe 1/f noise. Some developments have been made using GaN FET structures in order to benefit from the very high breakdown voltage, but 1/f is even higher than in GaAs FET devices and therefore seems to offset the gain due to the breakdown

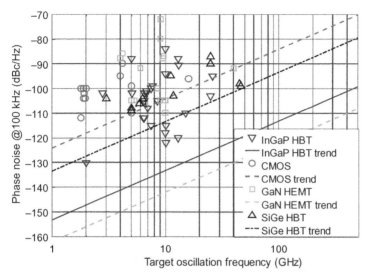

FIGURE 19.5

Phase noise versus oscillation frequency for oscillators in different semiconductor technologies [36].

voltage. Fig. 19.5 summarizes phase noise performance at 100 kHz offset vs oscillation frequency for different semiconductor technologies.

Last but not least, recent research reveals the impact of the LO noise floor on system performance [23]. This impact is insignificant if the symbol rate is low. When the rate increases, such as in 5G NR, the flat noise floor starts to increasingly affect the EVM of the modulated signal. Fig. 19.6 shows the measured EVM from a transmitter for different symbol rate and different noise floor level. The impact from receiver LO noise floor is similar. This observation may imply that it requires extra care when generating mm-wave LOs for wideband systems in terms of choice of technology, VCO topology, and multiplication factor, to maintain a reasonably low PN floor.

19.3 POWER AMPLIFIER EFFICIENCY IN RELATION TO UNWANTED EMISSION

Radio Frequency (RF) building block performance generally degrades with increasing frequency. The power capability of power amplifiers (PA) for a given integrated circuit technology roughly degrades by 20 dB per decade, as shown in Fig. 19.7 for a number of various semiconductor technologies. There is a fundamental cause for this degradation; increased power capability and increased frequency capability are conflicting requirements as observed from the so-called

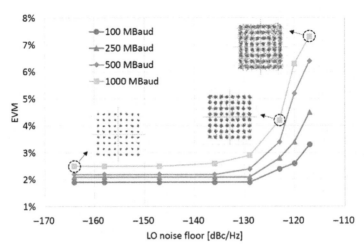

FIGURE 19.6

Measured EVM of a 64-QAM signal from a 7.5 GHz transmitter for different symbol rate and LO noise floor level [23].

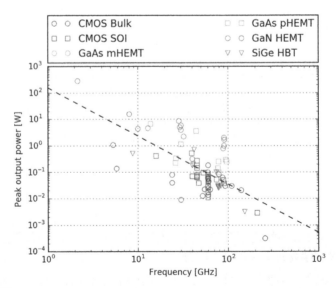

FIGURE 19.7

Power amplifier output power versus frequency for various semiconductor technologies. The dashed line illustrates the observed reduction in power capability versus frequency (−20 dB per decade). The data points are from a survey of published microwave and mm-wave power amplifier circuits.

Johnson limit [54]. In short, higher operational frequencies require smaller geometries, which subsequently result in lower operational power in order to prevent dielectric breakdown from the increased field strengths. To uphold Moore's law, the gate geometries are constantly shrunk and hence the intrinsic power capability is reduced.

A remedy is however found in the choice of integrated circuit material. mm-Wave integrated circuits have traditionally been manufactured using so-called III−V materials, that is a combination of elements from groups III and V of the periodic table, such as Gallium Arsenide (GaAs) and more recently Gallium Nitride (GaN). Integrated circuit technologies based on III−V materials are substantially more expensive than conventional silicon-based technologies and they cannot handle the integration complexity of, for example, digital circuits or radio modems for cellular handsets. Nevertheless, GaN-based technologies are now maturing rapidly and deliver power levels an order of magnitude higher compared to conventional technologies.

There are mainly three semiconductor material parameters that affect the efficiency of an amplifier: maximum operating voltage, maximum operating current density, and knee-voltage. Due to the knee-voltage, the maximum attainable efficiency is reduced by a factor that is proportional to:

$$\frac{1-k}{1+k}$$

where k is the ratio of knee-voltage to the maximum operating voltage. For most transistor technologies the ratio k is in the range of 0.05−0.01, resulting in an efficiency degradation of 10%−20%.

The maximum operating voltage and the current density limit the maximum output power from a single transistor cell. To further increase the output power, the output from multiple transistor cells must be combined. The most common combination techniques are stacking (voltage combining), paralleling (current combining), and corporate combiners (power combining). Either choice of combination technique will be associated with a certain combiner-efficiency. A lower power density requires more combination stages and will incur a lower overall combiner-efficiency. At mm-wave frequencies the voltage- and current-combining methods are limited due to the wave-length. The overall size of the transistor cell must be kept less than about 1/10th of the wavelength. Hence, paralleling and/or stacking are used to some extent and then corporate combining is used to get the wanted output power. The maximum power density of CMOS is about 100 mW/mm compared to 4000 mW/mm for GaN. Thus, GaN technology will require less aggressive combining strategies and hence give higher efficiency.

Fig. 19.8 shows the saturated power-added efficiency (PAE) as a function of frequency. The maximum reported PAE is approximately 40% and 25%, at 30 GHz and 77 GHz, respectively.

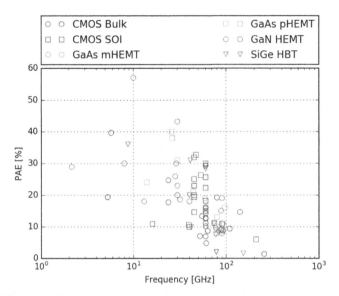

FIGURE 19.8

Saturated power-added efficiency versus frequency for various semiconductor technologies from a survey of published microwave and mm-wave power amplifier circuits.

PAE is expressed as

$$PAE = 100 * \frac{[P_{OUT}]_{RF} - [P_{IN}]_{RF}}{[P_{DC}]_{TOTAL}}.$$

At mm-wave frequencies, semiconductor technologies fundamentally limit the available output power. Furthermore, the efficiency is also degraded with higher frequency.

Considering the PAE characteristics in Fig. 19.8, and the non-linear behavior of the AM-AM/AM-PM characteristics of the power amplifier, significant power back-off may be necessary to reach linearity requirement such as the transmitter ACLR requirements (see Section 18.9). Considering the heat dissipation aspects and significantly reduced area/volume for mm-wave products, the complex inter-relation between linearity, PAE, and output power in the light of heat dissipation must be considered.

19.4 FILTERING ASPECTS

Using various types of filters in base station and device implementations is an essential component for meeting the overall RF requirements. This has been the case for all generations of mobile systems and will be essential also for NR, both below 6 GHz and in the new mm-wave bands. The filters mitigate the unwanted emissions arising from, for example, non-linearity in the transmitters generated due to intermodulation, noise, harmonics generation, LO leakage, and various unwanted mixing products. In the receiver chain, filters are used to handle either self-interference from own transmitter signal in paired bands, or to suppress the interferer at adjacent or other frequencies.

The RF requirements are differentiated in terms of levels for different scenarios. For base station spurious emission, there are general requirements across a very wide frequency range, coexistence requirements in the same geographical areas, and co-location requirements for dense deployments. Similar requirements are defined for devices.

Considering the limited size (area/volume) and level of integrations needed for mm-wave frequencies, the filtering can be challenging where discrete mm-wave filters are quite bulky and there is a challenge to embed such filters into highly integrated structures for mm-wave products.

19.4.1 POSSIBILITIES OF FILTERING AT THE ANALOG FRONT-END

Different implementations provide different possibilities for filtering. For the purpose of discussion, two main cases can be identified:

- Low-cost, monolithic integration with one or a few multi-chain CMOS/BiCMOS core-chips with built-in power amplifiers and built in downconverters. This case will give limited possibilities to include high-performance filters along the RF-chains since the Q-values for on chip filter resonators will be poor (5–20).
- High-performance, heterogeneous integration with several CMOS/BiCMOS core chips, combined with external amplifiers and external mixers. This implementation allows the inclusion of external filters along the RF-chains (at a higher complexity, size, and power consumption).

There are at least three places where it makes sense to put filters, depending on implementation, as shown in Fig 19.9:

- Behind or inside the antenna element (F1 or F0), where loss, size, cost, and wide-band suppression are important;

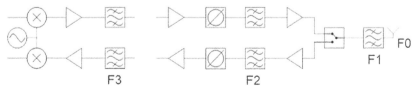

FIGURE 19.9

Possible filter locations.

- Behind the first amplifiers (looking from the antenna side), where low loss is less critical (F2);
- On the high-frequency side of mixers (F3), where signal have been combined (in the case of analog and hybrid beam forming).

The main purpose of F1/F0 is normally to suppress interference and emissions far from the desired channel across a wide frequency range (for example, DC to 60 GHz). There should not be any unintentional resonances or passbands in this wide frequency range. This filter will help relax the design challenge (bandwidth to consider, linearity requirements, etc.) of all following blocks. Insertion loss must be very low, and there are strict size and cost requirements since there must be one filter at each subarray (Figs. 19.9 and 19.10). In some cases, this filter must fulfill strict suppression requirements close to the passband, particularly for high output power close to sensitive bands.

The main purpose of F2 is suppression of LO-, image-, spurious-, and noise-emission, and suppression of incoming interferers relatively far from the desired frequency band. There are still strict size requirements, but more loss can be accepted (behind the first amplifiers) and even unintentional passbands (assuming F1/F0 will handle that). This allows better discrimination (more poles), and better frequency precision (for example, using half-wave resonators).

The main purpose of F3 is typically suppression of LO-, image-, spurious-, and noise-emission, and suppression of incoming interferers that accidentally fall in the IF-band after the mixer, and strong interferers that tend to block the mixers or ADCs. For analog (or hybrid) beam-forming it is enough to have just one (or a few) such filters. This relaxes requirements on size and cost, which opens the possibility to achieve sharp filters with multiple poles and zeroes, and with high Q-value and good frequency precision in the resonators.

The deeper into the RF-chain (starting from the antenna element), the better protected the circuits will get. For the monolithic integration case it is difficult to implement filters F2 and F3. One can expect performance penalties for this case, and output power per branch is lower. Furthermore, it is challenging to achieve good isolation across a wide frequency range, as microwaves tend to bypass filters by propagating in ground structures around them.

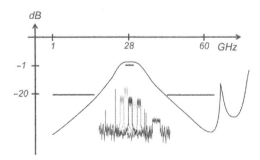

FIGURE 19.10

Filter example for the 28 GHz band.

19.4.2 INSERTION LOSS (IL) AND BANDWIDTH

Sharp filtering on each branch (at positions F1/F0) with narrow bandwidth leads to excessive loss at microwave and mm-wave frequencies. To get the insertion loss down to a reasonable level the passband can be made significantly larger than the signal bandwidth. A drawback of such an approach is that more unwanted signals will pass the filter. In choosing the best loss-bandwidth trade-off there are some basic dependencies to be aware of:

- IL decreases with increasing BW (for fixed fc);
- IL increases with increasing fc (for fixed BW);
- IL decreases with increasing Q-value;
- IL increases with increasing N.

To exemplify the trade-off, a three-pole LC-filter with $Q = 20$, 100, 500, and 5000, for 100 and 800 MHz 3 dB-bandwidth is studied, tuned to 15 dB return loss (with $Q = 5000$) is examined, as shown in Fig. 19.11.

From this study it is observed that:

- 800 MHz bandwidth or smaller, requires exotic filter technologies, with a Q-value around 500 or better to get an IL below 1.5 dB. Such Q-values are very challenging to achieve considering constraints on size, integration aspects, and cost;
- By relaxing the requirement on selectivity to 4×800 MHz, it is sufficient to have a Q-value around 100 to get 2 dB IL. This should be within reach with a low-loss printed circuit board (PCB). The increased bandwidth will also help to relax the tolerance requirements on the PCB.

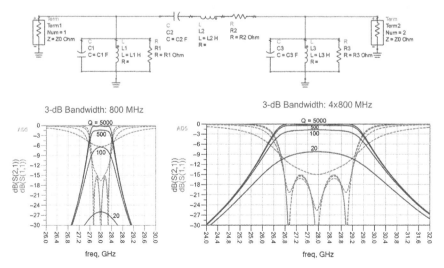

FIGURE 19.11

Example three-pole LC filter with 800 and 4×800 MHz bandwidth, for different Q values.

19.4.3 FILTER IMPLEMENTATION EXAMPLES

There are many ways to implement filters in a 5G array radio. Key aspects to compare are: Q-value, discrimination, size, and integration possibilities. Table 19.1 gives a rough comparison between different technologies and two specific examples are given below.

19.4.3.1 PCB Integrated Implementation Example

A simple and attractive way to implement antenna filters (F1) is to use strip-line or microstrip filters, embedded in a PCB close to each antenna element. This requires a low-loss PCB with good precision. Production tolerances (permittivity and patterning and via-positioning) will limit the performance, mainly though a shift in the pass-band and increased mismatch. In most implementations the pass-band must be set larger than the operating frequency band with a significant margin to account for this.

Typical characteristics of such filters can be illustrated by looking at the following design example, with the layout shown in Fig. 19.12:

- Five-pole, coupled line, strip-line filter;
- Dielectric permittivity: 3.4;
- Dielectric thickness: 500 μm (ground to ground);
- Unloaded resonator Q: 130 (assuming low-loss microwave dielectrics).

The filter is tuned to give 20 dB suppression at 24 GHz, while passing as much as possible of the band 24.25–27.5 GHz (with 17 dB return loss). Significant margins are added to make room for variations in the manufacturing processes of the PCB.

Table 19.1 Different Possible Technologies to Use for Filter Implementation

Technology	Q of Resonators	Size	Integration
On-chip (Si)	20	Small	Feasible
PCB (low-loss)	100	Medium	Feasible
Ceramic substrate	300	Medium	Difficult
Advanced miniature filters	500	Medium	Difficult
Waveguide (air-filled)	5000	Large	Extremely difficult

FIGURE 19.12

Layout of stripline filter on a PCB.

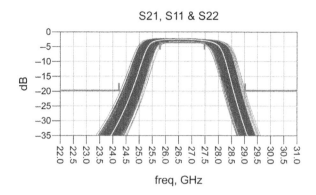

FIGURE 19.13

Simulated impact of manufacturing tolerances on the filter characteristics of a strip line filter in PCB.

A Monte Carlo analysis was performed to study the impact of variations in the manufacturing process on filter performance, using the following quite aggressive tolerance assumptions for the PCB:

- Permittivity standard deviation: 0.02;
- Line width standard deviation: 8 μm;
- Thickness of dielectric standard deviation: 15 μm.

With these distribution assumptions, 1000 instances of the filter were generated and simulated. Fig. 19.13 shows the filter performance (S21) for these 1000 instances (blue traces), together with the nominal performance (yellow trace). Red lines in the graph indicate possible requirement levels that could be met considering this filter.

From this design example, the following rough description of a PCB filter implementation is found:

- 3–4 dB insertion loss;
- 20 dB suppression (17 dB if IL is subtracted);
- 1.5 GHz transition region with margins included;
- Size: 25 mm², which can be difficult to fit in the case of individual feed and/ or dual polarized elements;
- If a 3 dB IL is targeted, there would be significant yield loss with the suggested requirement, in particular for channels close to the pass-band edges.

19.4.3.2 LTCC Filter Implementation Example

Another promising way to implement filters is to make components for Surface Mount Assembly (SMT), including both filters and antennas, for example based on Low-Temperature Cofired Ceramics (LTCC). One example of a prototype LTCC component was outlined in Ref. [31] and is also shown in Fig. 19.14.

FIGURE 19.14

Example of prototype of an LTCC-component containing both antenna elements and filters.

TDK Corporation, used with permission.

The measured performance of the corresponding filter is shown in Fig. 19.15 and it shows that the LTCC-filter adds about 2 dB of insertion loss for a 2 GHz passband, while providing 22 dB of additional attenuation 1 GHz from the pass-band edge.

Additional margins relative to this example should be considered to account for manufacturing tolerances and future adjustments of bandwidth, suppression level, guard bandwidth, antenna properties, integration aspects, etc. Accounting for such margins, the LTCC-filter shown could be assumed to add approximately 3 dB of insertion loss, for 17 dB suppression (IL subtracted) at 1.5 GHz from the pass-band edge.

Technology development, particularly regarding Q-values and manufacturing tolerances, will likely lead to improvements in these numbers.

19.5 RECEIVER NOISE FIGURE, DYNAMIC RANGE, AND BANDWIDTH DEPENDENCIES

19.5.1 RECEIVER AND NOISE FIGURE MODEL

A receiver model as shown in Fig. 19.16 is assumed here. The dynamic range (DR) of the receiver will in general be limited by the front-end insertion loss (IL), the receiver (RX) Low-noise Amplifier (LNA), and the ADC noise and linearity properties.

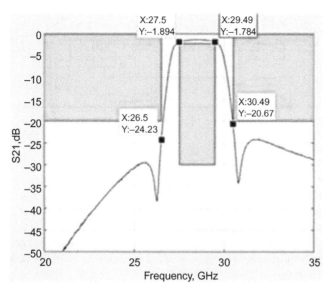

FIGURE 19.15

Measured performance of the corresponding filter without antenna.

TDK Corporation, used with permission.

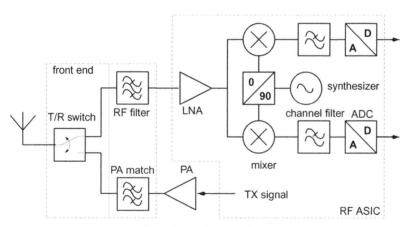

Typical zero-IF transceiver

FIGURE 19.16

Typical zero-IF transceiver schematic.

Typically $DR_{LNA} \gg DR_{ADC}$ so the RX use Automatic Gain Control (AGC) and selectivity (distributed) in-between the LNA and the ADC to optimize the mapping of the wanted signal and the interference to the DR_{ADC}. For simplicity, a fixed gain setting is considered here.

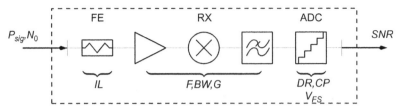

FIGURE 19.17

A simplified receiver model.

A further simplified receiver model can be derived by lumping the Front End (FE), RX, and ADC into three cascaded blocks, as shown in Fig. 19.17. This model cannot replace a more rigorous analysis but will demonstrate interdependencies between the main parameters.

Focusing on the small signal co-channel noise floor, the impact of various signal and linearity impairments can be studied to arrive at a simple noise factor, or noise figure, expression.

19.5.2 NOISE FACTOR AND NOISE FLOOR

Assuming matched conditions, Friis' formula can be used to find the noise factor at the receiver input as (linear units unless noted),

$$F_{RX} = 1 + (F_{LNA} - 1) + \frac{(F_{ADC} - 1)}{G}$$

The RX input referred small-signal co-channel noise floor will then equal

$$N_{RX} = F_{LNA} \cdot N_0 + \frac{N_{ADC}}{G}$$

where $N_0 = k \cdots T \cdots BW$ and N_{ADC} are the available noise power and the ADC effective noise floor in the channel bandwidth, respectively (k and T being Boltzmann's constant and absolute temperature, respectively). The ADC noise floor is typically set by a combination of quantization, thermal, and intermodulation noise, but here a flat noise floor is assumed as defined by the ADC effective number of bits.

The effective gain G from LNA input to ADC input depends on small-signal gain, AGC setting, selectivity, and desensitization (saturation), but here it is assumed that the gain is set such that the antenna referred input compression point (CP$_i$) corresponds to the ADC clipping level, that is the ADC full scale input voltage (V_{FS}).

For weak non-linearities, there is a direct mathematical relationship between CP and the third-order intercept point (IP$_3$), such that IP$_3 \approx$ CP $+$ 10 dB. For higher-order non-linearities, the difference can be larger than 10 dB, but then CP is still a good estimate of the maximum signal level while intermodulation for lower signal levels may be overestimated.

19.5.3 COMPRESSION POINT AND GAIN

Between the antenna and the RX there is the FE with its associated insertion loss (IL > 1), for example due to a T/R switch, a possible RF filter, and PCB/substrate losses. These losses have to be accounted for in the gain and noise expressions. Knowing IL, the CP_i can be found that corresponds to the ADC clipping as

$$CP_i = \frac{IL \cdot N_{ADC} \cdot DR_{ADC}}{G}$$

The antenna referred noise factor and noise figure will then become

$$F_i = IL \cdot F_{RX} = IL \cdot F_{LNA} + \frac{CP_i}{N_0 \cdot DR_{ADC}}$$

and

$$NF_i = 10 \cdot \log_{10}(F_i),$$

respectively.

When comparing two designs, for example, at 2 and 30 GHz, respectively, the 30 GHz IL will be significantly higher than that of the 2 GHz. From the F_i expression it can be seen that to maintain the same noise figure (NF_i) for the two carrier frequencies, the higher FE loss at 30 GHz needs to be compensated for by improving the RX noise factor. This can be accomplished by (1) using a better LNA, (2) relaxing the input compression point, that is increasing G, or (3) increasing the DR_{ADC}. Usually a good LNA is already used at 2 GHz to achieve a low NF_i, so this option is rarely possible. Relaxing CP_i is an option but this will reduce IP_3 and the linearity performance will degrade. Finally, increasing DR_{ADC} comes at a power consumption penalty (4 $\times$ per extra bit). Especially wideband ADCs may have a high power consumption, that is when BW is below some 100 MHz the $N_0 \cdots DR_{ADC}$ product (that is BW $\cdots DR_{ADC}$) is proportional to the ADC power consumption, but for higher bandwidths the ADC power consumption is proportional to $BW^2 \cdots DR_{ADC}$, thereby penalizing higher BW (see Section 19.1). Increasing DR_{ADC} is typically not an attractive option and it is inevitable that the 30 GHz receiver will have a significantly higher NF_i than that of the 2 GHz receiver.

19.5.4 POWER SPECTRAL DENSITY AND DYNAMIC RANGE

A signal consisting of many similar subcarriers will have a constant power-spectral density (PSD) over its bandwidth and the total signal power can then be found as $P = PSD \cdots BW$.

When signals of different bandwidths but similar power levels are received simultaneously, their PSDs will be inversely proportional to their BW. The antenna-referred noise floor will be proportional to BW and F_i, or $N_i = F_i \cdots k \cdots T \cdots BW$, as derived above. Since CP_i will be fixed, given by G and ADC clipping, the dynamic range, or maximum SNR, will decrease with signal bandwidth, that is $SNR_{max} \propto 1/BW$.

The above signal can be considered as additive white Gaussian noise (AWGN) with an antenna-referred mean power level (P_{sig}) and a standard deviation (σ). Based on this assumption the peak-to-average-power ratio can be approximated as $PAPR = 20 \cdots \log_{10}(k)$, where the peak signal power is defined as $P_{sig} + k \cdots \sigma$, that is there are k standard deviations between the mean power level and the clipping level. For OFDM an unclipped PAPR of 10 dB is often assumed (that is 3σ) and this margin must be subtracted from CP_i to avoid clipping of the received signal. An OFDM signal with an average power level, for example, 3σ below the clipping level will result in less than 0.2% clipping.

19.5.5 CARRIER FREQUENCY AND MM-WAVE TECHNOLOGY ASPECTS

Designing a receiver at, for example, 30 GHz with a 1 GHz signal bandwidth leaves much less design margin than what would be the case for a 2 GHz carrier frequency, $f_{carrier}$ with, for example, 50 MHz signal bandwidth. The IC technology speed is similar in both cases but the design margin and performance depend on the technology being much faster than the required signal processing, which means that the 2 GHz design will have better performance.

The graph shows expected evolution of some transistor parameters important for mm-wave IC design, as predicted by the International Technology Roadmap for Semiconductors (ITRS). Here f_t, f_{max}, and V_{dd}/BV_{ceo} data from the ITRS 2007 targets [39] for CMOS and bipolar RF technologies are plotted vs the calendar year when the technology is anticipated to become available. f_t is the transistor transit frequency (that is, where the RF device's current gain is 0 dB), and f_{max} is the maximum frequency of oscillation (that is, when the extrapolated power gain is 0 dB). V_{dd} is the RF/high-performance CMOS supply voltage and BV_{ceo} is the bipolar transistor's collector–emitter base open breakdown voltage limits. For example, an RF CMOS device is expected to have a maximum V_{dd} of 750 mV by 2020 (other supply voltages will be available as well, but at a lower speed).

The free space wavelength at 30 GHz is only 1 cm, which is one tenth of what is the case for existing 3GPP bands below 6 GHz. Antenna size and path loss are related to wavelength and carrier frequency, and to compensate the small physical size of a single antenna element multiple antennas, for example, array antennas will have to be used. When beam-forming is used the spacing between antenna elements will still be related to the wavelength, constraining the size of the FE and RX. Some of the implications of these frequency and size constraints are:

- The ratios $f_t/f_{carrier}$ and $f_{max}/f_{carrier}$ will be much lower at millimeter wave frequencies than for below 6 GHz applications. As receiver gain drops with operating frequency when this ratio is less than some $10 - 100 \times$, the available gain at millimeter waves will be lower and consequently the device noise factor, F_i, higher (similar to when Friis' formula was applied to a transistor's internal noise sources).

- The semiconductor material's electrical breakdown voltage (E_{br}) is inversely proportional to the charge carrier saturation velocity (V_{sat}) of the device due to the Johnson limit. This can be expressed as $V_{sat} \cdots E_{br} = $ constant or $f_{max} \cdots V_{dd} = $ constant. Consequently, the supply voltage will be lower for millimeter-wave devices compared to devices in the low GHz frequency range. This will limit the CP_i and the maximum available dynamic range.
- A higher level of transceiver integration is required to save space, either as system-on-chip (SoC) or system-in-package (SiP). This will limit the number of technologies suitable for the RF transceiver and limit F_{RX}.
- RF filters will have to be placed close to the antenna elements and fit into the array antenna. Consequently, they have to be small, resulting in higher physical tolerance requirements, possibly at the cost of insertion loss and stop-band attenuation. That is, IL and selectivity get worse. The filtering aspect for mm-wave frequencies is further elaborated on in Section 19.4.

Increasing the carrier frequency from 2 GHz to 30 GHz (that is $>10 \times$) has a significant impact on the circuit design and its RF performance. For example, modern high-speed CMOS devices are velocity saturated and their maximum operating frequency is inversely proportional to the minimum channel length, or feature size. This dimension halves roughly every 4 years, as per Moore's law (stating that complexity, that is transistor density, doubles every other year). With smaller feature sizes, internal voltages must also be lowered to limit electrical fields to safe levels. Thus, designing a 30 GHz RF receiver corresponds to designing a 2 GHz receiver using about 15-year-old low-voltage technology (that is today's breakdown voltage but 15 years old F_t (see Fig. 19.18) with ITRS device targets). With such a mismatch in device performance and design margin it is not to be expected to maintain both 2 GHz performance and power consumption at 30 GHz.

The signal bandwidth at mm-wave frequencies will also be significantly higher than at 2 GHz. For an active device, or circuit, the signal swing is limited by the supply voltage at one end and by thermal noise at the other. The available thermal noise power of a device is proportional to BW/g_m, where g_m is the intrinsic device gain (trans-conductance). As g_m is proportional to bias current it can be seen that the dynamic range becomes the ratio

$$\text{DR} \propto \frac{V_{dd}^2 \cdot I_{bias}}{BW} = \frac{V_{dd} \cdot P}{BW}$$

or

$$P \propto \frac{BW \cdot DR}{V_{dd}}$$

where P is the power dissipation.

Receivers for mm-wave frequencies will have increased power consumption due to their higher BW, aggravated by the low-voltage technology needed for

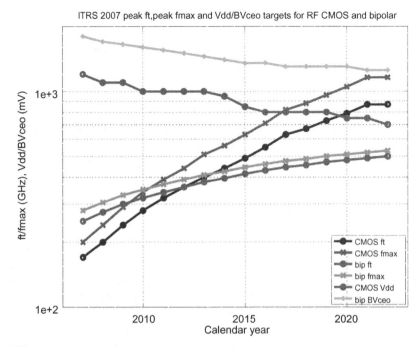

FIGURE 19.18

Expected evolution over time of some transistor parameters: f_t, f_{max}, and V_{dd}/BV_{ceo} [39].

speed, compared to typical 2 GHz receivers. Thus, considering the thermal challenges given the significantly reduced area/volume for mm-wave products, the complex interrelation between linearity, NF, bandwidth, and dynamic range in the light of power dissipation should be considered.

19.6 SUMMARY

This chapter gave an overview of what mm-wave technologies can offer and how to derive requirements. The need for highly integrated mm-wave systems with many transceivers and antennas will require careful and often complex consideration regarding the power efficiency and heat dissipation in small area/volume affecting the achievable performance.

Important areas presented were DA/AD converters, power amplifiers, and the achievable power versus efficiency as well as linearity. Receiver essential metrics are noise figure, bandwidth, dynamic range, and power dissipation and they all have complex dependencies. The mechanism for frequency generation as well as phase noise aspects were also covered. Filtering aspects for mm-wave frequencies were shown to have substantial impact in new NR bands and the achievable performance for various technologies and the feasibility of integrating such filters into NR implementations needs to be accounted for when defining RF requirements. All these aspects are accounted for throughout the process of developing the RF characteristics of NR in Frequency Range 2.

Beyond the First Release of 5G

20

The first release of NR, release 15, has focused on basic support for eMBB and, to some extent, URLLC.[1] Release 15 as described in the previous chapters is the foundation upon which the future evolution of NR will be built for the coming releases. The NR evolution will bring additional capabilities and further enhance the performance. Not only will the additional capabilities provide better performance in existing applications, they may also open for, or even be motivated by, new application areas.

In the following, some areas in which NR is likely to evolve are discussed. Studies in some of the areas are already ongoing in 3GPP, while other areas are more relevant for later releases.

20.1 INTEGRATED ACCESS-BACKHAUL

The use of wireless technology for backhaul has been used extensively for many years. In some regions of the world, wireless backhaul constitutes more than 50% of total backhaul. Current wireless-backhaul solutions are typically based on proprietary (non-standardized) technology operating as point-to-point line-of-sight links using special frequency bands above 10 GHz. The wireless backhaul is thus using different technology and operating in different spectra, compared to the access (base-station/device) links. Relaying, introduced in release 10 of LTE, is basically a wireless backhaul link, although with some restrictions. However, it has so far not been used in practice to any significant extent. One reason is that wirelessly connected small-cell deployments, for which relaying was designed, have not yet been extensively used in practice. Another reason is that operators prefer to use their precious low-frequency spectra for the access link. As already mentioned, current wireless backhauling relies on non-LTE technologies capable

---

[1]The first NR version primarily addressed the low-latency part of URLLC. Means to increase the reliability are worked upon in the latter parts of release 15, targeting the final NR release 15 in June 2018.

5G NR: The Next Generation Wireless Access Technology. DOI: https://doi.org/10.1016/B978-0-12-814323-0.00020-X

of exploiting significantly higher-frequency bands than LTE, thereby avoiding wasting valuable access spectra for backhaul purposes.

However, for NR, a convergence of backhaul and access can be expected for several reasons:

- The access link can exploit mm-wave frequencies—that is, the same frequency range that is currently used for wireless backhaul.
- The expected densification of the mobile networks, with many base stations located indoor and outdoor on street level, will require wireless backhaul capable of operating under non-line-of-sight conditions and, more generally, very similar propagation conditions as the access link.

The requirements and characteristics of the wireless backhaul link and the access link are thus converging. In essence, with reference to Fig. 20.1, there is, radio-wise, no major difference between the wireless backhaul link and the normal wireless link. Consequently, there are strong reasons to consider a convergence also in terms of technology and spectrum with a single radio-access technology that can be used for both access and wireless backhaul. There should preferably also be a common spectrum pool for both the access link and the wireless backhaul. It should also be noted that a common spectrum pool for access and wireless backhaul does not necessarily mean that the access link and the wireless backhaul link should operate on the same carrier frequency ("inband relaying"). In some cases, this will be possible. However, in other cases, having a frequency separation between the backhaul link and the access link is preferred. The key thing is that the separation of spectrum between backhaul and access should, as much as possible, not be a regulatory issue. Rather, an operator should have access to a single spectrum pool. It is then an operator decision how to use this spectrum in the best possible way and how to split it between access and backhaul.

To address backhaul scenarios, a study item on *integrated access-backhaul* [1] is part of release 15 to assess the possibilities and techniques for using NR for backhaul purposes. The NR radio access is well prepared to support the backhaul link and most of the necessary work is on higher-layer protocols.

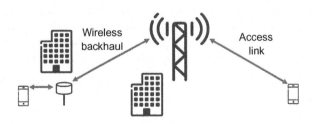

FIGURE 20.1

Wireless backhaul vs the access link.

20.2 OPERATION IN UNLICENSED SPECTRA

Spectrum is fundamental for wireless communication and there is a never-ending quest for more spectra to meet the ever-increasing demands of increased capacity and higher data rates. This is one of the reasons for supporting higher carrier frequencies in NR. The first release of NR was primarily designed for licensed spectra. Such spectra offer many benefits since the operator can plan the network and control the interference. Licensed spectrum is thus instrumental to providing quality-of-service guarantees and wide-area coverage. However, the amount of licensed spectra an operator has access to may not be sufficient and there is typically a cost associated with obtaining a spectrum license.

Unlicensed spectra, on the other hand, are open for anyone to use at no cost, subject to a set of rules, for example on maximum transmission power. Since anyone can use the spectra, the interference situation is typically much more unpredictable than for licensed spectra. Consequently, quality-of-service and availability cannot be guaranteed. Furthermore, the maximum transmission power is modest, making it unsuitable for wide-area coverage. Wi-Fi and Bluetooth are two examples of communication systems exploiting unlicensed spectra in the lower-frequency range: 2.4 GHz or 5 GHz. In addition, some of the higher-frequency bands which NR is likely to address are unlicensed.

From the discussion above, it can be seen that these two spectrum types have different benefits and drawbacks. An attractive option is to combine the two such that licensed spectra are used to provide wide-area coverage and quality-of-service guarantees, with unlicensed spectra used as a local-area complement to increase user data rates and overall capacity without compromising on overall coverage, availability, and reliability. This has been done as part of the LTE evolution, see *License-Assisted Access (LAA)* in Chapter 4 and Fig. 20.2. For NR, a study on *NR-based Access to Unlicensed Spectrum* [9] is part of release 15 with specification work targeting release 16.

Although NR release 15 does not support unlicensed spectra, it was considered in the development of the basic NR framework. One example hereof is the possibility to transmit over a fraction of a slot (see Chapter 7). Extending NR into an LAA-like operation is therefore relatively simple, using the existing flexibility and following the approach developed for LTE.

One important characteristic of operation in unlicensed spectra, which was accounted for in the LTE/LAA work, is fair sharing of unlicensed spectra with other operators and other systems, in particular Wi-Fi. There are several mechanisms that can be used to enable this. *Dynamic frequency selection* (DFS), where the network node searches and finds a part of the unlicensed spectra with low load, can be used to avoid other systems if possible. *Listen-before-talk* (LBT) mechanism, where the transmitter ensures there are no ongoing transmissions on the carrier frequency prior to transmitting, is another mechanism well proven at lower-frequency bands that could be added to NR. For higher-frequency bands,

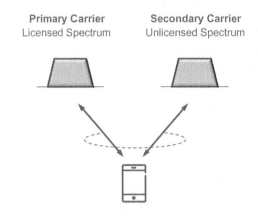

Primary Carrier
Licensed Spectrum

Secondary Carrier
Unlicensed Spectrum

FIGURE 20.2

License-assisted access.

where extensive beam-forming is typically used, the LBT mechanism may need some modifications.

Beyond license-assisted access to unlicensed spectra, a complete solution for standalone operation in unlicensed spectra can also be envisioned. This obviously requires mechanisms for system-information delivery and mobility capable of handling unlicensed spectra.

20.3 NON-ORTHOGONAL MULTIPLE ACCESS

NR primarily uses orthogonal multiple-access where different devices are separated in time and/or frequency. However, non-orthogonal access has the potential to increase capacity in some scenarios. During the early stages of NR development, *non-orthogonal multiple access* (NOMA) was briefly studied but down-prioritized. Nevertheless, studies on NOMA are ongoing in release 15 and may become relevant for NR in later releases.

20.4 MACHINE-TYPE COMMUNICATION

Machine-type communication is a very wide term, covering many different use cases and scenarios. It is common to divide machine-type communication into massive machine-type communication and ultra-reliable low-latency communication (URLLC), as already discussed at the beginning of this book.

Massive machine-type communication refers to scenarios where a device typically sends a very small amount of data, has relaxed latency requirements, but low power consumption and low cost are at premium. The number of devices is often very large. Such scenarios will be addressed by LTE and NB-IoT for the near- to mid-term perspective, in particular for the low-end massive MTC regime. Specific mechanisms such as the reserved resources discussed in Chapter 17 have been introduced to simplify the coexistence between NR and these access technologies. In the longer time perspective, NR is expected to evolve with improved native support of massive machine-type communication, primarily focusing on the mid-to-high-end massive MTC. Reduced bandwidth support, extended sleep-mode solutions, wake-up signaling, and non-orthogonal waveforms are examples of what could be relevant to study as part of such an evolution.

Factory automation is an example of an application area related to machine-type communication. In many cases, such applications are demanding in terms of reliability and latency and the URLLC aspects of NR are therefore highly relevant. Examples of possible enhancements to NR relevant for factory automation are higher-layer enhancements to support commonly used industrial protocols (other than TCP/IP) and local breakout from the core network.

20.5 DEVICE-TO-DEVICE COMMUNICATION

Support for direct device-to-device (D2D) connectivity (Fig. 20.3), also referred to as sidelink connectivity, using LTE was introduced in 3GPP release 12 with two main use cases in mind:

- Device-to-device communication, focusing on the public-safety use case;
- Device-to-device discovery, targeting public safety but also commercial use cases.

The D2D framework has also served as the basis for the V2V/V2X work in the LTE evolution in later releases as discussed in Chapter 4.

NR release 15 does not support direct device-to-device communication, but it is a likely candidate for a future release. Instead of focusing on a specific use case, device-to-device connectivity should be seen as a general tool to enhance

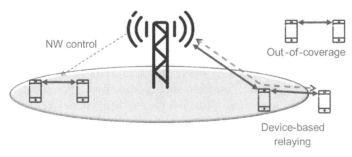

FIGURE 20.3

Device-to-device connectivity.

connectivity within the 5G network. In essence, direct data transfer between devices should be configured if the network concludes that this is more efficient (requires less resources) or provides better quality (higher data rates and/or lower latency) compared to indirect connectivity via the infrastructure. The network should also be able to configure device-based relay links to enhance the connectivity quality, for example for massive machine-type devices with bad or no coverage. The lower latency of NR could also prove valuable for some D2D applications, for example platooning, as mentioned in Chapter 4.

20.6 SPECTRUM AND DUPLEX FLEXIBILITY

Duplex flexibility is a wide area, aiming at improving the usage of the available spectrum. The tools part of NR from the start—for example bandwidth parts, a flexible slot structure, and carrier aggregation also across duplex schemes—provide a lot of flexibility and ensure NR can be deployed in a wide range of scenarios. Nevertheless, further enhancements in this area can be envisioned.

Currently, the FDD spectrum is split into a downlink part and an uplink part. However, what is relevant from a technical perspective is primarily not downlink vs uplink, but low power vs high power. The downlink typically uses high power and relatively high above-rooftop antennas, while the uplink uses significantly lower transmission power and antenna installations. Hence, from an interference perspective, a low-power downlink transmission in the uplink spectrum is not different from a low-power uplink transmission in the same spectrum. Consequently, there are ideas on allowing downlink transmission also in the uplink bands. To some extent, this is the FDD counterpart to dynamic TDD as it allows for a dynamic change to the "transmission direction." From a technical perspective, NR is well prepared to such enhancements because of the flexible slot structure. The potential issues are primarily regulatory.

Another area related to spectra and possible future enhancements is interference measurements and dynamic TDD. The TDD scheme in NR is built upon a dynamic framework and dynamic TDD is therefore part of release 15. However, in practice, such deployments are primarily limited to small cells. In larger cells, with a correspondingly higher downlink transmission power, the intercell interference typically calls for a more static duplex operation. One possibility to improve the number of scenarios where dynamic TDD is feasible could be to include various interference measurement mechanisms. For example, if the scheduler knows the interference situation for the different devices, it can schedule dynamically for some devices while taking a more static approach for other devices. Different intercell interference coordination mechanisms can also be thought of.

There have recently been different proposals for "true" full-duplex operation [53]. In this context, full-duplex operation means that transmission and reception

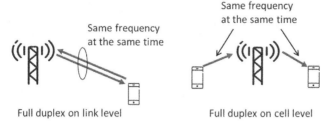

FIGURE 20.4

Full duplex on link level vs cell level.

are carried out *at the same frequency at the same time* (see also Fig. 20.4).[2] Full-duplex operation obviously leads to very strong "self" interference from the transmitter to the receiver, an interference that needs to be suppressed/canceled before the actual target signal can be detected.

In principle, such interference suppression/cancellation is straightforward, as the interfering signal is in principle completely known to the receiver. In practice, the suppression/cancellation is far from straightforward due to the enormous difference between the target signal and the interference in terms of received power. To handle this, current demonstrations of full-duplex operation rely on a combination of spatial separation (separate antennas for transmission and reception), analog suppression, and digital cancellation. The technology is still to a large degree at the research level and not mature enough for large-scale deployments. Implementation on the network-side only (see right part of Fig. 20.4) might be less complex than implementation on the device-side due to a higher degree of spatial separation of receive and transmit antennas on the network side.

Even if full duplex would be feasible in real implementation, its benefits should not be overestimated. Full duplex has the potential to double the link throughput by allowing for continuous transmission in both directions on the same frequency. However, there will then be two simultaneous transmissions, implying increased interference to other transmissions, something which will negatively impact the overall system gain. The largest gain from full duplex can therefore be expected to occur in scenarios with relatively isolated radio links.

20.7 CONCLUDING REMARKS

Above, some examples of technology areas relevant for NR evolution are outlined. Some of these are likely to be part of future NR releases, while other may not happen at all. However, as always, when trying to predict the future, there are

---

[2]Not to be up mixed with *full-duplex FDD* as used in LTE.

a lot of uncertainties and new, not-yet-known requirements or technologies, which may motivate evolutions into directions not discussed above. The emphasis on future compatibility in the basic NR design ensures that introduction of extension in most cases is relatively straightforward. It is clear though that NR is a very flexible platform, capable of evolving in a wide range of directions and an attractive path to future wireless communication.

References

[1] 3GPP RP-172290, New SID Proposal: Study on Integrated Access and Backhaul for NR.

[2] 3GPP TS 37.141, E-UTRA, UTRA and GSM/EDGE; Multi-Standard Radio (MSR) Base Station (BS) Conformance Testing.

[3] 3GPP R1-163961, Final Report of 3GPP TSG RAN WG1 #84bis.

[4] 3GPP TS 38.104, NR; Base Station (BS) Radio Transmission and Reception.

[5] 3GPP TS 38.101-1, NR; User Equipment (UE) Radio Transmission and Reception. Part 1. Range 1 Standalone.

[6] 3GPP TS 38.101-2, NR; User Equipment (UE) Radio Transmission and Reception. Part 2. Range 2 Standalone.

[7] 3GPP TS 38.101-3, NR; User Equipment (UE) Radio Transmission and Reception. Part 3. Range 1 and Range 2 Interworking Operation with Other Radios.

[8] 3GPP TS 38.101-4, NR; User Equipment (UE) Radio Transmission and Reception. Part 4. Performance Requirements.

[9] 3GPP RP-172021, Study on NR-Based Access to Unlicensed Spectrum.

[10] 3GPP TR 36.913, Requirements for Further Advancements for Evolved Universal Terrestrial Radio Access (E-UTRA) (LTE-Advanced) (Release 9).

[11] 3GPP TR 38.803, Study on New Radio Access Technology: Radio Frequency (RF) and Coexistence Aspects.

[12] 3GPP TS 23.402, Architecture Enhancements for Non-3GPP Accesses.

[13] 3GPP TS 23.501, System Architecture for the 5G System.

[14] 3GPP TS 36.211, Evolved Universal Terrestrial Radio Access (E-UTRA); Physical Channels and Modulation.

[15] 3GPP TS 38.331, NR; Radio Resource Control (RRC) Protocol Specification (Release 15).

[16] 3GPP TR 36.913, Requirements for Further Advancements for Evolved Universal Terrestrial Radio Access (E-UTRA) (LTE-Advanced).

[17] E. Arikan, Channel polarization: a method for constructing capacity-achieving codes for symmetric binary input memoryless channels, IEEE Trans. Inform. Theory 55 (7) (July 2009) 3051−3073.

[18] Roland E. Best, Phases Locked Loops: Design, Simulation and Applications, sixth ed., McGraw-Hill Professional, 2007.

[19] CEPT/ERC Recommendation 74-01 on unwanted emissions in the spurious domain, Cardiff 2011.

[20] CEPT, LS from to CEPT/ECC SE21, SE21(17)38, September 2017.

[21] T. Chapman, E. Larsson, P. von Wrycza, E. Dahlman, S. Parkvall, J. Sköld, HSPA Evolution: The Fundamentals for Mobile Broadband, Academic Press, 2014.

[22] D. Chase, Code combining—a maximum-likelihood decoding approach for combining and arbitrary number of noisy packets, IEEE Trans. Commun. 33 (May 1985) 385−393.

[23] J. Chen, Does LO noise floor limit performance in multi-Gigabit mm-wave communication? IEEE Microw. Compon. Lett. 27 (8) (2017) 769−771.

[24] J.-F. Cheng, Coding performance of hybrid ARQ schemes, IEEE Trans. Commun. 54 (June 2006) 1017−1029.

[25] D.C. Chu, Polyphase codes with good periodic correlation properties, IEEE Trans. Inform. Theory 18 (4) (July 1972) 531–532.

[26] S.T. Chung, A.J. Goldsmith, Degrees of freedom in adaptive modulation: a unified view, IEEE Trans. Commun. 49 (9) (September 2001) 1561–1571.

[27] D. Colombi, B. Thors, C. Törnevik, Implications of EMF exposure limits on output power levels for 5G devices above 6 GHz, IEEE Antennas Wirel. Propag. Lett. 14 (February 2015) 1247–1249.

[28] E. Dahlman, S. Parkvall, J. Sköld, 4G LTE-Advanced Pro and the Road to 5G, Elsevier, 2016.

[29] DIGITALEUROPE, 5G Spectrum Options for Europe, October 2017.

[30] Ericsson, Ericsson Mobility Report, November 2017. https://www.ericsson.com/assets/local/mobility-report/documents/2017/ericsson-mobility-report-november-2017.pdf.

[31] Ericsson, On mm-wave Filters and Requirement Impact, R4-1712718, 3GPP TSG-RAN WG4 Meeting #85, December 2017.

[32] Federal Communications Commission, Title 47 of the Code of Federal Regulations (CFR).

[33] P. Frenger, S. Parkvall, E. Dahlman, Performance comparison of HARQ with chase combining and incremental redundancy for HSDPA. In: Proceedings of the IEEE Vehicular Technology Conference, Atlantic City, NJ, USA, pp. 1829–1833. October 2001.

[34] R.G. Gallager, Low Density Parity Check Codes, Monograph, M.I.T. Press, 1963.

[35] Global mobile Suppliers Association (GSA), The future of IMT in the 3300–4200 MHz frequency range, June 2017.

[36] M. Hörberg, Low phase noise GaN HEMT oscillator design based on high-Q resonators (Ph.D. Thesis), Chalmers University of Technology, April 2017.

[37] IEEE, IEEE Standard for Local and metropolitan area networks Part 16: Air Interface for Broadband Wireless Access Systems Amendment 3: Advanced AirInterface, IEEE Std 802.16m-2011 (Amendment to IEEE Std 802.16-2009).

[38] IETF, Robust header compression (ROHC): framework and four profiles: RTP, UDP, ESP, and Uncompressed, RFC 3095.

[39] ITRS, Radio Frequency and Analog/Mixed-Signal Technologies for Wireless Communications, Edition International Technology Roadmap for Semiconductors (ITRS), 2007.

[40] ITU-R, Workplan, timeline, process and deliverables for the future development of IMT, ITU-R Document 5D/758, Attachment 2.12.

[41] ITU-R, Framework and overall objectives of the future development of IMT-2000 and sys- tems beyond IMT-2000. Recommendation ITU-R M.1645, June 2003.

[42] ITU-R, Unwanted emissions in the spurious domain. Recommendation ITU-R SM.329-12, September 2012.

[43] ITU-R, Future technology trends of terrestrial IMT systems. Report ITU-R M.2320, November 2014.

[44] ITU-R, Technical feasibility of IMT in bands above 6 GHz. Report ITU-R M.2376, November 2014.

[45] ITU-R, Detailed specifications of the terrestrial radio interfaces of International Mobile Telecommunications Advanced (IMT-Advanced). Recommendation ITU-R M.2012-2, September 2015.

[46] ITU-R, Frequency arrangements for implementation of the terrestrial component of International Mobile Telecommunications (IMT) in the bands identified for IMT in the Radio Regulations. Recommendation ITU-R M.1036-5, October 2015.

[47] ITU-R, IMT Vision—Framework and overall objectives of the future development of IMT for 2020 and beyond. Recommendation ITU-R M.2083, September 2015.

[48] ITU-R, Radio regulations, Edition of 2016.

[49] ITU-R, Detailed specifications of the terrestrial radio interfaces of International Mobile Telecommunications-2000 (IMT-2000). Recommendation ITU-R M.1457-13, February 2017.

[50] ITU-R, Guidelines for evaluation of radio interface technologies for IMT-2020. Report ITU-R M.2412 November 2017.

[51] ITU-R, Minimum requirements related to technical performance for IMT-2020 radio inter- face(s). Report ITU-R M.2410 November 2017.

[52] ITU-R, Requirements, evaluation criteria and submission templates for the develop-ment of IMT-2020. Report ITU-R M.2411 November 2017.

[53] M. Jain, et al., Practical, Real-Time, Full-duplex Wireless, MobiCom'11, Las Vegas, NV, USA, September 19–23, 2011.

[54] E.O. Johnson, Physical limitations on frequency and power parameters of transistors, RCA Rev. 26 (June, 1965) 163–177.

[55] E.G. Larsson, O. Edfors, F. Tufvesson, T.L. Marzetta, Massive MIMO for next gen-eration wireless systems, IEEE Commun. Mag. 52 (2) (February 2014) 186–195.

[56] J. Lee, et al., Spectrum for 5G: global status, challenges, and enabling technologies, IEEE Commun. Mag. (March 2018).

[57] D.B. Leeson, A simple model of feedback oscillator noise spectrum, Proc. IEEE 54 (2) (February 1966).

[58] O. Liberg, M. Sundberg, E. Wang, J. Bergman, J. Sachs, Cellular Internet of Things: Technologies, Standards, and Performance, Academic Press, 2017.

[59] D.J.C. MacKay, R.M. Neal, Near shannon limit performance of low density parity check codes, Electron. Lett. 33 (6) (July 1996).

[60] Motorola, Comparison of PAR and Cubic Metric for Power De-rating, R1-040642.

[61] B. Murmann, The race for the extra decibel: a brief review of current ADC perfor-mance trajectories, IEEE Sol. State Circ. Mag. 7 (3) (Summer 2015) 58–66.

[62] Murmann, B., ADC Performance Survey 1997–2017 [Online]. Available: http://web. stanford.edu/Bmurmann/adcsurvey.html.

[63] M. Olsson, S. Sultana, S. Rommer, L. Frid, C. Mulligan, SAE and the Evolved Packet Core—Driving the Mobile Broadband Revolution, Academic Press, 2009.

[64] E. Onggosanusi, et al., Modular and high-resolution channel state information and beam management for 5G new radio, IEEE Commun. Mag. 56 (3) (March 2018).

[65] J. Padhye, V. Firoiu, D.F. Towsley, J.F. Kurose, Modelling, TCP reno performance: a simple model and its empirical validation, ACM/IEEE Trans. Netw. 8 (2) (2000) 133–145.

[66] S. Parkvall, E. Dahlman, A. Furuskaär, M. Frenne, NR: the new 5G radio access technology, IEEE Commun. Stand. Mag. 1 (4) (December 2017) 24–30.

[67] M.B. Pursley, S.D. Sandberg, Incremental-redundancy transmission for meteor burst communications, IEEE Trans. Commun. 39 (May 1991) 689–702.

[68] T. Richardson, R. Urbanke, Modern Coding Theory, Cambridge University Press, 2008.

[69] C.E. Shannon, A mathematical theory of communication, Bell Syst. Tech. J. 27 (379−423) (July and October 1948) 623−656.

[70] Special Issue on Spread Spectrum, IEEE Trans. Commun. 25, 745−869. August 1977.

[71] S.B. Wicker, M. Bartz, Type-I hybrid ARQ protocols using punctured MDS codes, IEEE Trans. Commun 42 (April 1994) 1431−1440.

[72] Wozencraft, J.M., Horstein, M., Digitalised Communication Over Two-way Channels, Fourth London Symposium on Information Theory, London, UK, September 1960.

[73] C. Mollen, E.G. Larsson, U. Gustavsson, T. Eriksson, R.W. Heath, Out-of-band radiation from large antenna arrays, IEEE Commun. Mag. 56 (4) (April 2018).

[74] 3GPP, NR; General aspects for UE RF for NR, 3GPP TR 38.817-01.

[75] 3GPP, NR; General aspects for BS RF for NR, 3GPP TR 38.817-02.

Index

Note: Page numbers followed by "*f*" and "*t*" refer to figures and tables, respectively.

Made in the USA
Monee, IL
31 October 2020